Intelligent Systems, Control and Automation: Science and Engineering

Volume 81

Intelligent Systems, Control and Automation: Science and Engineering book series publishes books on scientific, engineering, and technological developments in this interesting field that borders on so many disciplines and has so many practical applications: human-like biomechanics, industrial robotics, mobile robotics, service and social robotics, humanoid robotics, mechatronics, intelligent control, industrial process control, power systems control, industrial and office automation, unmanned aviation systems, teleoperation systems, energy systems, transportation systems, driverless cars, human-robot interaction, computer and control engineering, but also computational intelligence, neural networks, fuzzy systems, genetic algorithms, neurofuzzy systems and control, nonlinear dynamics and control, and of course adaptive, complex and self-organizing systems. This wide range of topics, approaches, perspectives and applications is reflected in a large readership of researchers and practitioners in various fields, as well as graduate students who want to learn more on a given subject.

The series has received an enthusiastic acceptance by the scientific and engineering community, and is continuously receiving an increasing number of high-quality proposals from both academia and industry. The current Series Editor is Kimon Valavanis, University of Denver, Colorado, USA. He is assisted by an Editorial Advisory Board who help to select the most interesting and cutting edge manuscripts for the series:

Springer and Professor Valavanis welcome book ideas from authors. Potential authors who wish to submit a book proposal should contact Thomas Ditzinger (thomas.ditzinger@springer.com)

Indexed by SCOPUS, zbMATH, SCImago.

More information about this series at http://www.springer.com/series/6259

Maria Isabel Aldinhas Ferreira · Sarah R. Fletcher
Editors

The 21st Century Industrial Robot: When Tools Become Collaborators

Springer

Editors
Maria Isabel Aldinhas Ferreira
Centro de Filosofia da Universidade de Lisboa
University of Lisbon
Lisbon, Portugal

Sarah R. Fletcher
Industrial Psychology and Human Factors Group
Cranfield University
Cranfield, UK

ISSN 2213-8986 ISSN 2213-8994 (electronic)
Intelligent Systems, Control and Automation: Science and Engineering
ISBN 978-3-030-78512-3 ISBN 978-3-030-78513-0 (eBook)
https://doi.org/10.1007/978-3-030-78513-0

This Springer imprint is published by the registered company Springer Nature Switzerland AG
The registered company address is: Gewerbestrasse 11, 6330 Cham, Switzerland

Preface

I am C-3PO, human/cyborg relations. And you are?
in Star Wars

The disruption caused by the Covid-19 pandemics on economy and society as a whole has accentuated existing problems of an ongoing social and economic crisis,[1] mainly caused by the evolution of a global neo-liberal economic framework, with its speculative financial models, plus the destructive effects of a growing climate/environmental emergency. The dynamics of this very complex situation, that the sanitary crisis has deeply aggravated, calls for a fast, objective and ethically- guided strategic planning in order to reshape economies,[2] by placing equity and the well-being of individuals as their goals, and to restore the lost environmental equilibrium. To achieve these goals that, in a certain way, were already defined as a priority in the Development Sustainable Goals Framework,[3] the technological innovation brought about by the so-called 4th Industrial Revolution, namely the embodied and non-embodied artificial intelligent systems, will be essential tools. However, having intelligent systems producing tangible or non-tangible forms of work, and coexisting with human beings at the workplace or even co-operating, raises fundamental ethical and societal concerns. By addressing the complexities of this human/robot co-existence and co-operation in industrial settings, this book aims to contribute to their harmonious integration .

According to Deloitte[4], automation will remain a strategic priority for all countries over the next ten years. As 2014–2019 saw an estimated 85% global rise in factory deployment of industrial robots[5], it is clear that fast development and deployment of

[1] The so-called 2008 Great Recession.

[2] Cf. Emerging Stronger and Better: 12 Months, Twelve Lessons from the Pandemics: https://feature.undp.org/emerging-stronger-and-better/?utm_source=web&utm_medium=sdgs&utm_campaign=12lessonsfromCOVID19.

[3] https://www.undp.org/content/undp/en/home/sustainable-development-goals.html.

[4] Deloitte.uk- Robots are Coming. https://feature.undp.org/emerging-stronger-and-better/?utm_source=web&utm_medium=sdgs&utm_campaign=12lessonsfromCOVID19.

[5] International Federation of Robotics: World Robotics Report 2020. https://ifr.org/ifr-press-releases/news/record-2.7-million-robots-work-in-factories-around-the-globe.

ICTs and intelligent systems, either embodied or non-embodied, will remain a technological target of developed and developing societies. Despite early assumptions that industrial automation would just replace human operators, evidence has long shown that it simply cannot replace many complex human skills, particularly those involving flexible and intelligent responses.[6] The Global Partnership on Artificial Intelligence report on the Future of Work[7] refers to evidence that shows intelligent systems are not intended to automate entire processes but rather to improve the performance of human workers. This means that natural and artificial cognition/intelligence will not only coexist at the workplace but will co-operate, i.e. act in order to achieve a common productive outcome. This co-operation has been commonly referred to as human-robot 'collaboration'. However, artificial systems lack the true agency/free will features inherent to the semantics of that concept. Consequently, artificial systems need to be "made collaborative" through human-centric purposeful design. A human-centric approach to developing technology means applying design principles that respect human factors and follow strict ethical guidelines, so that systems are compatible for effective human-robot co-operation, work for human benefit and respect fundamental human values. Production processes cannot only be designed to achieve the expected productive outcome goals, but with respect to the nature, specificities and reasonable expectations of all the users involved. However, as the abovementioned GPAI report points out, putting the human at the centre is an incantatory discourse, it must not just be said, it must be accomplished.

A human-centric approach requires a robust understanding of, and reliable methods to examine, human behaviour and requirements. Social science, and particularly human factors (ergonomics), has a fundamental role to play in providing this knowledge, though its inputs have traditionally been neglected. But, in fact, until fairly recently, the size and payload of industrial robots was so hazardous, they were completely separated from the workforce in production processes and, understandably, there has been, until recently, no need to consider human issues (other than segregation protocols) in the design and engineering of systems. However, as customary boundaries between people and robots are progressively being removed and the deployment of collaborative systems advances at pace, this fundamental change brings a pressing need to gather and integrate human factors into system design. Consequently, engineering and social science disciplines are beginning to merge towards development of a rich field of research that not only seeks to develop technical systems but to also take account of human factors in order to promote worker well-being, both physical and mental.

The present book takes a multidisciplinary stance, where the insights and experience of academia and industry merge to highlight the need for an accurate vision of how the design, development and deployment of intelligent tools can incorporate fundamental knowledge about human behaviour and present societal values. In the

---

[6] de Winter, J. C. F., Dodou, D., 2014. Why the Fitts list has persisted throughout the history of function allocation. *Cognition, Technology & Work.* 16, 1, 1–11.

[7] The Global Partnership on Artificial Intelligence report on the Future of Work-November 2020-GPAI, the Future of Work https://gpai.ai/projects/future-of-work/.

following chapters, a range of contemporary issues and approaches demonstrate how emerging industrial human-robot collaboration challenges are being tackled.

In chapter "On Human Condition: The Status of Work", Ferreira refers both to the universal existential circumstances that human beings share with all the other life forms and to those that are human specific. The author claims that the capacity for work, in all its tangible and non-tangible forms, is a unique essential human attribute, responsible for the evolution of humankind as a species and for the development of their world. According to the author, being intrinsically human and not the result of a temporary condition or state, [work], i.e. [productive goal-oriented action] has been, throughout times, improved, enhanced and augmented by the creation of rudimentary, and then progressively more sophisticated, man-made tools. Ferreira refers that contemporary societies have become hybrid environments, where the physical is permeated by the digital, human interaction is mediated by advanced forms of virtual communication and work is being replaced, in many sectors, by task performance and decision making by artificial intelligent systems. This present context and its predictable development in the near future demands the emergence of a deep awareness on the part of different stakeholders: research and development institutions, policy makers and governance, business and on the part of society in general, so that intelligent technology remains a human tool for enhancing and augmenting [work], respecting its fundamental twofold dimension as: (i) a generative endowment for the creation of human reality and (ii) a means for the fulfilment and existential satisfaction of every human being.

Michalos, Karagiannis, Dimitropoulos, Andronas and Makris, in chapter "Human-Robot Collaboration in Industrial Environments", present a thorough exploration of the current state of *Human-Robot Collaboration in Industrial Environments*, arguing that the advancement of robotics technology in recent years and the parallel evolution of the AI, big data, Industry 4.0 and Internet of things (IoT) paradigms have paved the ground for applications that extend far beyond the use of robots as mindless repetitive machines. The number of technical configurations/solutions grows exponentially when considering factors such as (a) the particularities of the task to be performed (e.g. type of part, weight, dimensions, process to be carried out, etc.), (b) the type of robots that can address these requirements (fixed or mobile robots, high/low payload, exoskeletons, aerial robots, etc.), (c) the type of collaboration and interaction that would be appropriate for the task and (d) the special requirements of the production domain where such tasks are needed. The authors aim to identify the existing approaches on the implementation of human-robot collaborative applications and highlight the trends towards achieving seamless integration of humans and robots as co-workers in the factories of the future.

In chapter "Managed Systems Approach to Commissioning Collaborative Industrial Robot Systems", Quinlan-Smith describes how the field of collaborative robotics has expanded significantly over the past ten years, such that it is now the fastest growing segment of the global industrial robotics market with advances in robot software technology allowing robots and workers to work "hand-in-hand" to achieve higher levels of efficiency and productivity. This new relationship leverages the strength of the robot to perform dull, dirty and repetitive tasks (e.g. palletizing, painting, packaging, polishing, etc.) while combining the higher cognitive abilities and flexibility of the human colleague; a winning combination of brawn and brains that is clearly transforming traditional ways of working. Quinlan-Smith points out that, when properly executed, the partnership between the human and robot has the potential to improve safety while keeping up with forever changing customer requests and productivity demands, e.g. a robot that performs a repetitive task may help improve health by reducing injury incidence while also improving productivity at the same time. However, despite these claims, research has shown that this new technological introduction into our workplace will bring forth ethical issues. Sudden, coerced introductions of a robotic colleague into our work space, whom we have long been separated from by means of strategically placed "fencing", will threaten the adoption of such technology on the plant floor. Evidence presented in one study showed that improper attention to the "human aspects" is believed to be a primary cause of significant failures in the implementation of advanced manufacturing technology in the USA. Based on lessons learned in this study, organizations must learn to understand how changes in work tasks or the working environment that are made without consultation with, or involvement of, a worker can significantly impact the human experience and overall productivity.

Haninger, in chapter "Robot Inference of Human States: Performance and Transparency in Physical Collaboration", addresses the topic of *Robot Inference of Human States: Performance and Transparency in Physical Collaboration*. This refers to the need for a human-robot partnership to not only be designed for the human to monitor and anticipate forthcoming events but how, in order for a robot to flexibly collaborate towards a shared goal in human-robot interaction (HRI), it must also be appropriately designed to respond to changes in their human partner. The robot can realize this flexibility by responding to certain inputs or by inferring some aspect of their collaborator and using this to modify robot behaviour; these distinct approaches reflect design viewpoints of robots that regard robots respectively as tools or as collaborators. Independently of this design viewpoint, the robot's response to a change in the collaborator's state must also be designed. In this regard, HRI approaches can be distinguished according to the scope of their design objectives: whether the design goal depends on the behaviour of the individual agents or the coupled team. Haninger synthesizes work on physical HRI, largely in manufacturing tasks, according to the design viewpoint. HRI is posed as the coupling of two dynamic systems; a framework which allows a unified presentation of the various design approaches and, within which, common concepts in HRI can be posed (intent, authority, information flow). Special attention is paid to predictability at various stages of the design and deployment process: whether the designer can predict team performance, whether the designer can predict team performance, whether the human can predict robot behaviour, and to what degree the human behaviour can be modelled or learned.

Eimontaite, Chapter "Human-Robot Collaboration using Visual Cues for Communication" presents recent studies that have examined *Human-Robot Collaboration using Visual Cues for Communication*, in order to begin building a foundational knowledge of what type(s) of cues a robot should present to most effectively promote human awareness and anticipation of a robot's forthcoming actions. Thus, following on from chapter "Robot Inference of Human States: Performance and Transparency in Physical Collaboration" attention to robot capabilities, Eimontaite addresses another important aspect of communication and mutual awareness between robot and human partners. This chapter reviews how traditional industrial robots in the manufacturing sector have been used for repetitive and strenuous tasks for which they were segregated due to their hazardous size and strength and so are still perceived as threatening by operators in manufacturing. This means that successful introduction of new collaborative systems where robotic technology will be working alongside and directly with human operators depends on human acceptance and engagement. The chapter discusses the important reassuring role played by communication in human-robot interaction and how involving users in the design process increases not only the efficiency of communication, but provides a reassuring effect. After presenting findings to date, Eimontaite identifies remaining challenges that affect the development of productive and stimulating communication between manufacturing operators and robots, thereby highlighting future work needed in this area.

Chapter "Trust in Industrial Human-Robot Collaboration" explores the specific issue of workers' *Trust in Industrial Human-Robot Collaboration.* Here, Charalambous and Fletcher not only emphasize how trust is a vital component for successful co-operation within any team, regardless of the entities that form it, but how there is a need to identify and understand the specific characteristics of a system or context that influence trust. In the context of human-automation teaming, an ideal level of trust needs to be achieved to optimize performance because too much trust can cause over-reliance and confidence, but not enough trust can cause timidity and poor response. Therefore, the authors describe how the different attributes of robots, e.g. degree of autonomy, mobility, anthropomorphism, size, types of physical embodiments, etc., and the task for which the application is being used, may cause different impacts on trust. As these attributes have the potential to tease out different human/user responses, the authors emphasize the importance of designers and integrators being able to evaluate trust in relation to robot attributes. Having identified that no existing measure to evaluate trust in industrial human-robot collaboration, Charalambous and Fletcher describe research work they conducted to specifically identify relevant robot attributes and, at the same time, develop a new psychometric measure of Trust in Industrial Robots intended to aid design of future systems.

In chapter "Adapting Autonomy and Personalisation in Collaborative Human–Robot Systems", Marguglio, Caruso and Cantore point out why modern manufacturing systems need to be increasingly "adaptive" to an ever-changing environment, because evolving internal and external demands today require increasing flexibility, sustainability and human satisfaction. Adaptive automation is vital to meet these challenges but also so that systems can autonomously identify and apply best methods of employing the abilities offered by humans and automation, taking advantage of each other's strengths to balance flexibility and productivity requirements in an easy and cost effective way. The chapter describes a programme of work undertaken as part of the European Commission funded "A4BLUE" project to develop this "adaptive" robot architecture, and how a key objective was to integrate a degree of personalization and meet requirements of human operators to sustain their satisfaction.

The authors describe how this research fits into the wider context of the Europe 2020 strategy and its aims to promote market-oriented projects by bringing together private and public resources, and the role of the European Factories of the Future Research Association (EFFRA) which aims to promote development of new and innovative production technologies and pre-competitive research.

In chapter "Designing Robot Assistance to Optimize Operator Acceptance", Otero and Johnson present a real case study of *Designing Robot Assistance to Optimize Operator Acceptance* which was also conducted in Europe as part of the A4BLUE project. The chapter describes the main difficulties inherent to the implementation of an automated or robotic solution in a company, namely the doubts and concerns experienced by workers: "Will I be replaced by the robot? Is it safe to work collaboratively with an automated mechanism?" which reflect commonly held anxieties for workforces around the world as deployment of human-robot collaboration expands. At the same time, workers are asking these questions, there are usually economic concerns that the company are asking about the initially high investment and the expected benefits of such an innovative solution, where the risks can be higher than those traditional solutions which have been widely implemented and tested over the years. Otero and Johnson describe the process they undertook to involve operators in the development of the new system which was designed to both enhance design but also promote engagement and acceptance. The case study provides a real world example of successful implementation of a human-robot system in a Spanish aeronautical company in a production area where tasks are traditionally performed manually, including the assembly of complex aeronautical equipment and its auxiliary operations, and shows the benefits of involving the workforce in design and implementation of a new technology which will change their work methods.

In chapter The Role of Standards in Human-Robot Integration Safety", Franklin describes *The Role of Standards in Human-Robot Integration Safety* and specifically focuses on how standards have been developed for guiding the safe application of human-robot collaboration (HRC) in industrial settings standards. The chapter sets out the nature and purpose of standards, their utility and limitations, and how such standards are developed via voluntary industry consensus, i.e. how standards are constructed by international committees of volunteer experts. Franklin provides thoughts on how standards impact innovation in the marketplace and how standards provide safety requirements for industrial HRC and influence procedures, identifying ongoing challenges and limitations of voluntary industry consensus standards, both in general and specifically in the area of closer human-robot collaboration in industry. The history of industrial robot safety standards is also discussed, along with areas of potential future work.

In chapter "Engineering a Safe Collaborative Application", Dominguez provides a detailed account of risk assessment and design principles and procedures involved in the implementation or integration of a collaborative robot system. The chapter highlights the benefits brought by collaborative robotics when compared to the traditional robotic systems and points out that this new approach determines the priority of risk assessment when deploying collaborative systems at the workplace. In a collaborative context, risk assessment must identify all potential contact events, both intentional and unintentional. On the other hand, these contact events must be assessed in terms of probability of occurrence and the potential severity of injury, the definition of risk. If the risk level is not tolerable, then measures can be taken to either reduce the potential severity of injury or the probability of contact. The author identifies parameters involved, such as types of contact event and injury/pain threshold limits for force, and identifies the procedures respecting each of the steps concerning the assessment of risk in a collaborative context and in accordance with current standards.

Challenges in the Safety–Security Co-Assurance of Collaborative Industrial Robots explores existing approaches and best practices for ensuring the safety and security of collaborative robots; in chapter "Challenges in the Safety–Security Co-Assurance of Collaborative Industrial Robots", Gleirscher, Johnson, Karachristou, Calinescu, Law and Clark highlight the challenges posed by the complexity involved. In Sect. 1, the authors provide an overview of safety and security approaches applicable to collaborative robot systems, while in Sects. 2 and 3, they elaborate on particular methodologies in the context of a real industrial case study in the UK. Following the two perspectives—social and technical—of the Socio-Technical System (STS) design approach, Sect. 4 enumerates additional socio-technical and technical challenges arising from safety-security interactions. The chapter provides not only a thorough overview of current state of the art in safety and security assessment but also a real world example of application, presenting the foundations for a "preliminary research roadmap".

In chapter "Task Allocation: Contemporary Methods for Assigning Human-Robot Roles", Kousi, Dimosthenopoulos, Aivaliotis, Michalos and Makris propose that because human-robot collaboration (HRC) is now a major enabler for achieving flexibility and reconfigurability in modern production systems; it is vital to offer system designers an effective and systematic method of task and environment evaluation/optimization. The authors emphasize that the motivation for HRC applications arises from the potential of combining human operators' cognition and dexterity with the robot's precision, repeatability and strength that can increase system's adaptability and performance at the same time. The authors point out that to exploit this synergy effect on its full extent, production engineers must be equipped with the means for optimally allocating the tasks to the available resources as well as setting up appropriate workplaces to facilitate human-robot collaboration. The chapter discusses existing approaches and methods for task and process planning in collaborative work environments analysing the requirements for implementing such decision-making strategies, including modelling and visualization methods. The chapter also highlights future trends for progressing beyond the state of the art on this scientific field, exploiting the latest advances in artificial intelligence and digital twin techniques.

In chapter "Implementing Effective Speed and Separation Monitoring with Legacy Industrial Robots—State of the Art, Issues, and the Way Forward", Moel, Denenberg and Wartenberg present *Implementing Effective Speed and Separation Monitoring with Legacy Industrial Robots—State of the Art, Issues, and the Way Forward*, based on extensive work they have been conducting at VEO Robotics towards human-robot safety. The authors point out that collaborative applications using traditional industrial robots and Speed and Separation Monitoring (SSM per ISO/TS 15066) rely on safe stopping if a Protective Separation Distance (PSD per ISO/TS 15066) is violated. However, larger industrial robots have longer stopping times and their control architectures are not designed for flexible external interaction. Robot manufacturers also provide stopping time and distance data for calculating the PSD, but these data are often fragmented, hard to interpret and overly conservative. Hence, the "worst-case" PSD calculation for SSM is generally more conservative than warranted. The authors claim that truly fluid human-robot collaboration will be possible but will require a closer interlocking between the robot controller and the safety system and a more precise characterization of robot stopping times and distances. The present chapter describes techniques to improve latencies and response times using SSM with existing robot control architectures also proposing longer-term alternatives for consideration by the industry.

Chapter "Ethical Aspects of Human-Robot Collaboration in Industrial Work Settings" addresses the increasingly important topic of *Ethical Aspects of Human-Robot Collaboration in Industrial Work Settings*. Wynsberghe, Ley and Roeser review and expand upon the current ethical research on human-robot collaboration in industrial settings which primarily includes: job loss, reorganization of labour, informed consent and data collection, user-involvement in design, hierarchy in decision-making and coerced acceptance of robots. These wide-ranging issues are a useful starting point for discussion, yet as the number of robots designed and deployed as collaborators in industrial settings grows, ethical research must evolve to allow for more nuance in the previously listed issues as well as a recognition of novel concerns as they arise. The authors suggest a forthcoming emergence of new ethical aspects related to industrial human-robot collaboration, including: emotional impacts on workers; effects of limited movement; the potential effects of working with one's replacement; the "chilling effects" of performance monitoring; the possibility for disclosure of new and unintended information through data collection; and the inability to challenge computerized decisions. Individually these are all very concerning issues, and together they comprise a set of factors that will require new forms of moral learning for assessing the ethical acceptability of industrial human-robot collaborations.

Last but not at all least, the final chapter of this book, Chapter "Robots and the Workplace: The Contribution of Technology Assessment to Their Impact on Work and Employment", presents a comprehensive overview of *Robots and the Workplace: The contribution of Technology Assessment to their Impact on Work and Employment*. Here, Carvalho and Pereira address the challenges that automation/robotization poses for human labour and employment issues. The authors claim that technology assessment (TA) can provide a ground for both ethical reflection and social engagement towards participatory decision-making regarding the application of such technologies. The chapter also debates labour substitution as a dominant narrative in economic analysis, while also stressing the need to contextualize technological change and innovation regarding robots and automation in the concrete work processes or tasks, bringing narratives closer to the ground. This discussion leads to the second main theme of the chapter: the potential role of technology assessment in better exploiting the development and use of robots in the workplace, their unanticipated consequences and the ethical and social tensions arising therein. According to the authors, these approaches do not aim at complete or sound predictions but at building participatory and interdisciplinary processes being the chapter ultimately about how we ought to live and to relate to technology.

Lisbon, Portugal — Maria Isabel Aldinhas Ferreira
Cranfield, UK — Sarah R. Fletcher

Acknowledgements

The editors would like to thank to all the authors for their willingness and interest in reflecting on the technical and ethical challenges that the deployment of embodied and non-embodied artificial systems at the workplace poses, giving their best to significantly contribute to the state of the art at a very complex time.

To Thomas Ditzinger for his support and to all Springer's team for their patience and kindness and extreme professionalism throughout the making of this book.

Contents

On Human Condition: The Status of Work

Maria Isabel Aldinhas Ferreira

Who are we if not or when not productive?
Hanna Arendt

Abstract At a time we are experiencing the overspread deployment of artificial intelligent systems in all domains of life, the present paper addresses the topic of human condition questioning both the universal existential circumstances that human beings share with all the other life forms and those circumstances that are human specific, claiming, in this case, that the capacity for work, in all its tangible and non-tangible forms, is a unique essential human attribute, responsible for the evolution of humankind as a species and for the development of their world. Being intrinsically human and not the result of a temporary condition or state, [work], i.e. [productive goal-driven action] has been, throughout times, enhanced and augmented by the creation, at first, of rudimentary and then of progressively more and more sophisticated man-made tools. In the course of this developmental narrative contemporary societies have become hybrid environments, this means environments where the physical is permeated by the digital, where human interaction is mediated by advanced forms of virtual communication, where non-embodied and also embodied forms of artificial intelligence coexist with natural intelligence where ultimately [work] in its intrinsic humaneness is being replaced in many sectors by task performing and decision-making per artificial intelligent systems. This present context and its predictable development in the near future demand the emergence of a deep awareness on the part of different stakeholders: research and development institutions, policy makers and governance, business and on the part of society in general, so that intelligent technology remains a human tool for enhancing and augmenting work, respecting its fundamental twofold dimension as: (1) a generative endowment for the creation of human reality (2) a means for the fulfilment and existential satisfaction of every human being.

M. I. A. Ferreira (✉)
Centre of Philosophy, Faculdade de Letras, University of Lisbon, 1600-214 Lisboa, Portugal
e-mail: isabelferreira@letras.ulisboa.pt

Institute for Systems and Robotics, Instituto Superior Técnico, University of Lisbon, 1049-001 Lisboa, Portugal

M. I. Aldinhas Ferreira and S. R. Fletcher (eds.), *The 21st Century Industrial Robot: When Tools Become Collaborators*, Intelligent Systems, Control and Automation: Science and Engineering 81, https://doi.org/10.1007/978-3-030-78513-0_1

Keywords Human condition · Covid-19 crisis · Work · Intelligent tools-the ontological shift · Human dignity · Existential satisfaction

1 Introduction

The title of this chapter is primarily motivated by the consideration of a unique human dimension, that of productive goal-oriented action—the generative and transformative power of work- and the role tools play in this process, in their twofold symmetrical facets as the output of human work and also as work "coproducers".[1]

But this title is also certainly determined by the Covid-19 pandemics and its disruptive effects on the typical way human existence unfolded until March 2020. In fact, the massive disruption caused by this health emergency has made salient the essential fragility of humankind-a condition shared with all the other life forms, but for the sake of being, usually ignored by the self-conscious subject.[2]

Life requires a minimum of stability to unfold, however sometimes the essential parameters responsible for that stability are disrupted by internal or external circumstances: a health condition, a tsunami, a cyclone, a biological hazard- as in the case of plagues,- a war, an economic and social crisis... And it is at these times that human beings-the ones capable of creating worlds but also of destroying the planet-are brutally reminded of the existential frailty they share with their fellow species.

The pandemics has primarily made salient the fact that in spite of all the technological development achieved so far, in spite of having created a robotic engine capable of navigating a 290-million-mile journey and safely landing on the surface of Mars where it will collect samples to bring back home,[3] human beings are not all that mighty and can succumb by thousands when something gets out of control in their surrounding environment, like ants when something crushes their nest.

However, the pandemics has also made visible the wonder of the capacity, so many times challenged throughout history, of human beings to transcend and overcome critical situations thanks to the individual and collective effort of their intelligent productive goal-oriented action- their work capacity. We have seen this, we have all been and are being part of this common huge effort to overcome a crisis, an effort that initially caused the traditional existential private and work spheres to overlap, where most of the typical patterns of behaviour that defined the routines and lifestyles of our societies were erased or substantially changed, producing a time/space continuum where work status and offwork status merged in a nearly permanent working state, each of us trying to cope with the turbulence and instability caused by the pandemics in the personal- emotional/affective and social spheres of human existence and still

[1] We do not endorse here the distinction [labour]/[work] and their multiple interpretations throughout historical times and different ideological frameworks (c.f. Adam Smith, Marx, Arendt, among others) and refer in all circumstances only to [work].

[2] Those that are permanently or frequently tormented with these thoughts are diagnosed as having a mental condition that requires treatment.

[3] Nasa's Perseverance landed on the 18th of February 2021 on Mars.

trying to fulfill what was/is expected from the specificity of the roles previously assigned and assumed: health professionals giving their best in hospitals in order to save lives, thousands keeping services running, supply chains maintained, long-distance education and research going on… and last but not least scientists tracking the virus evolution in order to control its spreading, in order to understand its nature and behaviour producing vaccines in record time, vaccines capable of ending the threat, capable of restoring the normal patterns of the contemporary human social life.

And in all these situations, intelligent tools have come forward and have shown themselves as precious allies, proving that technological innovation and development, when ethically guided, pay off as their benefits for humanity can be immense.

In this context and bearing the heaviness of a strong philosophical tradition that coined and sustained the concept [human condition][4] and their multiple consecrated artistic expressions,[5] I have humbly accepted the challenge of its interpretation by assuming it as the grounding stone of an argument that tries to highlight the essential existential dimension that work plays in human life, simultaneously discussing the relationship between human beings and their tools, namely the new intelligent ones, within the transformative and generative dynamics triggered out by human productive goal-driven action.

2 The Key Elements of a Universal Existential Framework and the Specificity of Human Cognition

The human species shares with all the other fellow species three essential universal circumstances:

(i) It is endowed with a specific physical architecture that results from a long evolutionary and developmental individual and collective narrative, an architecture that has been shaped by its surrounding physical context and that has, on the other hand, shaped and given meaning to a particular world.[6]

(ii) It is embedded in a dynamic environment to which it is bound by a dialectic relationship of mutual influence and co-determination.

(iii) This physical architecture and its surrounding environment constitute a microcosm, a living unit.

It is thanks to the specificity of their physical bodies that living entities are capable of engaging in a particular way with their environment creating species specific

---

[4] cf. [1].

[5] cf. [20] and Magritte (1935).

[6] On the role of this physical architecture defining a species specific world cf. [3, 4, 6, 18, 25, 32].

worlds.[7] As Ref. [3] refers reality is not a unique and homogeneous thing, it is immensely diversified, having as many different schemes and patterns as there are different organisms. Every organism is according to that author a kind of monadic being as it has a world of its own, because it has an experience of its own.

However, though all life forms bear distinct physical architectures that determine the construction of specific worlds they are bound to their environments by the same dialectic relationship that characterizes the semiosic process inherent to all forms of cognition and on which the life of each individual entity and life in general depend.

Cognition, i.e., the intelligent capacity of every life form to guarantee their existence by interacting with its surrounding environment, is always an embodied, embedded and situated phenomenon.[8] This means the nature of this cognition depends on the specificity of the corporeal architecture involved and consequently on the type of interactions made available by it. It is embedded because it takes place within the boundaries of a typical environmental bubble which is fitted for that entity and has been extensively shaped by the cognitive architecture that has been evolving in it. This interaction is always situated as it happens along a bounded organic dynamics usually called lifespan, which is translated by a particular spatiotemporal flow according to the view of the human observer.

Though sharing the above mentioned key existential circumstances with all living entities, the human species distinguishes itself from all the others by having a set of specific attributes, namely: a conceptualizing and symbolic capacity that allows for the definition of an external objective reality, the definition of a sense of alterity, of Otherness, and the writing of an internal auto-narrative that permits the experiencer subject to recall past experiences, reflect on the present one and project, predict and eventually anticipate the modes of the future—a consciousness of their place and role in this lived social reality.

But the human being also distinguishes itself from the other species for its creative and transformative capacity to act on the surrounding environment, to model and shape it. It is the generative power of its productive goal-oriented action, of its work, that defines and develops the human world- a physical, economic, social, cultural and linguistic reality. Reflecting on the creative character of human action Marx points out: (1867:198): "A spider conducts operations that resemble those of a weaver, and the bee puts to shame many an architect in the construction of her cells. But what distinguishes the worst architect from the best of bees is that the architect raises his structure in imagination before he erects it in reality. At the end of every labour process, we get a result that already existed in the imagination of the laborer at its commencement". As Marx also points out, whereas animal is driven by natural impulses, man's specific form of activity is conscious life activity.

It is this conceptualizing capacity and its symbolic encoding—language- plus the power of productive-goal- oriented action that made possible the production

[7] cf. [31, 32]. According to Uexkull when we know the anatomical structure of a species we possess the necessary elements to reconstruct its experience. I would say we have the necessary elements to construct a rough approximation of that experience.

[8] cf. [6, 8].

of artifacts, namely of tools, the accumulation of knowledge and its transmission throughout generations allowing for the definition of a historic narrative that reflects the dialectics from which distinct modes of production, distinct economic, social and cultural models have emerged and that determined the scientific and technological development of humanity.

Given these assumptions, the concept of [human condition] stands in this text for [the set of universal objective circumstances-biological and developmental that determine or affect the being in the world of the humankind].[9] It comprehends on one hand the human fragility and the incomprehensiveness of human beings towards the existential phenomenon in all its dimensions. On the other hand it comprehends the generative power of the creative dimension overcoming the challenges of an ever evolving environment, constructing a human-sized world and using the power of love to hold it together. These circumstances define a condition to which every human being is intrinsically bound and from which cannot depart, as Ref. [1] refers: "The most radical change we can imagine in human condition would be an emigration of men from the earth to some other planet ... Yet even these hypothetical wanderers from the earth would still be human [...]".

As [8] points out, in a significant number of species, the evolutionary and developmental process has proceeded according to three interconnected axes: (1) interaction ability, (2) task performance ability, (3) tool making ability.[10] The abilities represented by these three axes are made possible by a set of innate endowments that, though exhibiting a degree of variability among species in what relates their level of sophistication and complexity, represent a continuum that is horizontal to all of them. This way the capacity for communication, that the interaction ability subsumes, attains its highest degree of sophistication in human conceptualizing and symbolic language capacity, the same happening with the capacities for distributed task performing and the tool making which are endogenous[11] and define a continuum of progressive complexity throughout different species.[12] Being innate as a potential, these abilities are in human beings also the result of a learning process that takes place in society, that goes on throughout the individual's lifetime and that relies substantially on the accumulated experience of precedent generations and on the intrinsic values of the prevalent economic, social and cultural models. One of the substantial differences we can immediately identify when contrasting tool making in humans

---

[9] To Ref. [1] the distinctive characteristic of human existence, its conditions, can be classified into two groups. The first group of human conditions consists of basic conditions under which life on earth has been given to man, and these are: life itself, worldliness, plurality, natality and mortality, and the earth itself. The second group of human conditions consists of conditions made by the men themselves, and these are man-made things and relations. Whatever enters the human world of its own accord or is drawn into it by human effort becomes part of human condition.

[10] The interaction ability, in fact, subsumes either (2) or (3). We make the distinction for purely analytical purposes.

[11] Greenfield [13].

[12] Although tool use has for long been assumed to be a uniquely human trait, there is now much evidence that other species such as mammals, namely primates, birds, cephalopods also use more or less rudimentary tools. cf. [2, 28].

and in other species is the fact that while human tools have been evolving exponentially—with some tools exhibiting nowadays a considerable degree of potential autonomy that will require little or no human intervention—tool making among other species has remained essentially rudimentary.

The creation of tools is consequently inherently associated to the biological, economic, social and cultural/scientific development of humankind. Throughout the ages, human beings have modified or updated the inventions of precedent generations or those of other communities of tool makers and have also created new ones in order to overcome difficulties, in order to achieve certain goals. From distinct social settings, distinct modes of production scientific and technological innovations have emerged determining specific working settings, specific working tools, specific divisions of labour.

Marx regarded work produced by means of tools as a feature differentiating human beings from the rest of the animal kingdom and referred to tool use as an extension of the laboring body,[13] viewing technologies as extensions of the human will domination over nature:

> Nature builds no machines, no locomotives, railways, electric telegraphs [...]These are products of human industry; natural material transformed into organs of the human will over nature...they are organs of the human brain created by the human hand (Marx 1993, p. 706).

Tool making and its natural evolution is inherently associated to the huge transformative and generative power resulting from the endogenous human capacity to define particular worlds, as can be observed by the way distinct artifacts have come to define distinct stages of human historical development, distinct civilization frameworks. Tools are in fact a particular subset of artifacts.[14] They share, with the broader category they belong to, the essential feature {function}, i.e. they are suited to a particular purpose, but their semantic specificity is also realized aby another fundamental feature. When looking at the definition of the concept [tool] in a language dictionary[15] we read:

1. A **tool**[16] is any instrument or piece of equipment that you hold in your hands **in order to help you to do a particular kind of work.** e.g., "workers downed tools in what soon became a general strike".
2. A **tool** is also any object, skill, idea etc., **that you use in your work** or that you need for a particular purpose.

By looking at these definitions we realize that the concept of [tool] is primarily associated to a working scenario and to the production/creation of a particular entity.

---

[13] Ref. [15].

[14] The definition of the concept of [tool] has been subject to different versions by researchers studying animal behavior. Ref. [16] defines [tool] as an object that has been modified to fit a purpose or an inanimate object that one uses or modifies in some way to cause a change in the environment, thereby facilitating one's achievement of a target goal.

[15] Collins Cobuild English Language Dictionary. Collins Publishers. University of Birmingham 1988 (1st edition).

[16] Emphases mine.

This means that inherent to the semantics grounding the concept of [tool] is the trait **{cause an object, an event or a state to come into being, through physical and/or mental activity}** which is the essence of the concept associated to the verb [work], weather the nature of this work is tangible or not.

This fact can be easily foreseen when we think not only of a shoemaker handling their tools to create or repair a pair of shoes, but also of a factory worker interacting with a machine to produce a particular piece, the farmer that drives a tractor to plough the field, the researcher that sits at the computer using a text processor to write a paper, the doctor that relies on machine learning to his medical procedures.

Tools can be seen primarily as extensions of the physical body[17] not only in the sense of being extensions of the "human hand" but essentially by being always somehow extensions of the human mind, providing a means for enhancing or augmenting the capacity for productive action that the bare human corporeal architecture finds difficult or is not able to attain by itself. This quasi prosthetic nature of tools is evident in the way human motor behaviour and its corresponding mental patterns are influenced by their handling. In fact, handling and/or operating any kind of artifact always requires the adoption of specific motor programmes[18] and the definition of new neural pathways that will allow particular patterns of behavior to become typical and routinary, being, this way, instantly triggered out by specific contexts of use without depending on a reflexive attitude.[19]

This prosthetic nature of the tool is also addressed by Ref. [26] that refers how tools are taken into ways into which human beings enroll and project themselves into work practices as they "withdraw" and become "ready-to-hand".

Perhaps because of this nearly physiological extension, this quasi-symbiotic process, between a human being and a specific instrumental artifact, through which a specific effect is produced causing a specific entity to come into being, there is frequently a link of affective attachment uniting workers to their tools and to their produced works. This frequent affective attachment reflects itself in the care often revealed by workers in the maintenance and keeping of their tools,[20] in the way artisans have always carved out or just signed their names on the created object or in the sense of achievement and even pride manifested by those that have contributed to the coming into being of important realizations, of particular endeavours. This feeling was evident, for instance, in a newspaper interview to the workers that participated

---

[17] It is particularly interesting how some technological tools are sometimes presented as extensions of the physical body. I recall on this purpose the sentence that opened up a small video that performed when I started my laptop computer produced by Texas Instruments in the early 90s—"Texas Extensa, an Extension of yourself".

[18] According to [32] the two fundamental human handgrips, first identified by J. R. Napier, and named 'precision grip' and 'power grip', represent a *throwing grip* and a *clubbing grip*, thereby providing an evolutionary explanation for the two unique grips, and the extensive anatomical remodelling of the hand that made them possible.

[19] cf. Ferreira [7].

[20] We recall on this purpose the particular attachment a hairdresser revealed towards her set of high specialized scissors, which she had acquired when becoming a professional and uses in her daily practice or the attachment and care a professional musician dedicates to his violin.

in the construction of the 25th of April Bridge (former Salazar Bridge) in Portugal, on the occasion of its 50 anniversary.[21] According to António Rosa, one of these workers, the construction of this bridge, the biggest in Europe at that time, was a real challenge to everyone and its construction site became a kind of second home to those deeply committed to their edification. With more than 40 years dedicated first to its building and then to its maintenance, this worker confessed that he still kept some of the tools he used then, namely a brush.

As Ref. [1] points out tools and implements have become an inalienable part of human existence and human beings have adapted to them from the moment they conceived, designed and produced them. Every tool is designed to make human life easier or more pleasant and to enhance human capacity or creativity in order to produce better work. The ontological dimension of tools, i.e., their nature and instrumentality can be understood exclusively in an anthropocentric sense that is historically determined.

3 Work as a Human Endowment

Different ideological perspectives, distinct epistemological frameworks[22] have converged on recognizing the uniqueness of [work] as a human endowment and its essential character in the definition of what it means to be human.

To Ref. [22], [work] is the unique means through which human beings objectify their existence and come into being, i.e., acquire an identity and a social role.[23] This essential objectivation is to him the essence of humanness. On the close connection between the role one plays as a worker and the definition of a psycho-social identity, Arendt writes [1]: "The moment we want to say who somebody is, our very vocabulary leads us astray into saying what he is […]".

Another fundamental perspective on the essential character of this endowment in the definition of humanness is the Encyclica Laborem Exercens (14 September 1981). This encyclical, written by Pope John Paul II, is part of the larger body of Catholic social teaching tracing its origin back to Pope Leo XIII's 1891 encyclical Rerum Novarum. The Encyclica Laborem Exercens highlights the fact that the capacity for work is an essential human feature that cannot be comparable by its intrinsic characteristics to the performing of certain tasks by other species in order to subsist.

---

[21] https://www.sabado.pt/portugal/detalhe/conhece-o-dono-da-ponte-25-de-abril.

[22] cf. on this purpose [17, 22].

[23] In "Is identity more than a name?" In On Meaning: Individuation and Identity, [6] refers that the construction of a psycho-social identity reinforces the already granted individual biological uniqueness. According to this author, identity formation starts at early infancy and develops throughout the individual's lifetime in a process identical to the formation of pearls. Beginning in the restricted early family circle it develops successively along other spheres of life- the enlarged circle of family and friends, the school and education circle, the work domain, giving this way substance to an identity that is generally actualized by a first name followed by a family name.

> Work is one of the characteristics that distinguishes man from the rest of creatures, whose activity for sustaining their lives cannot be called work [….] it bears a particular mark of man and of humanity, the mark of a person operating within a community of persons. (ibidem, p. 1).

As the encyclical points out [work] is universal in the sense that it embraces “all human beings, every generation, every phase of economic and cultural development” and it is simultaneously a process that takes place within each human being, a personal narrative acknowledged by the conscious subject. Consequently it develops along two fundamental inseparable and complementary dimensions:

1. An objective dimension
2. A subjective dimension

Its objective dimension relates to its generative and transformative power through which human beings act on the surrounding environment- “dominating nature”,[24] “subduing the earth”[25]- and by so doing creating with the effort of their bodies and the intelligence of their intellects the necessary conditions for their “being” throughout the dynamics of an existential historical time.

> [….]there thus emerges the meaning of *work in an objective sense,* which finds expression in the various epochs of culture and civilization (ibidem, 2).

This objective dimension is the tangible or non-tangible existential imprint registered not only by each society but by each of its individual members, from the most notorious to the most anonymous, since individual and collective existence and progress depend on the coordinated action and work of each and all in the different domains of human life.

John Paul II points out that [work] has an ethical value of its own, which clearly and directly remains linked to the fact that the one who carries it out is a person, a conscious and free subject.

> Working at any workbench, whether a relatively primitive or an ultramodern one, a man can easily see that through his work he enters into two inheritances: the inheritance of what is given to the whole of humanity in the resources of nature, and the inheritance of what others have already developed on the basis of those resources, primarily by developing technology, that is to say, by producing a whole collection of increasingly perfect instruments for work (ibidem, p. 6).

On the other hand, the subjective dimension relates to the consciousness every worker must acquire of their personal narrative and of the importance of their individual role, their contribution in a collective process to which all individual efforts converge. It is in its inherent humanity that resides the dignity of [work]:

> through work man not only transforms nature, adapting it to his own needs, but he also achieves fulfilment as a human being and indeed, in a sense, becomes “more a human being (ibidem, p. 9).

---

[24] Ref. [22].

[25] Laborem Exercens 1981.

4 When Tools Become Autonomous: The Ontological Shift

Technological development is the result of the physical and intellectual effort of millions, throughout multiple generations, the result of their creativity and accumulated experience/knowledge, aiming at producing the necessary conditions to liberate individuals from the toil frequently associated to hard work, promoting individual well-being and the society's development, improving life conditions, eradicating poverty and disease, assuring defence against eventual threats.

Laying aside the evident differences inherent to the distinct stages of development that characterize the momentum of the present and those of the past technological revolutions, perhaps the most important feature brought about by the present one is the ***ontological shift***[26] of the concept of [tool]. Until recently, either hand tools or machine tools have been manipulated or operated by human beings, depending on human skill and on their will. Language reflects this instrumental character assigned to the artifacts we handle or manipulate, to achieve a certain goal, in the syntax and semantics of most action verbs, e.g.,

X paints Y with Z
X: human – Agent. Y: a surface – Object. Z: artifact – Instrument.
Though we can have the Instrument in subject position it is never assigned agency, e.g.,
Z paints well.

This is interpreted as Z having intrinsic properties that allows it to be a good tool and not because Z is particularly skiillful or talented.

However, the present tools are becoming progressively more and more independent from human control. By introducing forms of artificial intelligence in the means of production, in work processes, by endowing machines with a form of intelligence that assigns them the capacity to operate and perform tasks independently, capable of decision making, technology is in fact not only introducing a factor of instability in the nature of the relationship between the human being and their tool but potentially interfering in the very nature of a fundamental human dimension- that of productive goal- oriented action, the generative capacity of work.

Martins [21] refers to this as "a technological mutation, that ceases to be instrumental and conceived as an extension of the human arm but merges with human being, producing the very arm and threatening to produce the whole being".[27]

[26] Refs. [8, 9].

[27] My translation.

More than the anticipated huge impact on employment[28]which is object of ongoing studies and monitoring[29] in order to reduce its negative consequences on the labour market,[30]and that in our opinion can be reverted or at least minimized by implementing the adequate social and political measures,[31]it is the potential expropriation of the generative and transformative power from human "hands/minds" that can become an existential problem.

Hal Varian, chief economist at Google, predicted the future in the following terms: "The future is simply what rich people have today. The rich have chauffeurs. In the future, we will have driverless cars that chauffeur us all around. The rich have private bankers. In the future, we will all have robo-bankers [...] One thing that we imagine that the rich have today are lives of leisure. So will our future be one in which we too have lives of leisure, and the machines are taking the sweat? We will be able to spend our time on more important things than simply feeding and housing ourselves?"[32]

These words come, in a way, nearly in line with the prediction made by John Maynard Keynes in *Economic Possibilities for our Grandchildren* (1930: I):

> My purpose in this essay [....] is not to examine the present or the near future, but to disembarrass myself of short views and take wings into the future. What can we reasonably expect the level of our economic life to be a hundred years hence? What are the economic possibilities for our grandchildren?
>
> [...] We are being afflicted with a new disease of which some readers may not yet have heard the name, but of which they will hear a great deal in the years to come–namely, technological unemployment. This means unemployment due to our discovery of means of economising the use of labour outrunning the pace at which we can find new uses for labour. [....]But this is only a temporary phase of maladjustment. All this means in the long run that mankind is solving its economic problem.
>
> I would predict that the standard of life in progressive countries one hundred years hence will be between four and eight times as high as it is to-day [....] Yet there is no country and

[28] Andy Haldane, chief economist at the Bank of England. Predicted, in 2015, that 15 million jobs in the UK, roughly half of all jobs, were under threat from automation. He pointed out that the first industrial revolution had occurred in the middle of the eighteenth century and the second in the latter half of the nineteenth century. The third industrial evolution—the era of information technology—appeared to have resulted in an intensification of trends seen in the first two: "a hollowing-out of employment, a widening distribution of wages and a fall in labour's income share".

https://www.theguardian.com/business/2015/nov/12/robots-threaten-low-paid-jobs-says-bank-of-england-chief-economist.

[29] cf. for instance the work going on at OEDE.AI observatory, GPAI working group on the Future of work. The Work of the Future. Report November 2020. Available at https://workofthefuture.mit.edu/.

[30] https://www.theguardian.com/business/2015/nov/12/robots-threaten-low-paid-jobs-says-bank-of-england-chief-economist.

[31] e.g., (i) involving all the stakeholders in this process of change, identifying the distinct needs determined by their respective functional contexts and tailoring technological solutions according to those specificities. (ii) Legally framing the hybrid working settings legislating in order to preserve not only the physical security, but also the mental health, the emotional stability of those working there; Last but probably first of all (iii) providing massive digital literacy and an ethical background that allows for a critical perspective on the new state of affairs and provides guidelines for human behaviour when acting/interaction with intelligent systems.

[32] https://www.ft.com/content/4329a987-9256-3059-b36f-1aba9338b800.

> no people, I think, who can look forward to the age of leisure and of abundance without a dread. For we have been trained too long to strive and not to enjoy. [...] For many ages to come the old Adam will be so strong in us that everybody will need to do some work if he is to be contented [...]Three-hour shifts or a fifteen-hour week may put off the problem for a great while. For three hours a day is quite enough to satisfy the old Adam in most of us!

Reflecting on the complex equilibrium demanded by the relationship between human beings and machines, Refs. [11, 12] refer to this relationship as a form of human empowerment. In his opinion machines do not limit human faculties as they are modelled on them but take them to a higher level. The fact that they can exalt human capacity to take physical action on our environment, even to the point of enabling new and unnatural functions (such as human flight), goes to show that machines are capable of forming part of a man–machine assembly for the purpose of going beyond boundaries previously believed impossible to overcome.

5 The Future of Work/The Work of the Future: The Way Forward

When one analyses the economic and social predictions and the expected growth estimates made before March 2020,[33] considering how ICTs and artificial intelligent systems would impact economy and society in the years to come, we realize that this type of predictions are exactly what they are—just hypotheses of further economic and social development to be verified if and only if the values of the variables involved don't change. But the reality is that the human environment in all its components—and economy is one of these components- is a complex dynamic system, an ever-evolving entity, whose evolution, that as it happens in the strict physical environment, can be disturbed by multiple factors. The Covid-19 pandemics is certainly one of the strongest causes of huge global economic and social turbulence in decades, causing a disruption whose consequences on the modes of production, on economy and on the social tissue cannot yet be fully anticipated. During the first lockdown and following the trend to introduce automation in management and delivery processes, a trend already initiated much before the pandemics, many companies deployed automation and AI in warehouses, grocery stores, call centers, and manufacturing plants to reduce workplace density and to cope with surges in demand.

According to the February 2021 McKinsey report,[34] Covid 19 has accelerated the already existing trends in remote work, e-commerce, and automation, with up to 25% more workers than previously estimated potentially needing to switch occupations (in the case of advanced economies). According to the evidence collected across the eight countries that were monitored, more than 100 million workers, or 1 in 16, will need to find a different occupation by 2030.

---

[33] cf. https://www.mckinsey.com/featured-insights/future-of-work/skill-shift-automation-and-the-future-of-the-workforce.

[34] https://www.mckinsey.com/featured-insights/future-of-work/the-future-of-work-after-covid-19.

The scale of workforce transitions set off by COVID-19 increases the urgency for policy makers to set up broad multidisciplinary education programmes across the curricula, starting up at the primary level and aiming at the digital literacy of all the students and the shaping of an ethical awareness relative to malicious uses of technology as well as the development of a critical perspective on the ways of their use. This educative action on the part of governance should also comprehend the development of programmes targeting those that are no longer either students or part of the workforce, providing this way for their digital literacy or its update.

On the other hand, business should not blindly start implementing technological transformations on site, but need to develop a systemic assessment of the structure of their organisations in order to:

1. identify the sectors that would, in the first hand, benefit most with the introduction of intelligent tools.
2. identify the adequate artificial intelligent systems for those particular contexts of use in order to have tailor-made solutions and not a general fits-all response.
3. assess the impact of those tools in the sectorial and global dynamics of the organisation
4. identify workers expectations and needs relatively to this structural and functional transformation and act in order to ensure trust and provide well-being
5. provide for the massive training of their workforce in the use of digital and intelligent tools so that each worker at their level of performance has:
    a. a clear insight of what is expected from them at each specific stage
    b. a clear view of how that tool performs
    c. a clear view of the procedures to be followed when operating with it.

A fundamental mindset for both employers and employees is to consider this education and training as an upskilling of previous competences and not as a reskilling. Though these two concepts may look identical, they are really completely distinct in epistemological and sociological terms. While upskilling considers how the set of competences an individual possesses– social, emotional … and those that are inherent to the performing of a certain function- can be even further enlarged through training and knowledge upgrading, reskilling presupposes that the skills the worker is endowed with are no longer suited for the job and have to be replaced by new ones. This is an error as the incorporation of skills and development of competences is a cumulative process that starts in infancy and goes on throughout the individual's lifetime defining the individual's know-how when interacting and performing in any context and circumstance. For instance the skills granted by one's experience as a parent may benefit extraordinarily their performance as a teacher, providing a better sensibility to the behavior and specificities of their students or improve their performance as a manager or executive by developing a better awareness of the difficulties working parents have to face to coordinate family life and work.

In the Future of Work [5] the authors point out that Artificial Intelligence is a disruptive technology that is currently changing, and will continue to radically change, all aspects of our life: the way we work, education and training, and the way we organize work and business models.

On the other hand, discussing the changes brought by robotics and automation in the workspace the IFR 2018 report highlights many cutting-edge technologies such as the concept of man–machine collaboration. According to Esben Østergaard collaborative robots—in its essence man–machine collaboration—are especially compelling as they work together with human workers instead of replacing them.

As the IFR 2018 report refers we can distinguish four distinct modes of robots and humans being in the workspace:

1. Co-existence: human and robot though working do not share the same workspace.
2. Sequential collaboration: human and robot share all or part of a workspace but do not work on a part or machine at the same time.
3. Co-operation: robot and human work on the same part or machine at the same time and the machine is able to adapt to the human.
4. Responsive collaboration: the robot responds in real-time to the worker's motion and/or conversation and is able to adapt to the human.

By the way they are formulated scenarios 3 and 4 raise particular concerns because both scenarios are grounded on concepts that do not apply to machines: that of cooperation and mainly the one of collaboration.

In fact, if we can think of co-operating as possible for a human and a machine simultaneously operating in a process, the same is hardly verifiable in what concerns the concept of collaboration.

When we think of the semantic structure that sustain this concept we realize that there is not only double agency involved, but that this agency has the semantic feature "human" and is involved in a symmetrical relation:

1. X collaborates with Y in order to attain Z
2. Y collaborates with X in order to attain Z
3. X and Y collaborate in order to attain Z

To collaborate involves a favorable disposition on the part of the agents (X and Y) translated in their **willingness** to be part of an effort that reveals itself in an intentional action. The intrinsic humanness that this double agency presupposes makes it unacceptable to apply this concept to a tool, as one cannot also say that the robot experiences a certain mental state, a certain disposition or feeling relatively to an event or circumstance.

This awareness relative to the concepts we use to talk about a new technological reality is important because concepts can be misleading. As machines can never be true collaborators, they are also never workers as there is no such thing as a new division of labour. These are concepts whose substance is intrinsically human, they relate not only to the human reality, but essentially to the nature of human condition.

Reflecting on the complexity of this relation human being/intelligent tool, Ref. [27] refers:

Ironman isn't ironman without the suit, but the suit has no power without the man. That is the future of robotic development, people and robots, working together hand-in-hand to accomplish more than we ever thought would be possible

We can conclude by saying that [work], in all its tangible and non-tangible forms, will continue to be a fundamental dimension of the human being of their humanness, a dimension where intelligent tools will play a fundamental role, empowering and enhancing action in order to produce a hybrid reality where human beings will hopefully not aim to subdue or dominate Nature or their peers, but will aim to be One with Nature and the rest of Humanity.

References

1. Arendt H (1958) The human condition. https://static1.squarespace.com/static/574dd51d62cd942085f12091/t/5eb57ba46186a7719bd8b924/1588951979399/Pages+from+hannah-arendt-the-human-condition-second-edition-2.pdf
2. Beck B (1980) Animal tool behaviour: the use and manufacture of tools by animals. Garland STPM Pub., New York
3. Cassirer E (1944) An essay on man: an introduction to a philosophy of human culture. Yale University Press, New Haven
4. Cassirer E (1996) The philosophy of symbolic forms, vols 1– 4. Yale University Press, New Haven
5. David M, Reynolds E (2020) The work of the future: building better jobs in an age of intelligent machines. MIT task force on the work of the future. https://workofthefuture.mit.edu/wp-content/uploads/2021/01/2020-Final-Report4.pdf
6. Ferreira MIA (2007, 2011) On Meaning: Individuation and Identity. Cambridge Scholars Publishing, England
7. Ferreira MIA (2013) Typical cyclical behavioral patterns: the case of routines, rituals and celebrations. Biosemiotics. Springer. https://doi.org/10.1007/s12304-013-9186-4
8. Ferreira MIA (2018) The concept of [work] in the age of autonomous machines. In Bringsjord S, Tokhi MS, Ferreira MIA, Govindarajulu NS (eds) Hybrid worlds: ethical and societal challenges: proceedings of the international conference on robot ethics and standards 2018. Clawar Association Book Series on Robot Ethics and Standards
9. Ferreira MIA (2021) The smart city: a civilization exponent in the context of a crisis. in how smart is your city? Technological innovation, ethics and inclusiveness, Maria Isabel Aldinhas Ferreira (ed) ISCA Series. Springer, New York
10. Gehlen A (1940) Der Mensch seine Natur und seiene Stellung in der Welt. Junker und Dünnhaupt, Berlin
11. Gehlen A (1957) Die seele im technischen Zeitalter. Hamburg: Rowohlt Taschenbuch Verlag. Jünger, Friedrich Georg. 1956. Die Perfektion der Technik . Frankfurt a. M: Klostermann
12. Global Partnership on AI (GPAI) The Future of Work Working Group. https://oecd.ai/wonk/contributors/gpai-working-group-on-the-future-of-work
13. Greenfield PM (1991) Language, tools, and brain: the development and evolution of hierarchically organized sequential behavior. Behav Brain Sci 14:531–595
14. Grigenti F (2016) Marx, Karl—from hand tool to machine tool, in existence and machine. In: The German Philosophy in the Age of Machines (1870–1960). Springer, Berlin. https://doi.org/10.1007/978-3-319-45366-8_2
15. Hauser M (2000) The evolution of communication. MIT Press, Cambridge, MA
16. Heidegger M (1962) Being and time (trans: Sein und Zeit), 7th ed. Max Niemeyer Verlag, Blackwell Publishing Limited. nternational Federation of Robotics

17. Kant I (1781, 2020) Critique of Pure Reason. The Project Gutenberg ebook of the Critique of Pure Reason. https://www.gutenberg.org/files/4280/4280-h/4280-h.htm
18. Keynes JM (1930) Economic possibilities for our grandchildren. http://www.econ.yale.edu/smith/econ116a/keynes1.pdf
19. Malraux A (1933) La Condition Humaine. Editions Gallimard
20. Martins ML (2011) Crise no Castelo da Cultura: Das Estrelas para os Écrans. https://repositorium.sdum.uminho.pt/bitstream/1822/29167/1/CriseCastelodaCultura.pdf
21. Marx K (1867) Das Kapital. https://www.marxists.org/archive/marx/works/download/pdf/Capital-Volume-I.pdf
22. Marx K (1973) On human nature in social research vol. 40, no. 3. Human Nature Reevaluation (Autumn 1973) John Hopkins University Press
23. Mckinsei (2020) The future of work in Europe. Discussion paper. https://www.mckinsey.com/~/media/mckinsey/featured%20insights/future%20of%20organizations/the%20future%20of%20work%20in%20europe/mgi-the-future-of-work-in-europe-discussion-paper.pdf
24. Merleau-Ponty M (1945, 2020) The phenomenology of perception. Taylor & Francis, London
25. OECD https://read.oecd-ilibrary.org/social-issues-migration-health/good-jobs-for-all-in-a-changing-world-of-work_9789264308817-en#page1
26. Paul II J (1981) Encyclica Laborem Exercens. http://www.vatican.va/content/john-paul-ii/en/encyclicals/documents/hf_jp-ii_enc_14091981_laborem-exercens.html
27. Shumaker RW, Walkup KR, Beck BB (2011) Animal tool behavior: the use and manufacture of tools by animals. Johns Hopkins University Press, Baltimore
28. Sikka S (2018) Heidegger moral and politics, questioning the shepherd of being. Cambridge University Press, New York
29. Smith A (1976) The wealth of nations. In: Campbell RH, Skinner AS (eds) The Glasgow edition of the Works and Correspondence of Adam Smith
30. von Uexküll J (1926) Theoretical biology (trans: Mackinnon DL). Harcourt, Brace, New York
31. von Uexkuhll J (2010, 1940) A Foray into the Worlds of Animals and Humans. University of Minnisota Press, London
32. Young RW (2003) Evolution of the human hand: the role of throwing and clubbing. J Anat 202(1):165–174

Human Robot Collaboration in Industrial Environments

George Michalos, Panagiotis Karagiannis, Nikos Dimitropoulos, Dionisis Andronas, and Sotiris Makris

Abstract The advancement of robotics technology over the last years and the parallel evolution of the AI, Big Data, Industry 4.0 and Internet of Things (IoT) paradigms have paved the ground for applications that extend far beyond the use of robots as mindless repetitive machines. The number of technical configurations/solutions grows exponentially when considering factors such as (a) the particularities of the task to be performed (e.g. type of part, weight, dimensions, process to be carried out etc.) (b) the type of robots that can address these requirements (fixed or mobile robots, high/low payload, exoskeletons, aerial robots etc.), (c) the type of collaboration and interaction that would be appropriate for the task and (d) the special requirements of the production domain where such tasks are needed. This chapter aims to present the existing approaches on the implementation of human robot collaborative applications and highlight the trends towards achieving seamless integration of humans and robots as co-workers in the factories of the future.

1 Introduction

Industrial robots have greatly demonstrated their fitness in serial production lines where they have efficiently undertaken a multitude of production operations thanks to their ability to be reprogrammed and repurposed [1]. Their high power, accuracy and repeatability allows them to reliably carry out small tasks which are key for reducing the overall cycle time and achieving high technical availability [2].

When comparing traditional industrial robots such as fixed manipulators, gantry robots or parallel kinematic machines to the most flexible production resource, i.e. a human operator, the capabilities for autonomous and adjustable operations are quite in favour of the latter (Table 1).

G. Michalos · P. Karagiannis · N. Dimitropoulos · D. Andronas · S. Makris (✉)
Laboratory for Manufacturing Systems and Automation, University of Patras, Patras, Greece
e-mail: makris@lms.mech.upatras.gr

M. I. Aldinhas Ferreira and S. R. Fletcher (eds.), *The 21st Century Industrial Robot: When Tools Become Collaborators*, Intelligent Systems, Control and Automation: Science and Engineering 81, https://doi.org/10.1007/978-3-030-78513-0_2

Table 1 Qualitative comparison between industrial robots and human operator attributes

	Industrial robots	Human operator
Configurability for different tasks	Medium—Usually enabled by adding tooling, peripherals and re-programming	High—May require training on the task and equipment. Can be supported by information systems
Mobility	Low—Fixed robots are most common in industrial practice with the exception of AGVs	High—Can freely move between stations, use stairs, access narrow spaces etc.
Sensing	Medium—Low: Specific to the sensors used the application. Cannot use multiple sensors at once but excellent performance	Medium—Simultaneous use of multiple senses directly linked to motions. Limited number and range of each sense
Manipulation ability—dexterity	Dependent on the tooling (non-universal) and integrated sensing/control systems	Very high—Broad range of parts. Effective for small sized complex geometries and deformable objects
Perception	Low—Tailored to the executed process. High bandwidth performance for focused tasks (e.g. visual servoing)	Very high—Combining multiple senses to perceive process/environment—low accuracy
Cognition abilities and decisional autonomy	Low—Limited to pre-programmed scenarios and behaviours that must be repeatable for certification	Very high—Can identify new conditions/states, reason over them and act or adapt behaviour
Interaction capability	Restricted to HMIs and operation in fenced environments. Several collaborative robot applications	Very high—Can communicate with other operators in multiple modalities (verbal, gestures etc.)
Dependability	Very high—Mean time between failure in the range of decades Very robust error/quality detection systems	Medium—Low: Higher error probability, absenteeism and external environment influence
Power	Very high—Industrial robots exhibit payloads up to 1 ton	Low—depending on the physical attire of the individual
Accuracy	Very high—depending on the payload and feedback devices can achieve up to micron level accuracy	Low—In the mm range, can be enhanced with the use of supporting devices engineered for the process
Repeatability/Speed of execution	Very high—especially for simple pick and place operations—can be sustained throughout all operation	Low—Cannot be sustained for long periods. More efficient for complex/deformable objects

The paradigm shift from mass production to mass personalization indicates that the high speed and technical performance of robots is not enough to guarantee sustainability of robots in all sectors. Simply relying on the programming of new trajectories or equipping manipulators with new tooling is not enough for operating in a dynamic production environment. Further enhancement of the robotic systems is required, to exhibit more of the human operators qualities.

Under this notion the principle of Human Robot Collaboration and the concept of Robotic Co-workers (also referred to as cobots, collaborative robotics etc.) [3] has been researched, aimed at enabling robots to work alongside human operators and seamlessly cooperate with them. However, the target of creating a single robot type that is able to efficiently work in any type of environment, carrying out any task and behaving at the same autonomy levels as a human is still quite distant. For this purpose, step changes are pursuit to achieve the following conditions which are considered critical for true HRC to take place:

1. Development of safety related technologies that will allow industrial robots to operate in fenceless environments and/or robots that are intrinsically safe to operate around humans.
2. Deployment of appropriate robotic systems depending on the requirements of the production environments (fixed high payload robots, mobile manipulators, drones, exoskeletons).
3. Integration of Human Centred Interfaces which allow direct, natural and efficient 'human to robot' and 'robot to human' communication. The interface should be "invisible" during the interaction with the humans and ensure that physiological aspects are not negatively compromised [4].
4. Enablement of robotic perception to enable awareness of the tasks, environment and humans and cognition capabilities that allow them to reason upon the perceived states and adjust their behaviour for balancing task execution and interaction efficiency as a human operator would do.

Figure 1 provides a visual representation of the key differences between the

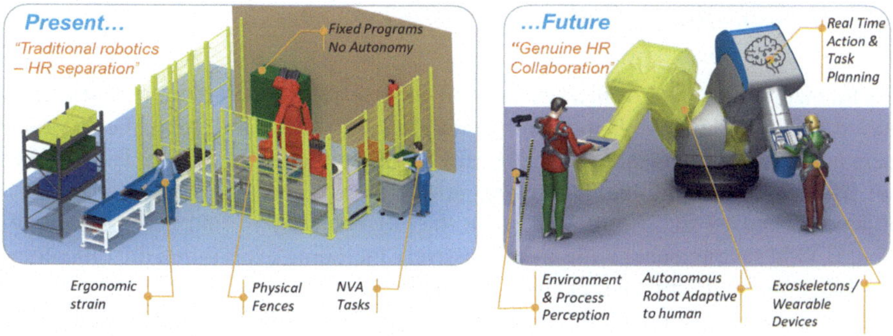

Fig. 1 Transition from industrial robotic system to human robot collaboration

existing industrial practice and the vision of Human Robot Collaborative production paradigm.

Already the first applications for introducing collaborative robots in different industrial sectors have been identified and are reviewed in Sect. 2 of this chapter. The aim is to provide an overview of the different approaches in terms of the deployed robotic systems and their collaboration capabilities as well as the diversified industrial requirements they are addressing. Based on this analysis and considering the latest scientific and technological advancements, Sect. 3 provides an outlook on how HRC systems are expected to evolve and which are the key enablers for an efficient and sustainable uptake of this technology. Finally, Sect. 4 summarizes the key findings and outlines areas of future work for this very promising field of industrial robotics.

2 Current State

In the previous section a brief introduction has been provided, showing the strengths and weakness among the two main resources that play key role in the manufacturing systems: the industrial robots and human operators. Researchers are working on creating a new industrial paradigm in which the benefits from both the aforementioned resource types would be, ideally, combined. The challenges that should be overcame to do this are numerous. They depend on the type of task that needs to be executed, the type of hardware required for a specific operation, the level and type of collaboration among the two resources required to achieve better results etc.

This chapter aims to discuss about the level at which technology has reached until now, in terms of key concepts, principles, methods and tools. Using the experience that researchers have so far, what are the best practices, the outstanding gaps and the new challenges that have emerged with the progress so far and the real-world problems that exist nowadays.

2.1 Human–Robot Collaborative Tasks

Starting with an analysis from a task point of view, several parameters are used to categorize a task as a Human–Robot Collaborative (HRC) one or not. Also, these parameters help the designers decide whether the human–robot collaboration is needed or not. The two most important parameters are space and time. If those two factors can be ameliorated by the use of HRC then the engineers try to find a way to implement it. Other parameters, without exhaustively including everything, are efficiency, flexibility, productivity, human stress and workload [5–7].

Firstly, focusing more on time and space parameters, HRC tasks are based on the principle of having two types of resources, namely human and robots, sharing the workspace and/or the time to perform a specific task, depending on the case [7, 8]. As a result, four cases of human robot collaboration are foreseen and visualized in

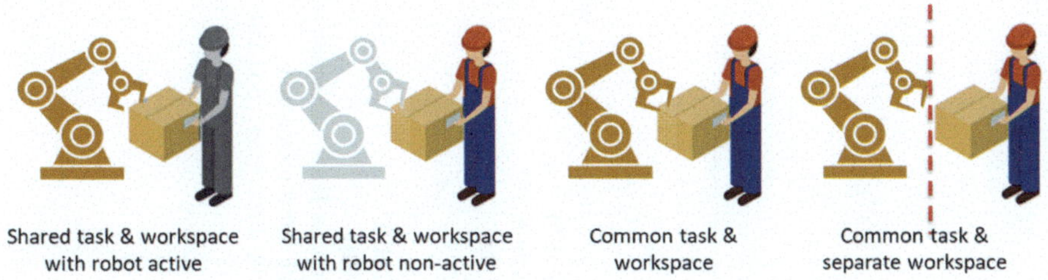

Fig. 2 Human robot collaborative task types

Fig. 2: The robot and the human operator could have a common task and workspace, a shared task and workspace, or a common task and a separate workspace, with either the robot or human active. Nevertheless, it should be mentioned that in all cases both human and robots are dedicated only on one task, allowing the human having multiple roles, acting as supervisor, operator, teammate, mechanic/programmer and bystander, depending on what he/she is supposed to do [9].

Moving on to the other parameters, namely efficiency, flexibility, productivity, human stress and workload, hybrid production systems aim to improve them to ameliorate the working conditions and quality of products. The manual assembly lines, are human-centred, causing a lot of stress to the operators and fatigue in the long run. Despite the high flexibility and cognitive capabilities human operators possess, in the long run this affects their physical and mental health, making them prone to errors [10, 11]. In order to avoid such cases, or at least restrict them, using robots as assistance in the production seemed the next sensible step [11]. Nevertheless, using just robots is part of the solution. Designing a hybrid cell by selecting and planning the tasks that can be considered collaborative, based on the aforementioned parameters, is of high importance [12].

2.2 Human Robot Interaction Types

The expansion of HRC applications in robotics has initiated an attempt of normalizing and standardizing the interaction levels between humans and robots. According to ISO 10218-1:2011, "collaborative workspace" is a "workspace within the safeguarded space where the robot and a human can perform tasks simultaneously during production operation". The types of HRI are therefore depended by the methodology on: (a) how the collaborative workspace is allocated or shared and (b) how operators and cobots work or collaborate during operation. In recent years, academia has published reports where interaction levels are categorized. The outcomes of some indicative works are summarized on the following Table 2. Some other approaches classify the interaction types based on "workspace" and "time sharing" aspects [7, 13].

Table 2 Human robot interaction classification types

	Level classification considerations						
Method reference	Robot status	Shared workspace	Shared tasks	Physical interaction	Safety measures	Mobility	Types of interaction
Michalos et al. [14]	✔	✔	✔	✔			i. Common task and workspace ii. Common task and separate workspace iii. Shared task and workspace with active robot iv. Shared task and workspace with non-active robot
Bdiwi et al. [15]		✔	✔	✔			i. Level 1: Shared workspace (WS) without shared task (ST) ii. Level 2: Shared WS with ST without physical interaction (PI) iii. Level 3: Shared WS with ST with Handing over of part iv. Level 4: Shared WS with ST with PI
KUKA Robotics [16]		✔	✔	✔	✔	✔	i. Fixed safety fence ii. Laser or virtual guarding separation with occasional HRC iii. Laser or virtual guarding separation with intended HRC iv. Shared workspace with no laser or virtual barrier without direct interaction v. Shared workspace with no laser or virtual barrier with direct interaction vi. Shared workspace with no laser or virtual barrier with direct interaction and mobile robot presence

In 2016, the International Organization for Standardization has published TS/ISO-15066 and classified the interaction levels into four categories: (a) safety-rated monitored stop (b) hand guiding (c) speed and separation monitoring (d) power and force limiting. Each industrial application applies different interaction levels and human robot collaboration strategies. The selection of those levels is clearly depending on the manufacturing process itself and the semi-automation concept that is implemented. However, the types of interaction have a huge effect on the sensorial system as whole and more specifically on the safety devices. Aiming industrial implementation, the safety system needs to comply with standards and regulations. According to ISO 10218-2:2011, all safety related parts the system should conform with "Performance Level d, Category 3" specifications. Devices of this category ensure that any kind of fault or error is detectable by the system itself and does not cause any loss of safety function. Safety devices and tools for monitoring of the "collaborative workspace" are mostly optoelectronic devices, radars, safety mats, etc. Regarding, power and force limiting, force sensors or safety skins are mostly applied in cases where such features are not directly integrated on the robot arm. In terms of interaction, commercial cobots come with standard features including soft edges, intuitive HMIs and buttons and in some cases protective skins. Starting from the touch sense, the most direct HRI method that exists is the one at which the operator uses his/her bare hands to move the robot freely in space. This can be achieved either by using a Force/Torque (F/T) sensor or the current values from the motors. More specifically, in the first case, the operator grabs the F/T sensor installed at the robot flange or at the gripper and pushes it towards the desired direction. In contrary, following a sensorless approach, the current of the robot motors is constantly monitored and the values are compared with the expected ones, based on the robot pose, speed, payload. In case of significant deviation, a collision is implied. This technique is called power and force limiting and is described in [19]. More information about pioneer and future human robot interaction interfaces is provided in Sect. 3.

2.3 Collaborative Robots and HRC

Robotic systems that have been engineered to serve hybrid production system include a wide range of machines that can either operate on their own or passively support the operators. This section provides an overview of the robotic solutions encountered so far, especially in industrial settings, as well as more innovative ones that have been developed for hybrid environments.

The following Table 3 provides a summary of highlighted products in the domain of collaborative robots, when considering mobility, number of arms and payload as the main variation factors. It is not aimed to be an exhaustive list, but to serve as a quick example finder under each category.

As it will be described in Sect. 2.3.4, apart from the aforementioned, there are other types of robots that can be used for HRC purposes as well. Such case are the drones which present different characteristics compared and their development is still under

Table 3 Commercial collaborative robots

Mobility	Brand and model	Number of arms	Degrees of freedom	Payload per arm	Safety Skin
Fixed	Universal Robots UR3/UR5/UR10/UR16	1	6	3/5/10/16	
	KUKA LBR iiwa 7 R800/14 R830	1	7	7/14	
	FANUC CRX-10iA	1	6	10	
	FANUC CR-35Ia	1	6	35	✔
	ABB Single-arm YuMi—IRB 14,050	1	7	0.5	
	YASKAWA MOTOMAN HC10/HC20XP	1	6	10/20	
	Franka Emika Panda	1	7	3	
	Rethink Robotics Sawyer	1	7	4	
	COMAU AURA	1	6	170	✔
	ABB YuMi—IRB 14,000	2	14	0.5	
	Rethink Robotics Baxter	2	14	2.2	
	EPSON Bertie Dual Arm	2	14	3	
	Robotnik MRP	2	14	10	
	KUKA KMR iiwa	1	7	14	
	KUKA flexFELLOW	1	7	14	
Mobile	MIR iAM-R	1	6	10	
	HIROTEC OTTO	2	6	3	

research. Thus, there are no commercial solutions for HRC scenarios. Nevertheless, there are commercial products that are used for other purposes inside and outside the manufacturing lines such as survey and mapping purposes. Indicative examples are the DJI Matrice 300/200 series as well as the Yuneec H520/H920 models.

2.3.1 Fixed Manipulators

Starting from the most standardized and well-known technology, that of stationary robotic arms. These machines are fixed on the ground with a certain range, with multiple degrees of freedom (usually 5–7Kgs) and payloads spanning from few kilograms up to one ton. They have dominated production environments due to the high levels of speed, repeatability and accuracy they can achieve in their tasks. It would make sense that these manipulators have inspired the starting point that robots can act as co-workers. Nevertheless, the safety requirements for allowing them to operate outside fences or close to human operators have posed great challenges to robotics OEMs and system integrators [14]. These challenges originate from (a) the fact that an industrial manipulator itself is a high powered machine that can exert high amounts of force/torque and is not equipped with appropriate, safety certified hardware to limit them when human injury is probable, (b) the tooling of the robot which may also contain high power actuation (pneumatic, electrical, hydraulic, mechanic) and elements that may potentially be hazardous (blunt edges) (c) the lack of safety rated monitoring systems that can track the working environment of the robot under any circumstances and ensure that no harmful action for humans is allowed and d) the existence of appropriate legislation and standards to cover the certification and liability aspects of such systems.

In order to overcome the aforementioned challenges, robot makers have followed two different paths: (a) the creation of new robots and (b) the enhancement of the existing ones. Thus, when discussing articulated manipulators, there are two main lines of development: low payload collaborative arms and medium—high payload manipulators with external sensorial and mechatronic enhancements.

The first category involves low payload robots that employ direct drive motors and/or embedded force/torque sensors inside each joint allowing them to measure/identify contact with objects or humans in the vicinity. Unlike the case of higher payload robots, the absence of gearboxes and reduced masses of the links allows direct identification of external contacts more efficient adaptation of the robot control that minimizes the probability of injury through high force application (clamping, high velocity impact or other). The approach however comes at a cost of payload (between 1 and 35 kgs) and speed of operations which is not comparable to the capabilities of large industrial robots (Fig. 3).

To overcome such limitations and truly take advantage of the benefits of industrial robots, recent research and technology development activities have focused on ways to enable safe HRC either by equipping these robotic arms with extra sensing modules or by supplementing their control systems with external safety related data [20]. The most typical examples of such applications are (a) the development of robot skins that cover the main robot surface and are able to identify human presence through

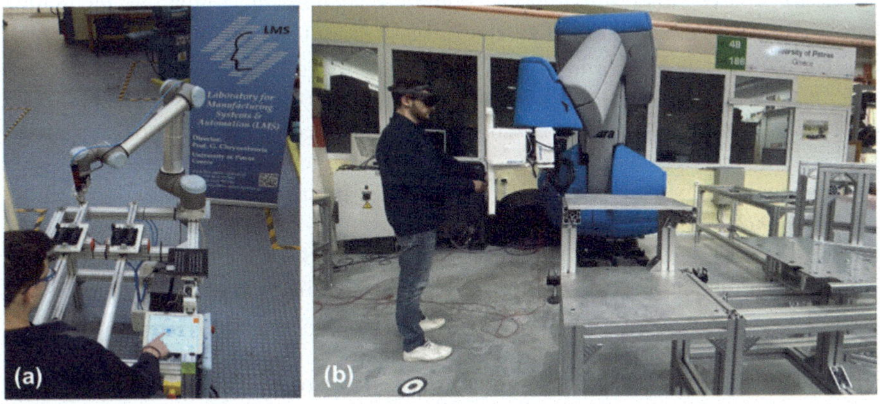

Fig. 3 **a** Low-payload collaborative arm, **b** High-payload collaborative manipulator

contact (pressure sensitive), capacitance, heat or other principles [21, 22] and (b) the introduction of cameras, laser based sensors and pressure sensitive floor around the robot that undertake the task of identifying humans and notifying the robot control system.

A further distinction of robots aimed at HRC come from the number of arms that they are equipped with. By observing the flexibility of humans to perform bimanual operations, research has tried to mimic this function by combining two or more manipulators under a common control scheme [23].

As discussed in Sect. 2.1, HRC can be implemented in multiple forms (coexistence, collaboration etc.) and thus the notion of collaborative robotics does not strictly involve robots in direct physical contact with operators. In that sense, and provided that the safety requirements foreseen by the international standards are met [14], HRC applications can be carried out in industrial shopfloors where fences can be eliminated and operators may enter the workspace of the robot which is supervised by means of external sensors such as overhead cameras, laser scanners etc. Indicative examples are the collaboration between industrial dual arm robots and humans where static virtual fences are implemented [24] as well as the case of robot workspace segmentation and dynamic monitoring through overlaying real robot position and human detection data from safety certified sensors [20].

2.3.2 Mobile Robots

The main flexibility constraint of industrial robots has been the ability to be easily relocated and maintain their accurate programming in a short time period. Nevertheless, concepts of highly reconfigurable shopfloors using mobile robotic resources have been investigated quite extensively in the recent past [25].

Humanoid robots, unmanned rovers, entertainment pets, and so on are great examples of mobile robots. Nevertheless, in the industrial paradigm the most commonly used is the mobile platform [26]. It consists of a surface equipped with wheels and sensors to achieve either autonomous or guided, such as line-following, navigation in its working area. Commercial products that belong to this category are the, MIR [27], KUKA [28] and ARTI3 [29].

A more advanced concept includes the introduction of mobile platforms equipped with robotic manipulators to the assembly lines and execute a number of operations, previously done by fixed robots (Fig. 4). The advantage of such approach is the fact that the robot can be easily replaced with others, in case of breakdown or failure, as well as avoiding using fixed conveyors and fixtures to hold the parts. Additionally, multiple robots can collaborate together, by working on the same part or by exchanging tools and parts using tool changers. Commercial products that belong to this category are KUKA KMR iiwa [30], KUKA KMR QUANTEC [31] and Robotino with arm from FESTO [32].

Another concept is based on a similar idea as described in the previous section of the fixed industrial robots. Following the introduction of a single arm manipulator on a mobile platform, the next logical step would be the creation of a mobile platform

Fig. 4 Mobile robots concept: (1) A group of robots work on the same part (2) Two robots exchange the part (3) Welding spot robot ready to work on the part

equipped with a multi-arm robot. This will give higher flexibility to the resource since this idea increase its capabilities. More specifically, the new concept allows the robot to carry heavier parts using two or more arms, equip more sensors such as a camera in the one arm while the other arm is equipped with a tool, as well as work on two or more different parts at the same time using one arm per part [33]. Commercial products that belong to this category are KUKA YouBot in the respective configuration [34] as well the TIAGo++ from PAL robotics [35] (Fig. 5).

Moving to a different concept of mobile robots to that of crawl and snake robots. The advantage of these robots is that they can easily be controlled remotely in very narrow areas and be equipped with sensors providing streams of data for processing. Such robots are mainly used in the Inspection and Maintenance domain, where small areas should be inspected, such as water or drainage pipes. Commercial example of snake arm robots has been provided by OCRobotics [36].

Fig. 5 Mobile platform equipped with two robot arms and sensors

2.3.3 Exoskeletons

Another interesting application of industrial robotics is that of wearable robotics, namely exoskeletons. This invention emerged from the fact that human operators can be assisted by robotic solutions in a more direct way, enhancing their strength and minimizing their physical risks. Worldwide, a significant interest in industrial exoskeletons does exist, but a lack of specific safety standards and several technical issues hinder mainstay practical use of exoskeletons in industry. Specific issues include discomfort (for passive, semi-active and active exoskeletons), weight of device, alignment with human anatomy and kinematics, and detection of human intention to enable smooth movement (for active exoskeletons) [37].

Nowadays, human operators are exposed in a lot of hazardous situations during the execution of production operations in an industrial environment, such as performing repetitive assembly tasks, carrying heavy parts and tools, having non-neutral postures, undergo hand, arm or even whole-body vibrations etc. Taking into consideration the ageing of the working population in the European countries, exoskeletons are considered to offer a huge support to the working population minimizing the work related Musculoskeletal disorders (MSDs) [38].

Nowadays, commercial exoskeleton solution exists for multiple purposes, apart from operator support in industrial lines, such as rehabilitation, walking assistance, entertainment etc. [39].

2.3.4 Other HRC Capable Robot Types

Based on the specific circumstances, more types of robots have been created, which adapt and cover the special conditions such as underwater robots, flying robots, in large working areas etc. These robots do not belong to the category of industrial robotics, nevertheless they could be used in a similar way to form an HRC application.

More specifically, drones are a relatively new technology that is gaining popularity [40]. The evolution of control systems enables the better control of the drones as well as having bigger payloads [41]. Thus, they can be used to carry tools to the operators in big heights or to be used as operators to assist the humans on the worksite. Additionally, drones can be controlled remotely by humans to navigate in difficult areas or places that are dangerous for humans such as radioactive areas. Last but not least, the combination of multiple drones, namely swarm of drones, can enhance their capabilities and be more effective in their activities [42].

Furthermore, cable robots are an interesting application of robots that can be easily installed and used in large areas. This invention has been used initially for entertainment purposes and it is not so popular in the industrial area. Nevertheless, although cable robots are hardly used in practical industrial applications, there is a number of prototypes with promising results [43–45].

2.3.5 Industrial Applications

In current industrial practice several examples of collaborative robots can be found. An exhaustive list would not be efficient to provide however the following examples provide a good overview on existing deployments and applications.

Automotive: Indicative applications include mainly low payload robots for handling of small components (e.g. differential case handling at BMW [46]), tightening (e.g. screwing on car motors at CSVW [47]) while humans work on it or adjustment (e.g. fog lights tuning at FORD [48]). The automotive is also a pioneer in the use of exoskeletons for supporting work positions with ergonomic strains such as the assembly of underbody panels [49].

Electrical/Electronics: Assembly of electronic components covering large families of products with different specifications and dimensions has been covered by collaborative robotics in the QISDA group [50] while dual arm SCARA type robots working without fences next to humans have also been deployed [49]. This was the domain where collaborative robots debuted as in the case of ASM where the Sawyer robot was installed for packaging and manufacturing of electronic components [51].

Plastics: Indicative examples include the use of collaborative robots for machine tending applications as in the case of Swedish polymer manufacturer Trelleborg [52] and the execution of trimming in blow moulding operations at BBM, Germany [53].

Food processing: applications mainly focus on packaging as in the case of UR robots installed in Atria facilities [54] or food preparation cobots such as the one installed at Huis Ten Bosch in Japan [55].

Aerospace: in aerospace a plethora of applications are now served by collaborative robots such as the inspection of large composite parts [56], assembly of subcomponents (e.g. actuation systems at Whipanny [57]) or even largest aircraft assemblies as in the case of BAE systems [58].

Metal production: Reported applications in this area include metal parts rinsing and handling of metallic parts (e.g. in Tool Gauge company based in the US [59]) all the way to assisting in sheet metal production as in the case of SFEG company where 20% productivity gains have been observed [60].

In addition to the aforementioned already operating at the shopfloor scenarios, other sectors have already started the investigation of collaborative robotics potential for their production in the form of research projects:

White goods: recent applications showcased two operators working in close collaboration with two medium payload industrial arms performing pre-assembly and sealing operations on refrigerator's cabinet [61].

Elevators production: experimentation with utilization of a high-payload collaborative robot (COMAU AURA) lifting, handling and presenting to a human operator heavy metal plates, while the operator focuses on installing various smaller metal parts is ongoing [56].

Industrial modules: such application involve the utilization of low payload collaborative arm, seamlessly coexisting with humans, performing repetitive riveting tasks, while the operators are focusing on actions that require human dexterity [56].

Heavy machinery production: recent applications explore the utilization of upper limb smart exoskeletons assisting operators while lifting up-high, positioning and assembling large and heavy metal parts for machine tools production [56].

3 Future Trends and Transformations

Over the last years both Research and Industry have tried to address the requirement for flexible production by introducing technologies that allow humans and robots to coexist and share production tasks. The focus has been to ensure the safety of humans while interacting with robots. It was revealed though that HRC applications present drawbacks that limit industrial adoption:

- **Diversified needs for robotics by the production tasks**. Robot arms are not able to cover all applications scenarios (rigidity, adaptability, payload, working envelope etc.) and soft robotics may seem preferable.[1]
- **Low performance of collaborative operations**. In the scenarios demonstrated so far, humans can perform tasks faster when working on their own as are robots when they function independently. This is due to:
    - o **Complexity of safety systems** that create more separation: Resolution, response time and robustness in uncertain conditions cannot be handled by existing safety/sensing/perception technologies and hardware/software implementations.
    - o **Inefficient design of the system** as HRC hazards are not systematically evaluated in the design process.
    - o **Operator acceptance and efficient integration in the workflow**: People are not used to work cooperatively with robots as they would with other humans. Workers need to develop familiarity and trust in robots through intuitive interaction and accurate anticipation of robot actions/intentions.
- **Lack of cognition for autonomy**: the robot cannot adjust its behaviour to shape its operation around the human behaviour. There are no structured methods to validate implementations of autonomous behaviour.
- **Programming of the robots is not user friendly, requiring programming skills knowledge** and is detached from the method of interacting with it during execution.

[1] Liyu Wang, Surya G. Nurzaman and Fumiya Iida (2017), "Soft-Material Robotics", Foundations and Trends in Robotics, 5(3), 191–259.

Cyber physical systems integration

Nowadays, manufacturing systems have to comply with the arising need for customized products, while maintaining fast and low-cost production as well as high level of flexibility and adaptability to changing production requirements. Cyber-physical systems (CPS) are being used in multiple ways, aiming to achieve a seamless and efficient HRC.

One application of CPS aims the estimation of the high-level status of the resources within the cell. Data from sensors installed around the shop floor as well as output from various runtime modules are collected, fused and analysed, providing an accurate representation of the cell status. The analysis results can then be utilized to trigger other processes at coordination or business level [62].

Another useful application of CPS is focused on estimation of the low-level actions that the operator is performing. Artificial intelligence (AI) enhanced vision-based recognition is used to detect the various objects of interest, namely the parts that are involved in the assembly as well as the hands of the operator. By continuously tracking the hands of the operator and detecting the parts that he/she is interacting with, the following can be achieved: (i) estimation of current assembly step, (ii) verification of assembly correctness, (iii) prediction and prevention of assembly errors prior happening. This information can be fed to task and action planners to efficiently plan the next course of actions as well as to AR based operator support tools providing visual indication to the operator in case of assembly errors or automatically updating the assembly instructions in case a of correct task execution [63].

Besides increasing the efficiency of the HRC workplaces, CPS are being used to boost their safety which is a major concern as the same workplace is shared between humans, static robots and auto-guided vehicle. Data from distributed sensors network installed around the shop floor are collected, fused and analysed in terms of safety distance between operators and robotic resources, allowing closed loop control algorithms to perform collision preventative actions [64].

Flexibility through HR task planning

Having multiple resources working at the same work cell, a new challenge arises for the workplace engineers related to the allocation of tasks among the resources as well as the design of the workplace to facilitate the HRC nature of interaction. Multi-criteria decision-making algorithms are being used to formulate the alternative layouts as well as task allocations among the resources, utilizing analytical models and simulations to estimate the criteria values and evaluate the various alternatives [65, 66]. The evaluation criteria are usually set based on the targets of the production line including, among others, cycle time, ergonomic values such as operator walking distance and RULA based metrics, non-value adding activities time, resources utilization percentage, safety focused criteria such as HR separation distance [67]. Various approaches for alternative generation are being used either based on analytic programming or heuristics producing similar results, differentiating a lot in alternative generation speed though.

Despite producing an efficient initial task allocation though, manufacturing systems need to be as responsive as possible to unexpected changes occurring in the shop floor. The task planning solutions should be able to continuously adapt to the competencies and preferences of operators, unforeseen machine breakdowns and changes to the production plan. This means that they should run continuously through the whole shift, analyse current production data, exploit data from sensors installed around the shop floor, analyse human presence and intention and model the robot reaction to it. In such way, efficient utilization of the available resources will be achieved.

3.1 Novel Interaction Means

In the future, in a hybrid cell a human and a robot are considered co-workers. They are sharing tasks, workspace, parts, even in some cases tools. In order to ensure a smooth collaboration, an intuitive mean of communication should be established. For this reason, a new research subject has emerged, where a lot of technical developments have been performed, that of Human Robot Interaction (HRI). There could be many types of HRI that are basically human-centred, like in HRC, and more specifically take full advantage of human senses and capabilities, such as hear, touch, speech, vision etc. The key target of HRI is to digitalize for the robot the information that comes from the human and the opposite, meaning give to the human, any digital information that is produced by the robot. The more intuitively and fluently this transition takes place, better the communication and collaboration among the partners that occurs [66].

Starting from the touch sense, the most direct HRI method that exists is the one at which the operator uses his/her bare hands to move the robot freely in space. This can be achieved either by using a Force/Torque (F/T) sensor or the values of the current from the motors. More specifically, in the first case, the operator grabs the F/T sensor and pushes it towards the desired direction. The sensor understands the forces and torques applied on the sensor and translates them to movement for the robots' end effector, leading to the movement of the arm [17, 18]. On the other side, in the sensor less approach, smart algorithms are used to track the values of the current in the robot's motors. Any deviation from the expected values mean that an impact occurs, leading to robot emergency stop. This technique is called power and force limiting and is described in [19].

Moving on to the other human senses, using speech and hearing senses is another way to interact with a robotic system. The operator can use a headset to give verbal instructions to the robot and receive feedback in a similar way. Usually, in a simpler form, the operator can use a predefined list of instructions that form a dictionary, while in more advanced and smart algorithms the software can recognize more complex instructions from the operator [61].

Concluding with the sense-based interaction methods, one of the most intuitive HRI technologies is the Augmented Reality (AR). The operator is using a wearable device that projects digital information in his/her field of view. Based on the stereoscopy effect, the user can see 3D objects such as parts, safety zones and robot's trajectory. These functionalities aim to support operators in the assembly lines as well as increase his/her safety awareness for potential hazards [68–70]. Additionally, the operator can see messages in the form of text that provide alerts and information regarding the production. This technology is considered the most immersive, since it blends virtual and real-world components, targeting one of the strongest human sense, that of vision.

Last but not least, another form of HRI, as described in the introduction, is by using human movement and posture to perceive his/her activity or receive commands. More specifically, the operator can use his body and create specific signs in front of a camera and then a smart algorithm can translate those signs to robot commands [71, 72]. Apart from this, similar to the previous approaches which understand only a predefined set of instructions, there are more sophisticated algorithms that can understand human activity, such as passing nearby, performing a process such as assembly or screwing, or even idling. Having this understanding, the robot and the system in general, can adapt its activity based on what the human operator is about to do [73].

All the aforementioned HRI methods are still in the lab level. Technical as well as legislation constraints prevent them from being widely used in the manufacturing lines. Ones the technology is mature enough to ensure human safety, as it is ensured right now by the standards as described in Sect. 2.2, it would enter the assembly lines of the future.

4 Conclusion

This chapter has provided an overview of robotic applications aimed at redefining the role of industrial robots as co-workers and collaborators in realistic production environments. The review has covered attempts to enable traditional robotic systems to act in a collaborative way as well as the conception and development of robots that are particularly built for the purpose. Although great leaps have been observed in individual technologies such as robotic cognition, machine vision, interfaces etc. there is still a lot of ground to be covered before seamless human robot collaboration can be achieved. From the industrial perspective and considering the technological state presented in this chapter, there are several requirements that need to be met, including but not necessarily limited to:

- **Easiness in designing, deploying and operating HRC systems that are inherently safe**: The integration of technological enablers for HRC has proven to be very complex across the whole lifecycle of such project. Different tools, protocols, vendor specific functionalities etc. limit to a great extent the capabilities of HRC systems which are, in most case, programmed as standard robots which are limited to tolerate the existence of humans around them. When considering the safety certification aspects, AI and similar technologies that would make a robot behave more like its human counterparts, are automatically discarded from the options. In that sense proper tools that can integrate the autonomous behaviour of robots in the design stage and allow to program behaviours rather than routines are becoming more and more important.

- **Handling non-expected situations—Adjusting uncertainty in real time**: Apart from the complexity in designing and implementing autonomous behaviours, environments where humans operate are in inherently dynamic and unstructured. While humans may vary their actions (sequence of execution, way of performing a task etc.) depending on their physical and psychological status at that time they can easily adjust to the new state of play by using their intuition and senses and by naturally communicating with their co-workers. Such behaviours are crucial to be achieved if robots aspire to work equally efficiently at the side of human workers.
- **Improvement in operator's working conditions**: When discussing collaborative robotics, amelioration of human working environment is of the highest importance. HRC equipment need to adjust their operation so as to match the behaviour and physical qualification of individual operators (e.g. exoskeleton regulating the assistance it offers in lifting weights or high payload robot adjusting its position for different operator heights). By developing systems that will assist and augment human performance across skills diversity, HRC systems will improve workers' chances of maintaining employment regardless of personal characteristics, capabilities and experience. Granting human workers the ability to operate in a stress free, adapting environment, without compromising their performance eliminates the uncertainty of losing their skills/qualification for the job (Labour market security). Ideally a balance between demands which brings pressure and resources that enable workers to perform tasks effectively is sought.
- **Flexibility to combine different type of resources in the same workspace**. When considering genuine and smooth collaboration between humans and robots, the robot type should not be a limiting factor. This chapter has reviewed several types of robots aimed at collaboration, demonstrating different geometrical, kinematic and performance characteristics. If modern production systems are to benefit from their introduction, ways to ensure that they can be homogeneously integrated with the shopfloor infrastructure as well as in their interaction with humans need to be established. A unified integration and interaction scheme for all technologies will allow to reduce the technical complexity in deployment and maintenance as well as the cognitive load (and efficiency) of humans that will not have to become accustomed to the particularities of their robotic co-workers.

Setting aside the technical capabilities of HRC systems, cost is also among the most significant factors for adoption and sustainability of technology. Robot costs have been significantly reduced up to now becoming affordable for even smaller firms. The overall costs including the peripherals etc. will continue to decrease and thus the decision making will eventually be regulated by the maintenance costs and the ability to use the full flexibility of the robot [74, 75].

As a closing statement, it can be said that a great investment for creating robotic co-workers is taking place around the globe. The evolution of computer science and AI together with the benefits of world-scale communication through the Internet of Things paradigm has set the perfect ground for achieving this vision which was technologically impossible before. Industrial applications are at the forefront of developments and great benefits in terms of working conditions for humans can be expected. These can hopefully be migrated to other areas outside of the manufacturing/business domains (e.g. service and healthcare robotics) and further enhance quality of life for all humans.

References

1. Michalos G, Makris S, Papakostas N, Mourtzis D, Chryssolouris G (2010) Automotive assembly technologies review: challenges and outlook for a flexible and adaptive approach. CIRP J Manuf Sci Technol 2:81–91. https://doi.org/10.1016/j.cirpj.2009.12.001
2. Chryssolouris G (2006) Manufacturing systems: theory and practice. Springer, New York
3. Galin R, Meshcheryakov R (2019) Review on human–robot interaction during collaboration in a shared workspace. Lecture notes in computer science (including subseries lecture notes in artificial intelligence and lecture notes in bioinformatics). Springer
4. Paletta L, Brijacak I, Reiterer B, Pszeida M, Ganster H, Fuhrmann F, Weiss W, Ladstatter S, Dini A, Murg S, Mayer H (2019) Gaze-based human factors measurements for the evaluation of intuitive human-robot collaboration in real-time. In: 2019 24th IEEE international conference on emerging technologies and factory automation (ETFA). IEEE, Zaragoza, Spain, pp 1528–1531
5. Heyer C (2010) Human-robot interaction and future industrial robotics applications. In: 2010 IEEE/RSJ international conference on intelligent robots and systems. IEEE, Taipei, pp 4749–4754
6. Gleeson B, MacLean K, Haddadi A, Croft E, Alcazar J (2013) Gestures for industry Intuitive human-robot communication from human observation. In: 2013 8th ACM/IEEE international conference on human-robot interaction (HRI). IEEE, Tokyo, Japan, pp 349–356
7. Krüger J, Lien TK, Verl A (2009) Cooperation of human and machines in assembly lines. CIRP Ann 58:628–646. https://doi.org/10.1016/j.cirp.2009.09.009
8. Bannat A, Gast J, Rehrl T, Rösel W, Rigoll G, Wallhoff F (2009) A multimodal human-robot-interaction scenario: working together with an industrial robot. In: Jacko JA (ed) Human-computer interaction. Novel interaction methods and techniques. Springer Berlin Heidelberg, Berlin, Heidelberg, pp 303–311
9. Scholtz J (2003) Theory and evaluation of human robot interactions. In: Proceedings of the 36th annual Hawaii international conference on system sciences, 2003. IEEE, Big Island, HI, USA, p 10
10. Kosuge K, Yoshida H, Taguchi D, Fukuda T, Hariki K, Kanitani K, Sakai M (1994) Robot-human collaboration for new robotic applications. In: Proceedings of IECON'94—20th annual conference of IEEE industrial electronics. IEEE, Bologna, Italy, pp 713–718
11. Bauer A, Wollherr D, Buss M (2008) Human-robot collaboration: a survey. Int J Human Robot 5(1):47–66
12. Tan JTC, Duan F, Zhang Y, Watanabe K, Kato R, Arai T (2009) Human-robot collaboration in cellular manufacturing: Design and development. In: 2009 IEEE/RSJ international conference on intelligent robots and systems. IEEE, St. Louis, MO, USA, pp 29–34
13. Yanco HA, Drury J (2004) Classifying human-robot interaction: an updated taxonomy. In: 2004 IEEE international conference on systems, man and cybernetics (IEEE Cat. No.04CH37583). IEEE, The Hague, Netherlands, pp 2841–2846

14. Michalos G, Makris S, Tsarouchi P, Guasch T, Kontovrakis D, Chryssolouris G (2015) Design considerations for safe human-robot collaborative workplaces. Procedia CIRP 37:248–253. https://doi.org/10.1016/j.procir.2015.08.014
15. Bdiwi M, Pfeifer M, Sterzing A (2017) A new strategy for ensuring human safety during various levels of interaction with industrial robots. CIRP Ann 66:453–456. https://doi.org/10.1016/j.cirp.2017.04.009
16. Stages of human-robot collaboration. https://www.kuka.com/en-us/future-production/human-robot-collaboration/6-stages-of-human-robot-collaboration. Accessed 12 Feb 2020
17. Mousavi Mohammadi A, Akbarzadeh A (2017) A real-time impedance-based singularity and joint-limits avoidance approach for manual guidance of industrial robots. Adv Robot 31:1016–1028. https://doi.org/10.1080/01691864.2017.1352536
18. Michalos G, Kousi N, Karagiannis P, Gkournelos C, Dimoulas K, Koukas S, Mparis K, Papavasileiou A, Makris S (2018) Seamless human robot collaborative assembly—an automotive case study. Mechatronics 55:194–211. https://doi.org/10.1016/j.mechatronics.2018.08.006
19. Kokkalis K, Michalos G, Aivaliotis P, Makris S (2018) An approach for implementing power and force limiting in sensorless industrial robots. Procedia CIRP 76:138–143. https://doi.org/10.1016/j.procir.2018.01.028
20. Michalos G, Makris S, Spiliotopoulos J, Misios I, Tsarouchi P, Chryssolouris G (2014) ROBO-PARTNER: seamless human-robot cooperation for intelligent, flexible and safe operations in the assembly factories of the future. Procedia CIRP 23:71–76. https://doi.org/10.1016/j.procir.2014.10.079
21. Robot safety skin. https://www.koris-fs.de/en/products/robot-safety-skin/ (2020). Accessed 12 Feb 2020
22. Robot safety skin. https://www.bluedanuberobotics.com/airskin/ (2020). Accessed 12 Feb 2020
23. Krüger J, Schreck G, Surdilovic D (2011) Dual arm robot for flexible and cooperative assembly. CIRP Ann 60:5–8. https://doi.org/10.1016/j.cirp.2011.03.017
24. Makris S, Tsarouchi P, Matthaiakis A-S, Athanasatos A, Chatzigeorgiou X, Stefos M, Giavridis K, Aivaliotis S (2017) Dual arm robot in cooperation with humans for flexible assembly. CIRP Ann 66:13–16. https://doi.org/10.1016/j.cirp.2017.04.097
25. Michalos G, Makris S, Chryssolouris G (2014) The new assembly system paradigm. Int J Comput Integr Manufact, Available Online
26. Rubio F, Valero F, Llopis-Albert C (2019) A review of mobile robots: Concepts, methods, theoretical framework, and applications. Int J Adv Rob Syst 16:172988141983959. https://doi.org/10.1177/1729881419839596
27. Mobile robots. https://www.mobile-industrial-robots.com/en/solutions/ (2020). Accessed 12 Feb 2020
28. Mobile robot. https://www.kuka.com/en-us/products/mobility/mobile-platforms (2020). Accessed 12 Feb 2020
29. Mobile platform. https://www.solvelight.com/product/arti3-mobile-robot-platform/ (2020). Accessed 12 Feb 2020
30. KMR IIWA mobile robot. https://www.kuka.com/en-us/products/mobility/mobile-robot-systems/kmr-iiwa (2020). Accessed 12 Feb 2020
31. KMR QUANTEC mobile robot. https://www.kuka.com/en-us/products/mobility/mobile-robot-systems/kmr-quantec (2020). Accessed 12 Feb 2020
32. FESTO mobile robot. https://www.festo.com/group/en/cms/10239.htm (2020). Accessed 12 Feb 2020
33. Kousi N, Michalos G, Aivaliotis S, Makris S (2018) An outlook on future assembly systems introducing robotic mobile dual arm workers. Procedia CIRP 72:33–38. https://doi.org/10.1016/j.procir.2018.03.130
34. Mobile robot app competition offers $25,000 prize. http://linuxgizmos.com/kuka-youbot-robot-app-competition/ (2020). Accessed 12 Feb 2020
35. TIAGo++, the robot you need for bi-manual tasks. http://blog.pal-robotics.com/tiago-bi-manual-robot-research/ (2020). Accessed 12 Feb 2020

36. Snake-arm robots. http://www.ocrobotics.com/technology-/snakearm-robots/ (2020). Accessed 12 Feb 2020
37. de Looze MP, Bosch T, Krause F, Stadler KS, O'Sullivan LW (2016) Exoskeletons for industrial application and their potential effects on physical workload. Ergonomics 59:671–681. https://doi.org/10.1080/00140139.2015.1081988
38. Karvouniari A, Michalos G, Dimitropoulos N, Makris S (2018) An approach for exoskeleton integration in manufacturing lines using Virtual Reality techniques. Procedia CIRP 78:103–108. https://doi.org/10.1016/j.procir.2018.08.315
39. Commercial exoskeletons in 2015. https://exoskeletonreport.com/2015/04/12-commercial-exoskeletons-in-2015/ (2020). Accessed 12 Feb 2020
40. Giones F, Brem A (2017) From toys to tools: the co-evolution of technological and entrepreneurial developments in the drone industry. Bus Horiz 60:875–884. https://doi.org/10.1016/j.bushor.2017.08.001
41. Hassanalian M, Abdelkefi A (2017) Classifications, applications, and design challenges of drones: a review. Prog Aerosp Sci 91:99–131. https://doi.org/10.1016/j.paerosci.2017.04.003
42. Tosato P, Facinelli D, Prada M, Gemma L, Rossi M, Brunelli D (2019) An autonomous swarm of drones for industrial gas sensing applications. In: 2019 IEEE 20th international symposium on "a world of wireless, mobile and multimedia networks" (WoWMoM). IEEE, Washington, DC, USA, pp 1–6
43. Pott A, Mütherich H, Kraus W, Schmidt V, Miermeister P, Verl A (2013) IPAnema: a family of cable-driven parallel robots for industrial applications. In: Bruckmann T, Pott A (eds) Cable-driven parallel robots. Springer Berlin Heidelberg, Berlin, Heidelberg, pp 119–134
44. Kraus W, Schmidt V, Rajendra P, Pott A (2014) System identification and cable force control for a cable-driven parallel robot with industrial servo drives. In: 2014 IEEE international conference on robotics and automation (ICRA). IEEE, Hong Kong, China, pp 5921–5926
45. Izard J-B, Gouttefarde M, Michelin M, Tempier O, Baradat C (2013) A reconfigurable robot for cable-driven parallel robotic research and industrial scenario proofing. In: Bruckmann T, Pott A (eds) Cable-driven parallel robots. Springer Berlin Heidelberg, Berlin, Heidelberg, pp 135–148
46. HRC in the production of the BMW Group. https://www.kuka.com/en-de/industries/solutions-database/2017/06/solution-systems-bmw-dingolfing (2020). Accessed 12 Feb 2020
47. SIASUN collaborative robot helps the automobile industry to change its manufacturing mode. https://ifr.org/ifr-press-releases/news/siasun-collaborative-robot-helps-the-automobile-industry-to-chan (2020). Accessed 12 Feb 2020
48. Human-robot collaboration during headlight adjustment. https://www.kuka.com/en-de/industries/solutions-database/2019/01/hrc-headlight-adjustment (2020). Accessed 12 Feb 2020
49. Collaborative robotics puts people first. https://www.robotics.org/content-detail.cfm/Industrial-Robotics-Industry-Insights/Collaborative-Robotics-Puts-People-First/content_id/7021 (2020). Accessed 12 Feb 2020
50. Collaborative robots improve Qisda production efficiency https://www.digitimes.com/news/a20190514PD210.html (2020). Accessed 12 Feb 2020
51. Electronics industry increasingly relies on collaborative robots to boost productivity. https://www.prnewswire.com/news-releases/electronics-industry-increasingly-relies-on-collaborative-robots-to-boost-productivity-300606226.html (2020). Accessed 12 Feb 2020
52. Collaborative robots in plastic and polymer production. https://www.digitalconnectmag.com/collaborative-robots-in-plastic-and-polymer-production/ (2020). Accessed 12 Feb 2020
53. What 'Cobots' can do for Blow Molders. https://www.ptonline.com/blog/post/what-cobots-can-do-for-blow-molders (2020). Accessed 12 Feb 2020
54. Collaborative industrial robots improve the food and agriculture industries. https://www.universal-robots.com/industries/food-and-agriculture/ (2020). Accessed 12 Feb 2020
55. First dumpling-making robot optimize production, entertain guests. https://www.universal-robots.com/case-stories/huis-ten-bosch/ (2020). Accessed 12 Feb 2020
56. Dimitropoulos N, Michalos G, Makris S, (2020) An outlook on future assembly systems—the SHERLOCK approach. In: 8th CIRP conference on assembly technologies and systems, (CATS 2020), Procedia CIRP, Athens, Greece (2020)

57. Whippany actuation systems. https://www.universal-robots.com/case-stories/whippany-actuations-systems/ (2020). Accessed 12 Feb 2020
58. BAE unveils smart factory for Tempest aircraft. https://www.wearefinn.com/topics/posts/bae-unveils-smart-factory-for-tempest-aircraft/ (2020). Accessed 12 Feb 2020
59. Universal robots doubles production of plastic and metal aerospace components despite labor shortage. https://www.assemblymag.com/articles/95566-universal-robots-doubles-production-of-plastic-and-metal-aerospace-components-despite-labor-shortage (2020). Accessed 12 Feb 2020
60. Collaborative robots in high mix, low volume shops. https://www.fabricatingandmetalworking.com/2016/02/collaborative-robots-in-high-mix-low-volume-shops/ (2020). Accessed 12 Feb 2020
61. Papanastasiou S, Kousi N, Karagiannis P, Gkournelos C, Papavasileiou A, Dimoulas K, Baris K, Koukas S, Michalos G, Makris S (2019) Towards seamless human robot collaboration: integrating multimodal interaction. Int J Adv Manuf Technol 105:3881–3897. https://doi.org/10.1007/s00170-019-03790-3
62. Argyrou A, Giannoulis C, Sardelis A, Karagiannis P, Michalos G, Makris S (2018) A data fusion system for controlling the execution status in human-robot collaborative cells. Procedia CIRP 76:193–198. https://doi.org/10.1016/j.procir.2018.01.012
63. Andrianakos G, Dimitropoulos N, Michalos G, Makris S (2019) An approach for monitoring the execution of human based assembly operations using machine learning. Procedia CIRP 86:198–203. https://doi.org/10.1016/j.procir.2020.01.040
64. Nikolakis N, Maratos V, Makris S (2019) A cyber physical system (CPS) approach for safe human-robot collaboration in a shared workplace. Robot Comput-Integr Manufact 56:233–243. https://doi.org/10.1016/j.rcim.2018.10.003
65. Tsarouchi P, Michalos G, Makris S, Athanasatos T, Dimoulas K, Chryssolouris G (2017) On a human–robot workplace design and task allocation system. Int J Comput Integr Manuf 30:1272–1279. https://doi.org/10.1080/0951192X.2017.1307524
66. Tsarouchi P, Makris S, Chryssolouris G (2016) Human—robot interaction review and challenges on task planning and programming. Int J Comput Integr Manuf 29:916–931. https://doi.org/10.1080/0951192X.2015.1130251
67. Evangelou G, Dimitropoulos N, Michalos G, Makris S (2020) An approach for task and action planning in human-robot collaborative cells using AI. In: 8th CIRP conference on assembly technologies and systems (CATS 2020). Procedia CIRP. Athens, Greece.
68. Michalos G, Karagiannis P, Makris S, Tokçalar Ö, Chryssolouris G (2016) Augmented reality (AR) applications for supporting human-robot interactive cooperation. Procedia CIRP 41:370–375. https://doi.org/10.1016/j.procir.2015.12.005
69. Makris S, Karagiannis P, Koukas S, Matthaiakis A-S (2016) Augmented reality system for operator support in human–robot collaborative assembly. CIRP Ann 65:61–64. https://doi.org/10.1016/j.cirp.2016.04.038
70. Gkournelos C, Karagiannis P, Kousi N, Michalos G, Koukas S, Makris S (2018) Application of wearable devices for supporting operators in human-robot cooperative assembly tasks. Procedia CIRP 76:177–182. https://doi.org/10.1016/j.procir.2018.01.019
71. Tsarouchi P, Athanasatos A, Makris S, Chatzigeorgiou X, Chryssolouris G (2016) High level robot programming using body and hand gestures. Procedia CIRP 55:1–5. https://doi.org/10.1016/j.procir.2016.09.020
72. Xu Y, Chen J, Yang Q, Guo Q (2019) Human posture recognition and fall detection using kinect V2 camera. In: 2019 Chinese control conference (CCC). IEEE, Guangzhou, China, pp 8488–8493
73. Mainprice J, Berenson D (2013) Human-robot collaborative manipulation planning using early prediction of human motion. In: 2013 IEEE/RSJ international conference on intelligent robots and systems. IEEE, Tokyo, pp 299–306

74. Boston Consulting Group, 2015, How a take-off in advanced robotics will power the next productivity surge. http://www.automationsmaland.se/dokument/BCG_The_Robotics_Revolution_Sep_2015.pdf (2020). Accessed 12 Feb 2020
75. McKinsey report automation, robotics, and the factory of the future (2017). https://www.mckinsey.it/file/7736/download?token=WJBccDzU (2020). Accessed 12 Feb 2020

Participatory Approach to Commissioning Collaborative Industrial Robot Systems

Carolann Quinlan-Smith

Abstract This chapter describes how the field of collaborative robotics has expanded significantly over the past ten years, such that it is now the fastest growing segment of the global industrial robotics market with advances in robot software technology allowing robots and workers to work "hand-in-hand" to achieve higher levels of efficiency and productivity. This new relationship leverages the strength of the robot to perform dull, dirty and repetitive tasks (e.g., palletizing, painting, packaging, polishing, etc.) whilst combining the higher cognitive abilities and flexibility of the human colleague; a winning combination of brawn and brains that is clearly transforming traditional ways of working. However, this new technological introduction into our workplace will bring forth ethical issues. Sudden, coerced introductions of a robotic colleague into our work space, whom we have long been separated from by means of strategically placed "fencing", will threaten the adoption of such technology on the plant floor. Evidence presented showed that improper attention to the "human aspects" is believed to be a primary cause of significant failures in the implementation of advanced manufacturing technology in the United States of America. Based on lessons learned in this study, organizations must learn to understand how changes in work tasks or the working environment that are made without consultation with, or involvement of, a worker can significantly impact the human experience and overall productivity.

1 Introduction

The field of collaborative robotics has expanded significantly over the past ten years, and it is now the fastest growing segment of the global industrial robotics market. A recent report from *Interact Analysis* predicts the collaborative robot market to be worth $5.6 billion by 2027, accounting for approximately 30% of the total robot market [1]. A consistently decreasing price tag and simple set-up are the main reasons

C. Quinlan-Smith (✉)
CRSP, Workplace Safety and Prevention Services, St. Thomas, ON, Canada
e-mail: Carolann.QuinlanSmith@wsps.ca

M. I. Aldinhas Ferreira and S. R. Fletcher (eds.), *The 21st Century Industrial Robot: When Tools Become Collaborators*, Intelligent Systems, Control and Automation: Science and Engineering 81, https://doi.org/10.1007/978-3-030-78513-0_3

why collaborative robots are experiencing market growth on a global level. Advances in robot software technology allow robots and workers to work "hand-in-hand" to achieve higher levels of efficiency and productivity. This new relationship leverages the strength of the robot to perform dull, dirty and repetitive tasks (e.g., palletizing, painting, packaging, polishing, etc.) with the cognitive abilities and flexibility of the human colleague; a winning combination of brawn and brains.

Collaborative robot technology is clearly changing our traditional ways of working. When properly executed, the partnership between the human and robot has the potential to improve safety while keeping up with forever changing customer requests and productivity demands, e.g., a robot that performs a repetitive task may help improve health by reducing injury incidence while improving productivity. However, despite these claims, research has shown that this new technological introduction into our workplace will bring forth ethical issues, as discussed in previous chapters. Sudden, coerced introductions of a robotic colleague into our work space, whom we have long been separated from by means of strategically placed "fencing", will threaten the adoption of such technology on the plant floor. Evidence presented in one study showed that improper attention to the "human aspects" is believed to be a primary cause of significant failures in the implementation of advanced manufacturing technology in the United States of America [2]. Based on lessons learned in this study, organizations must learn to understand that changes in work tasks or the working environment, which were made without the consultation or involvement of a worker, can significantly impact the human experience and overall productivity.

This chapter begins by discussing current findings in research on how and why these psychological health and safety aspects, or ethical issues, are likely to impact the worker's experience and acceptance of new technology. We move on to highlight opportunities for operator consultation and participation in the planning, design and implementation phases of a collaborative robot system as a potential solution to the human issues influencing technology adoption. In the absence of specific industry standards or guidelines on how to address the human factors associated with human and robot collaboration, the author presents solutions that are within the responsibility, control or influence of the workplace. CAN/CSA Z1003-13, *Psychological health and safety in the workplace—Prevention, promotion, and guidance to staged implementation*, Canada's first national standard for mental health in the workplace, was referenced as it provides a "systematic approach to develop and sustain a psychologically healthy and safe workplace" [3]. Additionally, the following key elements of an occupational health and safety management system standard (e.g., ISO 45001:2018) are considered for managing the health and safety risks identified in collaborative applications: leadership and worker participation, e.g., leadership commitment, consultation and participation of workers; planning, e.g., hazard identification and risk assessment; and support, e.g., resources, communication, and training.

2 Psychological Health and Safety Aspects Inhibiting Technology Acceptance

2.1 Physical and Psychological Protection

An employer has a legal and ethical responsibility for the health and safety of its workers, and those affected by its activities. This responsibility includes protecting a workers physical safety, and now with greater concern, a workers psychological safety. According to CAN/CSA Z1003-13, psychological safety is defined as "*the absence of harm and/or threat of harm to mental well-being that a worker might experience*" [3].

In recent years, worker physical and psychological safety issues are becoming an increasing concern as more robots are used in collaborative applications and with varying degrees of human and robot collaboration—from limited collaboration (e.g., speed and separation monitoring, safety-rated monitored stop, hand guiding) to full collaboration (power and force limiting). The robot speed, movement (planned or unplanned) and the operator's proximity to the robot [4, 5] are all factors that may cause stress to the human engaged in a collaborative partnership with a robot. This is in part due to fact that the traditional "physical" measures (e.g. perimeter guarding) used to protect the operator from the robots movement and other associated risks has now been replaced with inherently safe design measures built into the robot (e.g., vision systems, collision detection, etc.) This new, advanced technology is unfamiliar and presents a challenge for operators, both the experienced and unexperienced with robotics alike, to trust the efficacy of technology that they cannot see or feel.

The ongoing introduction of collaborative robots will require organizations to take a proactive and participatory approach in the design of collaborative industrial robot systems in order to reduce stress and build a workers knowledge and trust in the functionality and reliability of advanced robot safety technologies. Each phase of the collaborative robot system implementation, or as referred to herein as the "commissioning process", provides an opportunity for worker participation in discussions pertaining to worker safety and well-being.

2.2 Lack of Participation

According to CAN/CSA Z1003-13, "*Risks to mental health are more likely to arise and contribute to a psychologically unsafe workplace when discretion over the means, manner, and methods of their work (including "voice" or the perceived freedom to express views or feelings appropriate to the situation or context) is withheld from workers…*" [3].

Robots will continue to take over tasks that are undesirable; however, it is important for an organization to understand work preferences before tasks are distributed between the human and robot worker. Workers want to engage in meaningful work that facilitates a positive work experience and reduce negative ones. However, the positives may not always be evident to management as highlighted in this study [6] where management deemed a delivery job as their biggest pain point from a production sense, when in fact, it was a job that workers preferred because it involved several positive work attributes such as autonomy, exercise/movement, etc.

Several studies [2, 7–10] have shown that having influence over how work is organized (e.g., identifying and retaining meaningful work) will give workers a sense of ownership and control regarding upcoming changes, such as the introduction of collaborative robots, and will increase the likelihood of their willingness to adopt the changes and of their understanding of the technology. Based on the work of [2], we can propose that the use of a human-centered philosophy and worker involvement during each stage of the integration of advanced technology, such as collaborative robot systems, will have a significant impact on the overall success of the implementation. This concept aligns with an occupational health and safety (OH&S) management systems approach, which emphasizes worker consultation and participation in temporary or permanent changes related to their work.

2.3 *Lack of Communication*

A lack of communication regarding impending organizational change, such as the introduction of robotic technology, can cause a level of uncertainty and anxiety about the future of one's job. This fear is further manifested by the plethora of articles that can be found online and in our local newspapers touting an impending risk of losing our jobs to robots. Furthermore, a fear of the unknown or unfamiliarity (e.g., lack of training) of advanced robot technology can result in operator resistance or acceptance to work with the new technology.

Managing fear in the workplace is important during times of change; and organizations can influence the level of fear through effective communication [11]. According to point 8 "drive out fear", of Dr. W. Edwards Deming's 14 points for total quality management (or the "Deming Model of Quality Management"), organizations should "encourage effective two way communication and other means to drive out fear throughout the organization so that everybody may work effectively and more productively for the company" [12].

2.4 *Lack of Training*

Technology is being introduced into the workplace at a record speed, however, engaging and upskilling the workforce is lagging [13]. Moreover, robot technology has historically been thrust upon workers without adequate training, leading to technology adoption and acceptance issues [6]. According to a national survey by Eagle Hill Consulting [14], which polled a random sample of 675 healthcare industry employees from across the United States of America on the impact of technology (e.g., robotics) on their current and future work environment, companies are not soliciting worker feedback when making decisions about introducing new technology in the workplace (17% of healthcare workers were asked for their input) and they are not preparing them to work with the new technology. A lack of worker input, support and training will have a significant impact on the human experience and can undermine the investment in technology [14].

3 Opportunities for Worker Consultation and Participation

3.1 Leadership, Communication and Participation of Workers

According to CAN/CSA Z1003-13, "***Clear leadership and expectations****is present in an environment in which leadership is effective and provides sufficient support that helps workers know what they need to do, explains how their work contributes to the organization, and discusses the nature and expected outcomes of impending changes. An organization with clear leadership and explicit expectations would be able to state that workers are informed about important changes at work in a timely manner*" [3].

Change has become prevalent in the workplace, with organizations becoming more nimble in order to stay competitive in today's marketplace. Failure of organizational changes can have a significant impact on stakeholders in an organization and on the ultimate longevity of an organization [15]. A key factor in the successful outcome of organizational change is to ensure that there are clear lines of communication at all levels within the organization, and consultation and participation of workers within the change process.

When implementing a change, such as the integration of a collaborative robot system, organizations will need to convince stakeholders to alter practices, processes, procedures, work arrangements, and often beliefs and values as well [15]. In change implementation, communication is a means by which an organization can increase a worker's contextual understanding of the reasons for the change and their ability to adapt and adjust during the implementation phase of a change process [16]. Similar findings in this project [7], where communication to the operators regarding the reasons for the introduction of the automation and associated benefits was a major enabler.

Prior to the introduction of a collaborative robot system, senior leadership should clearly communicate their intent, rationale and goals for the change, as well as the potential impact on roles and specific jobs (if any), their commitment and support to the commissioning process, and solicit worker feedback and participation. This type of transparent leadership and ongoing, open dialogue between leadership and workers throughout the duration of the commissioning process will provide a platform for raising such issues while managing rumours, feelings and concerns about the impending change (e.g., job loss) and it will promote employee engagement. In an interview with AJung Moon, Assistant Professor in the Department of Electrical and Computer Engineering at McGill University, and the founder and director of the Open Roboethics Institute, Moon recommends allowing workers to voice any concerns that they may have about working alongside collaborative robots. These concerns may "raise important issues that hasn't been thought of at the design or deployment stage of the system, and that can make the difference between successful and failed deployment of cobots" [17].

Implementation of change can often compete for time, attention and resources that might otherwise be devoted to other things, e.g., maintaining production rates [15]. According to ISO 45001, top management can demonstrate commitment and leadership with respect to health and safety activities by providing adequate time, support, training, information and resources, such as ensuring that there is adequate labour assigned to support production and to participate in the collaborative robot system commissioning process. In [7], participants felt that senior management's visible support and commitment to the project throughout its development was a major enabler and enhanced its credibility.

A key requirement in ISO 45001:2018 is the involvement of workers in the OH&S management system, as outlined in Clause 5.4 "Consultation and participation of workers". "Consultation" involves seeking worker feedback and considering it before making a decision; whereas, "participation" involves the contribution of workers in the decision-making process. An organization should establish, implement, and maintain a process for worker consultation and participation in the decision-making process and the change management process, which includes technological changes, e.g., collaborative industrial robot systems. Opportunities for active and ongoing worker participation and involvement in the collaborative robot system commissioning process will be presented herein.

3.2 Task Selection and Planning

According to CAN/CSA Z1003-13, "***Involvement and influence****is present in a work environment where workers are included in discussions about how their work is done and how important decisions are made. Opportunities for involvement can relate to a worker's specific job, the activities of a team or department, or issues involving the organization as a whole.*"

Numerous case studies and research suggest that early and significant worker involvement in decisions pertaining to changes to their work gives them a sense of ownership and control over upcoming changes [7, 18], and this ownership can be the key to a successful outcome. Welfare et al. states, "If the operators want the system to work, they try everything they can to ensure that the system does work. Conversely, if they do not feel part of the project and do not have any ownership of the system, they do not actively support resolution of any problems, instead waiting for others to address any issues" [18]. Worker participation by those impacted by the technology should begin in the concept stage of an automation project. Participation from production (e.g., those with knowledge and experience of current production, familiarity with equipment) maintenance (e.g., those with experience servicing the specific equipment), vendors (e.g., automation), and other key stakeholders (e.g., engineering staff, health and safety professionals, trades union, etc.) is essential, especially if an organization is considering integrating a collaborative robot into an existing system.

One case study showed the benefits of having a "process champion" who understands the technology and common applications and can cascade this knowledge to the rest of the team [2, 7]. As a starting point, the organization should designate an individual tasked with leading workshops where key stakeholders from across the organization can meet to identify simple processes for automation, with a primary focus on identifying tasks that operators dislike and preserving work preferences. Work preferences are important because they link to job performance and long-term success via motivation and satisfaction [18]. This understanding of work preference is important because, as collaborative industrial robot technology is considered for integration, work preferences can be targeted for preservation, which can positively affect a worker's psychological health and technology adoption and implementation [18].

Next, relevant data should be gathered that will assist the organization in identifying tasks or jobs which are physically and/or psychologically demanding, e.g., accident reports, worker feedback (discomfort surveys, formal complaints), claims data (short-term disability, benefits utilization, etc.), organizational audit/inspection reports, administrative data (absenteeism reports, employee turnover, etc.), job descriptions, physical demands description, etc. Ergonomic assessments (e.g., physical demands analysis), using industry standards and tools (e.g., RULA, SNOOK, etc.) can also assist the organization in identifying ergonomic risks (e.g., highly repetitive tasks) which could result in the development of work-related musculoskeletal disorders. The ergonomic assessment process involves observing workers performing their jobs and engaging them in the conversation about the aspects of their job that they consider physically and/or psychologically demanding; simply put, "what hurts?".

Following the selection of a job or task for automation, the next step in the planning process is providing workers with the opportunity to visualize the job sequence and the degree of human and robot collaboration using computer-generated simulation software or virtual reality technology. Simulation and visualization provides a good visual representation of the system as it is operating, which allows the user to validate the design of the system prior to procurement [19]. Most robot manufacturers offer free robot simulation software that allows the user to "build" their collaborative robot system online and view a 3D simulation of the system running. This software, and other commercial virtual software platforms (e.g., virtual reality technology), provides the visual learner with a better understanding of the technology and its breadth of capabilities as well as limitations, and potentially easing the workers fear or anxiety about the proposed new working environment or new "ways of working". Alternatively, most (if not all) robot manufacturers offer "demo" robots which gives workers an opportunity to gain practical "hands-on" experience using the technology in the workplace or on the manufacturer's showroom floor, which is ideal for the visual and kinesthetic learner.

3.3 Risk Assessment

According to CAN/CSA Z1003-13, "***Protection of physical safetyis** present when a worker's psychological, as well as physical safety, is protected from hazards and risks related to the worker's physical environment*" [3].

In collaborative robot operations, humans work in close proximity to the robot while power is available to the robot's actuators. ISO 10218 and the technical specification for collaborative robots (ISO/TS 15066) outline four different types of collaborative operations, with power and force limiting by design or control causing the greatest concern for physical safety. In such operations, the design of the a robot system and the associated cell layout [20]. is of major importance; and a key process in the design of a collaborative operation is the elimination of hazards and the reduction of risk. A risk assessment is a normative or "mandatory" requirement in ISO 10218 and a necessary process for identifying hazards and determining the appropriate risk reduction measures to ensure the operator's safety during collaborative operation.

Hazard identification is fundamental in the planning process to prioritize actions to address risks and opportunities [21]. An organization must consider the hazards associated with the entire collaborative robot system application, including the robot end-effector (shape, type, etc.), appliation (e.g., painting, drilling, capping, etc.), workpiece, workspace (e.g. clearances, access points, etc.) and the potential impact type and bodily contact area [22]. There are two types of impacts, with different pain thresholds, outlined in ISO/TS 15066: transient and quasi-static. Transient contact is where there is contact between an operator and part of the robot system (without restriction or clamping), whereas, quasi-static refers to situations where a body part can be crushed against an object. Each of these hazards should be addressed on an individual basis through a risk assessment for the specific collaborative application [20].

A risk assessment "enables the systematic analysis and evaluation of the risks associated with the robot system over its whole life cycle" [23], from design to decommissioning. According to ISO 10218, the integrator (the company contracted to design and build the robot system) is responsible for carrying out a task-based risk assessment in the equipment design phase to determine the appropriate safeguarding measures, and update it as the design matures. It also states that the "user" shall be "consulted" during the hazard identification and risk assessment process to ensure that the integrator has identified all reasonably foreseeable hazards and hazardous situations (task and hazard combinations) associated with the robot system. Annex A, in ISO 10218:2 contains a list a significant hazards associated with a robot system for use as a reference when carrying out a risk assessment; however, this list is not inclusive of all hazards, and is limited to the physical hazards (e.g., mechanical, electrical, thermal, etc.) associated with the robot and robot system. Although indirectly implied, additional hazards can be created by a specific robot application and should be addressed in the risk assessment, e.g., psychosocial hazards arising from working in close proximity to a robot in a collaborative operation.

As mentioned earlier, the ISO 45001 OH&S management system standard places a great emphasis on leadership to enhance participation and engagement with workers in the implementation of an effective OH&S management system, such as the participation of non-managerial works in identifying hazards and assessing risk and opportunities [21]. Leadership should ensure the availability of key stakeholders within the organization to participate in the hazard identification and risk assessment and reduction process. The "process champion" should lead the risk assessment process and as such, should be knowledgeable in the risk assessment process and the safety requirements for the design and integration of collaborative robot systems, including permissible limit values for force and pressure outlined in ISO/TS 15066 to prevent biomechanical load created by quasi-static and transient contact. According to ISO/TS 15066, "The robot system shall be designed to adequately reduce risks to an operator by not exceeding the applicable threshold limit values for quasi-static and transient contacts, as defined by the risk assessment" [20]. Therefore, the risk assessment for power and force limited collaborative applications must consider the potential contact between parts of the collaborative robot system and the operator, and the specific part of the body where contact may occur, in order to determine these threshold limit values and the appropriate protective devices (e.g., limiting devices).

The organization should also establish, implement and maintain a procedure for the hazard identification, risk assessment and determination of adequate risk reduction measures. The hazard identification and risk assessment procedure should take into account changes or proposed changes in the workplace and the manner in which the risk assessment will be carried out and by whom. The ISO 10218:2 and TS/ISO 15066 standards do not identify a recommended or prescribed risk assessment methodology, however, the ISO 12100:2010 (*Safety of machinery—General principles for design—Risk assessment and risk reduction*) recommends using a task-based risk assessment methodology that provides direction on the safety circuit performance level required to mitigate the risks identified. Since technological solutions and their associated reliability cannot address psychosocial risk factors, a psychosocial risk assessment methodology or industry best practice, with similar guidance (as outlined in Annex A, ISO 10218:2) on the sources of psychosocial hazards should be used.

3.4 Knowledge and Skill Development

According to CSA Z1003-13, "***Growth and development****is present in a work environment where workers receive encouragement and support in the development of their interpersonal, emotional, and job skills. Such workplaces provide a range of* *internal and external opportunities for workers to build their repertoire of competencies, which will not only help with their current jobs, but will also prepare them for possible future positions*".

Changes in technology typically brings forth a change or increased demand for skills and as such, workers may be reluctant to accept changes if they are anxious about their ability to perform their job after the change. Research has shown that adequate training can strengthen a worker's confidence in their abilities to accommodate workplace change and empowers them to learn new skills [6, 24]. In this study [2], worker orientation, training, and education was found to reduce workers fears (e.g., change, job loss, etc.) and their resistance to adopt new technology. The training can also spur positive work attributes such as social interaction if workers are training their colleagues [6].

Various sources of information on collaborative robot technology are available from robot manufacturers (e.g., website, brochures, e-books), robot system integrators, automation specialists and industry standards. Most robot manufacturers offer support and training in a setting (e.g., training center) that allows the user to experience the robot, such as in-person training that is customized to their specific application and needs. Additionally, a variety of online training options is available for operators and those programming and maintaining the systems. To minimize worker resistance to technology adoption generated by fear and anxiety, it is critical that affected workers receive training prior to handing over the fully commissioned system to production [25].

Lastly, robot system testing and debug of the collaborative robot system at the integrators site provides an excellent opportunity to introduce operators to the system, enhancing their knowledge, familiarity and confidence with the technology [20]. Operator involvement in the final installation on the factory floor can also provide hands-on learning and encourage ownership of the system, which is critical to a successful deployment [25].

3.5 Robot System Verification and Validation

The ISO 10218-2 standard places the responsibility for the final verification and validation of the design and construction of the robot system on the robot system integrator. Worker consultation and participation is a requirement for successful planning and implementation of changes affecting worker health and safety, and as such, worker participation in the verification and validation process should be encouraged.

The "process champion" should organize a cross functional team comprised of individuals in the organization who interact with the robot system (e.g., maintenance, operators, set-up personnel, supervisors, etc.) in the verification and validation process as outlined in Annex G, Table G.1 of ISO 10218-2. The verification and validation process employs various methods essential to the safety of the robot system including, but not limited to, visual inspection, measurement and practical testing, e.g., testing pressure and force outputs in power and force limited collaborative robot applications [20]. A reliable test method should be used by the team for measuring the pressures and forces associated with quasi-static and transient contact events of power and force limiting by inherent design or control collaborative applications. With this information, the team can verify that the collaborative robot is functioning within its permissible limits, or make adjustments to the system to ensure the safety for all users of the system.

According to CAN/CSA Z1003-13, "*An organization with good organizational culture would be able to state that workers and management trust one another*" [3]. Lastly, the inclusion of workers in the verification and validation process may also lend to a building of trust in the systems that they will work with and the organization's commitment to a psychologically healthy and safe workplace.

3.6 Conclusion

Collaborative robots are experiencing a rapid growth due to the plethora of benefits due to a decreasing price tag, flexibility and minimal integration costs and time [26]. Organizations seeking to reap the benefits of this technology must understand the psychological impact of sharing a workspace with a robot that was once "caged" for our protection. Forcing such change upon workers can generate fear (e.g., job loss, safety, lack of trust in technology) which can affect technology acceptance and adoption. A systematic and structured approach, based on a management systems model, that involves worker consultation and participation in the planning, design and implementation phases of a collaborative robot system and where leadership is effective and provides sufficient support throughout the robot commissioning process, will positively affect the human experience and develop and sustain a psychologically healthy and safe workplace.

References

1. Xiao M (2019) The collaborative robot market—2019. Interact Analysis. https://www.interactanalysis.com/the-collaborative-robot-market-2019-infographic/
2. Chung CA (1996) Human issues influencing the successful implementation of advanced manufacturing technology. J Eng Tech Manage 13(3):283–299
3. CAN/CSA Z1003-13, Psychological health and safety in the workplace—prevention, promotion, and guidance to staged implementation
4. Koppenborg M, Nickel P, Naber B, Lungfiel A, Huelke M (2017) Effects of movement speed and predictability in human–robot collaboration. Human Fact Ergonom Manufact Ser Indus 27(4):197–209. https://doi.org/10.1002/hfm.20703
5. Arai T, Kato R, Fujita M (2010) Assessment of operator stress induced by robot collaboration in assembly. CIRP Ann Manuf Technol 59(2010):5–8. https://doi.org/10.1016/j.cirp.2010.03.043

6. Welfare KS, Hallowell MR, Shah JA, Riek LD (2019) Consider the human work experience when integrating robotics in the workplace. In: 2019 14th ACM/IEEE international conference on human-robot interaction (HRI), pp 75–84. https://doi.org/10.1109/HRI.2019.8673139
7. Charalambous G, Fletcher S, Webb P (2015) Identifying the key organisational human factors for introducing human-robot collaboration in industry: an exploratory study. Int J Adv Manufact Technol 81(9–12). https://doi.org/10.1007/s00170-015-7335-4
8. Armenakis AA, Harris SG, Mossholder KW (1993) Creating readiness for organisational change. Human Relat 46(6):681–703
9. Wagner SH, Parker CP, Christiansen ND (2003) Employees that think and act like owners: effects of ownership beliefs and behaviors on organisational effectiveness. Pers Psychol 56(4):847–887
10. Pierce JL, O'Driscoll MP, Coghlan AM (2004) Work environment structure and psychological ownership: the mediating effects of control. J Soc Psychol 144(5):507–534
11. Stanleigh S (2011) Diminishing fear in the workplace. Business Improvement Architects. https://bia.ca/diminishing-fear-in-the-workplace/
12. Deming WE (1986) Out of the crisis. Massachusetts Institute of Technology, Center for Advanced Engineering Study.
13. Nealon S (2020) As health companies double down on technology investment, employees out of loop on decisions according to new eagle hill research: lack of employee input & training puts technology and automation ROI at risk. PR Newswire
14. The missing factor sabotaging technology change in healthcare: employee input. https://www.eaglehillconsulting.com/wp-content/uploads/2020/05/EHC_Healthcare_Technology_Thought_Leadership.pdf
15. Lewis L (2011) Organizational change: creating change through strategic communication. Wiley, Incorporated
16. Parsells R (2017) Addressing uncertainty during workplace change: Communication and sense-making. Administ Issues J Educat Pract Res 7(2):47
17. Cole E (2020) The Cobot experience: AJung Moon & resolving human-cobot resource conflicts. Robotiq. https://blog.robotiq.com/the-cobot-experience-ajung-moon-the-ethics-of-industrial-robotics
18. Welfare KS, Hallowell MR, Shah JA, Riek LD (2019) Consider the human work experience when integrating robotics in the workplace. Paper presented at the 75–84. https://doi.org/10.1109/HRI.2019.8673139
19. Yap HJ, Taha Z, Md Dawal SZ, Chang S-W (2014) Virtual reality based support system for layout planning and programming of an industrial robotic work cell. PLoS ONE 9(10):e109692. https://doi.org/10.1371/journal.pone.0109692
20. International Organization for Standardization (2016) Technical Specification Robots and robot devices—collaborative robots (ISO/TS 15066:2016). https://www.iso.org/standard/62996.html
21. International Organization for Standardization (2018) Occupational health and safety management systems—requirements with guidance for use (ISO 45001:2018). https://www.iso.org/standard/63787.html
22. Braman R (2019) The basics of designing for safety with collaborative robots a primer on collaborative robots, how to conduct safety assessments, and how to design safe implementations for robot/human coworking spaces. Machine Design. https://www.machinedesign.com/mechanical-motion-systems/article/21837445/the--of-designing-for-safety-with-collaborative-robots
23. International Organization for Standardization (2011) Robots and robotic devices—safety requirements for industrial robots—part 2: robot systems and integration (ISO 10218-2:2011). https://www.iso.org/standard/41571.html

24. Wanberg CR, Banas JT (2000) Predictors and outcomes of openness to changes in a reorganizing workplace. J Appl Psychol 85(1):132–142
25. Wilson M (2014) Implementation of robot systems: an introduction to robotics, automation, and successful systems integration in manufacturing. Butterworth Heinemann, US
26. Robotics Industries Association (RIA) Collaborative robots market experiencing exponential growth. https://www.robotics.org/Collaborative-Robots

Robot Inference of Human States: Performance and Transparency in Physical Collaboration

Kevin Haninger

Abstract To flexibly collaborate towards a shared goal in human–robot interaction (HRI), a robot must appropriately respond to changes in their human partner. The robot can realize this flexibility by responding to certain inputs, or by inferring some aspect of their collaborator and using this to modify robot behavior—approaches which reflect design viewpoints of robots as tools and collaborators, respectively. Independent of this design viewpoint, the robot's response to a change in collaborator state must also be designed. In this regard, HRI approaches can be distinguished according to the scope of their design objectives: whether the design goal depends on the behavior of the individual agents or the coupled team. This chapter synthesizes work on physical HRI, largely in manufacturing tasks, according to the design viewpoint and scope of objective used. HRI is posed as the coupling of two dynamic systems; a framework which allows a unified presentation of the various design approaches and, within which, common concepts in HRI can be posed (intent, authority, information flow). Special attention is paid to predictability at various stages of the design and deployment process: whether the designer can predict team performance, whether the human can predict robot behavior, and to what degree the human behavior can be modelled or learned.

1 Introduction

Factory workstations are designed to achieve favorable outcomes in safety, ergonomy, and productivity when operated by a typical employee. This performance depends on the human's needs and behavior in situ, which are predicted by established human factors principles, thus allowing appropriate design of the workstation. To extend

This project has received funding from the European Union's Horizon 2020 research and innovation programme under grant agreement No 820689—SHERLOCK.

K. Haninger (✉)
Fraunhofer Institute for Production Systems and Design Technology (IPK), Berlin, Germany
e-mail: kevin.haninger@ipk.fraunhofer.de

M. I. Aldinhas Ferreira and S. R. Fletcher (eds.), *The 21st Century Industrial Robot: When Tools Become Collaborators*, Intelligent Systems, Control and Automation: Science and Engineering 81, https://doi.org/10.1007/978-3-030-78513-0_4

factory workstations with robotic collaborators inherits the structure of this design problem—performance depends on how the human interacts with the robot—but the novelty and complexity of robots compounds the design difficulty. As a result, human–robot interaction (HRI) today is highly application specific, relying heavily on the robot programmer's intuition to predict the human collaborator's needs and behavior in situ.

HRI promises new opportunities for operators and factories [1, 2], especially in agile manufacturing [3]. Specifically, flexible robot collaborators can adapt themselves to an operator, meeting their individual preferences (robot speed, key points for handover, task timing) or current state (fatigue, attention, intent). This flexibility has the potential to improve task performance (e.g. cycle time), ergonomics, and acceptance.

To realize these benefits, HRI faces a challenge beyond just novelty: complexity. If the robot responds to the operator and vice versa, they form a coupled system where the behavior of one affects the other. The coupling of any two systems can lead to unexpected behavior [4]. As HRI performance often depends on the coupled behavior [5], the potential for unexpected behavior limits the ability to design high-performance robotic collaborators.

To a degree, HRI is similar to robot-environment interaction: the robot must perceive, plan, and act in response to external changes. Classical robotics, with a well-modelled and deterministic environment, have mature methods for relating robot design (sensors, actuators, and control) to performance and safety. A direct application of such methods to HRI is not possible. These methods are largely model-based, whereas engineering models for humans are limited, application-specific, and susceptible to any number of external factors.

Modern robotics often uses learning methods and explicit treatment of uncertainty in less-structured environments. Such methods have potential in HRI, but face similar fundamental challenges: data-driven methods generalize from historical data based on certain assumptions, such as similarity in training and application data. Human behavior is not stationary and can also depend on the robot behavior, making data-driven optimization of robot behavior difficult. Furthermore, human data collection is often relatively expensive, limiting the use of techniques such as neural networks which still require large datasets.

In spite of these design challenges, a large breadth of HRI approaches have been developed, and this chapter considers those in industrial, physical HRI. As a number of thorough review papers cover the techniques and applications of modern HRI [6–8], the focus here is rather on the design frameworks employed. Here, HRI is posed as the coupling of two systems, each of which with internal states which evolve over time and respond to the state of the other. A taxonomy for physical HRI is proposed, delineated by two questions:

- Does the robot form an explicit representation of the collaborator's state?
- Is the interaction of the agents considered in the design/learning of robot behavior?

These questions can be well-formulated in the coupled systems framework: what information informs each system's dynamic behavior, and what informs the design

Table 1 Parallel terminology used for humans and robots

Human	Robot	Explanation
Behavior	Dynamics	How the system's state evolves over time
Intent	Objective	Preferred state of the world, may vary between task iterations
Preference	Parameter	Partly defines objective or dynamics, may vary between iterations

objectives of the robot. This framework is also used to formalize common concepts in HRI (e.g., intent and authority), with connections made to adjacent fields of haptics, human–computer interaction, and control theory.

Terminology

A *task* is defined by its outcome: a desired state of the world, which requires a series of coordinated actions to attain. A *system* is a set of relations between quantities which can be separated from its environment by well-defined inputs and outputs. A dynamic system has an internal *state* which evolves over time and fully characterizes the current system configuration (typically a vector $\times$ R^n which renders future outputs independent of previous inputs). An *agent* is a system with an objective, where its outputs are directed at achieving a certain external outcome. Parallel terms are often used to refer to properties of humans and robots, the definitions in Table 1 will be used here.

Intent, in particular, is widely used and somewhat ambiguous [6]. Colloquially, intent is often used as an explanation for someone's behavior over a period of time, typically posed from a desired outcome. Here, *intent* refers to a human's desired state of the world, and can be considered (in a factory setting) to be defined by the current task. Quite often in physical HRI, the human's intent is a desired configuration of the environment—either a goal position for certain objects, or the instantaneous desired motion [9].

HRI can be considered as the coupling of two systems, each of which has internal states, dynamics, and possibly objectives. Interaction couples these subsystems, where the state of one influences the state of the other, represented in Fig. 1 with input/output arrows. In physical interaction, these inputs/outputs will include forces and motion [10]. Relating properties of a coupled system to those of constituent subsystems is a fundamental concern of control theory a basis for much work in classical robotics. Other frameworks, such as behavioral systems formalize coupled systems as a constraint over the states of the two systems, thus avoiding the need to assign causality to each system, as causality in HRI can often be meaningfully defined in either direction (e.g., does a human produce a force in response to motion, or vice versa?).

Scope

HRI has a wide variety of applications, modalities of interaction, and objectives. Here, the principle concern is the application of HRI to manufacturing, where tasks are iterative (with possible variation between iterations), and the human's intent is

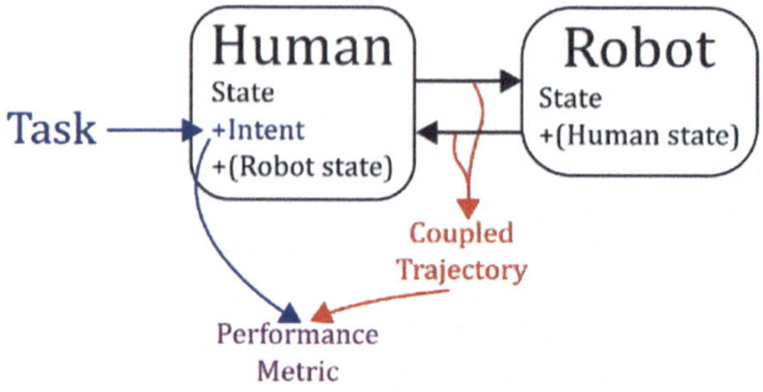

Fig. 1 A human and robot are coupled, where the evolution of one's state depends on the state of the other. The human's state includes intent and potentially an estimate of the robot state. The robot's state can include an estimate of the human state (robot as collaborator) or not (robot as tool)

fully defined by the current task. Furthermore, physical interaction is considered: the intentional exchange of mechanical power between human and robot. Such applications typically involve some sort of compliant behavior from the robot, such as impedance control [11], where the robot's motion is affected by human force [12].

In this chapter, the focus is on direct collaboration, rather than indirect collaboration where the robot detects and responds to human-caused changes in the workspace (e.g. workspace sharing). Such problems are important, but can largely be considered from a general robotics perspective—they do not require special consideration of the human (except for safety), and are thus outside the scope here.

Substantial progress in HRI has been made in safety, where new hardware, sensors, and ISO norms have been introduced to quantify and manage the risk of physical injury, which is dominated by collision concerns [13, 14]. While certification of safe HRI applications still requires substantial analysis and risk management; it is outside the scope here.

2 Interactive Robots

This section distinguishes two approaches to interactivity (Fig. 2), based on whether the robot makes an explicit estimate of the human state.

2.1 Robot as Tools

When is a robot a tool? A tool is an object with a predictable ability to transform the input that the operator applies to it. A motorized screwdriver transforms a button press to rotational torque in an obvious and repeatable manner, thus allowing an operator to choose actions which, when mediated by the tool, achieve desired outcomes. When a robot is a tool, the human applies inputs to the robot (typically via an explicit

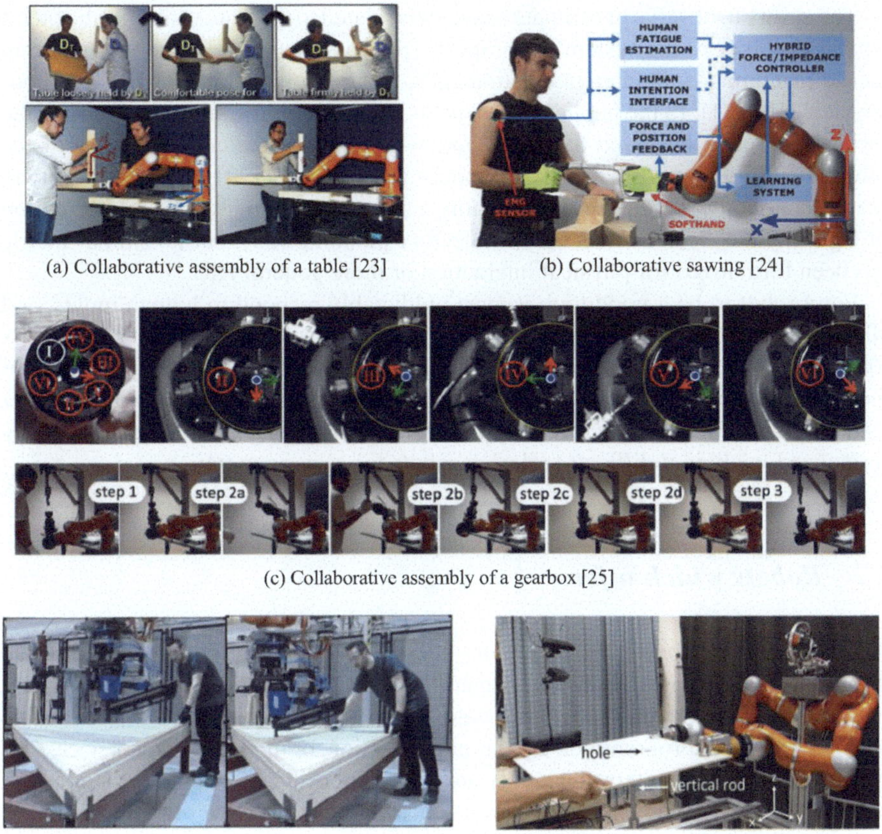

(a) Collaborative assembly of a table [23]

(b) Collaborative sawing [24]

(c) Collaborative assembly of a gearbox [25]

(d) Assembly of pre-fabricated building elements [26]

(e) Collaborative manipulation and assembly [27]

Fig. 2 Various HRI applications aiming at manufacturing and assembly tasks

interface such as with buttons), such that the robot responds as appropriate for the current task.

A robot as a tool should be designed such that the human can understand and control the robot. Formal conditions for the ability to estimate and control a system are established in control theory (observability and controllability [15]), but these are binary conditions which do not characterize the relative difficulty of a human to operate a system. A human's ability to control complex systems has been studied from a controls perspective [16], but more substantially in teleoperation and haptics. There, it has been shown that humans have the ability to control systems with a wide range of dynamics [17], even when unstable [18]. This suggests that humans can learn how to operate complex, unnatural systems—under certain experimental conditions. Characterizing the ability of humans to operate robots as tools has been further developed in workspace-sharing HRI [19].

The ability to observe or estimate a system is related to the state visibility, the literal optical visibility of the system's condition. This has been proposed as an important metric in human–computer interaction [20], programming for HRI [3], and non-physical HRI [21]. In the haptics experiments above, the position and velocity of the robot are directly observed by the operator—this is also the complete state of the robot. Visibility is an important condition for predictability [22], where the ability to see the state improves the ease of learning a system's dynamics. The related notion of legibility has been formalized in trajectory planning [23], but no similar metric has been formalized for physical interaction or more general HRI.

For a robot to be a good tool, it must predictably respond to human inputs, and allow the human to observe the robot state. This is directly jeopardized by complex robots, such as those which learn and adapt their behavior, or infer properties of the human—these add additional, non-visible internal states to the robot. A further challenge is that the relative difficulty for a human to operate a robot is not generally formalized, making it difficult to design robots for easy operation.

2.2 *Robots which infer*

An alternative approach to interactivity is where the robot estimates an aspect of the human state: a low-dimensional quantity which explains (some of) the observed behavior of the operator. Quite often these approaches estimate a quantity which is intuitive to a robot programmer, e.g. fatigue, attention, or intent. While not necessary (the semantics of the inferred quantity are of little relevance for the robot), inferring intuitive states does allow the connection with existing bodies of research and suggest sensors/methods for inference. The act of inferring the state of the human collaborator can be viewed as a type of information flow.

2.2.1 What to Infer

What should a robot infer? Anything that requires a change in the robot behavior, according to the application goals and requirements. Answering this question for classical robots in assembly lines can be done by exhaustive enumeration of the operating conditions of the robotic cell. For each possible condition, the need for a change in robot behavior can be evaluated, and if necessary, appropriate sensors can be added to detect these changes. The structured environment allows a high degree of predictability, which in turn allows the evaluation and selection of robot design.

Unfortunately (from the perspective of HRI design), humans 'contain multitudes'. The observable behavior of a human—e.g. their motion—can be informed by innumerable and unquantifiable internal states. Even humans, who benefit from shared experience and empathy, can sometimes struggle to understand the actions of others.

At the same time, not all human states are of equal importance. Is it important to for a robot to infer how hungry the operator is? Probably not, but this may cause other states of fatigue or a tendency to be distracted, which *are* potentially relevant. Other operator states, e.g., an injury, are better handled by the infrastructure of the factory—the worker notifying the floor manager, appropriate modifications being made to their schedule and duties.

What human states should inform the robot's behavior in HRI? The robot should respond to human states which impact the safety, performance, and ergonomic requirements. Several examples are established: the desired motion of the robot/payload [8, 24–26], fatigue [27], and preferred interaction force range [28]. Human intent can also reflect discrete changes in the task [29]. These human states are closely related to the human's physical behavior which can be directly measured; making their observation models (the relation between human state and sensor measurements) more concrete.

More abstract mental states, such as anxiety [30], have been explored in laboratory settings, but are not widely deployed. Much more work has been done in inference outside physical HRI, for example in detecting whether a driver is attentive or distracted [31].

Human intent is by far the most common state to be inferred, and several types of intention recognition can be distinguished [32]: keyhole, intended, and obstructed recognition. Keyhole recognition is when an agent is unaware of being observed, intended recognition is when the agent is aware their intentions are being estimated and seeks to facilitate this, and obstructed is an agent seeks to obfuscate their intentions. In factory physical HRI, it can be assumed that all intention recognition is intended recognition.

2.2.2 How to Infer It

To estimate a human state online requires a means of inference from sensor values. The relationship between the human state and the sensor values rarely has an analytical or first-principles expression, and must be informed by either domain knowledge or learned from labeled data (where the sensor measurement is paired with correct values of the human state).

The operator's desired motion is fundamental for physical collaboration, and has been inferred in several ways. In [9], the operator's intended position in a positioning task is assumed to be inferred through the velocity of the operator via ad hoc rules. Many other examples of intent estimation are reviewed in [8]. EMG sensors can be employed to detect the muscular activity (and therefore, desired motion) in collaborative tasks [33, 34].

In physical interaction, the physical dynamics of the human can also be relevant. Humans can modulate the stiffness of many body joints through antagonistic contraction of opposing muscles, a property which is suggested to be fundamental to human motor control [35]. Estimating the stiffness of the human arm has been explored as a useful human state in HRI. In [36], the desired motion is estimated from the dynamics that a human renders. Similarly, [37] estimates the experience of a welder by the stiffness which they present at their hand. In [38], the opera-tor's intent is considered indirectly via the directional stiffness that the operator presents at their hand. Estimation of human dynamics presents challenges to traditional identification techniques, largely due to the unmeasured input which is internal to the human [39], and typically only the low-frequency dynamics (i.e. stiffness) can be reliably estimated.

Learning or statistical techniques can be used to infer human states, using historical labelled data instead of an observation model. In [24], a neural network is used and in [40] a recurrent neural network was used to model and infer human motion. A recurrent network learns an internal state representation, allowing for historical observations to be incorporated into the current estimate. The use of Bayesian inference to human intent has also been developed, where methods such as Gaussian Processes can be used to give an estimate to the posterior distribution of human intent [41]. This allows the uncertainty in the robot's belief about the human to be characterized, which can allow unusual operating conditions to be detected (i.e. where the model breaks down).

In learning approaches, a major concern is the means of data collection and the ensuing quality. Often, labeled data—which pairs raw observations and the true value—is required to train an inference method based on regression. This data can be collected from human–human collaboration, offering a means for transferring established collaborative behaviors [42]. Controlled experimental conditions often allow the collection of a corresponding 'true' value; e.g. the configuration of the task for each iteration.

2.3 Dangers of Inference

While the benefits of inferring a human state have been shown in toy problems [43], the robot's belief of the human state becomes an additional state which impacts the robot's behavior—one which is not immediately visible to the human, potentially reducing the transparency and predictability of the robot's behavior. An example from human–human interaction can be seen in Fig. 3, where two courteous people are trying to pass, but neither knows which direction the other plans to take, resulting in a stalemate. If one makes a decision and communicates their plan, the other can accommodate. Similarly; if one is discourteous or distracted (ignored the state of the other), their behavior would be more predictable and the situation would be resolved.

3 From Agent to Team

Regardless of whether a human's state is inferred or inputs are taken directly, the robot's response to this information must be designed. The human-related application objectives in HRI—ergonomic, physiological—are often difficult to quantify, much less relate to the behavior of the robot. Technical design objectives which approximate

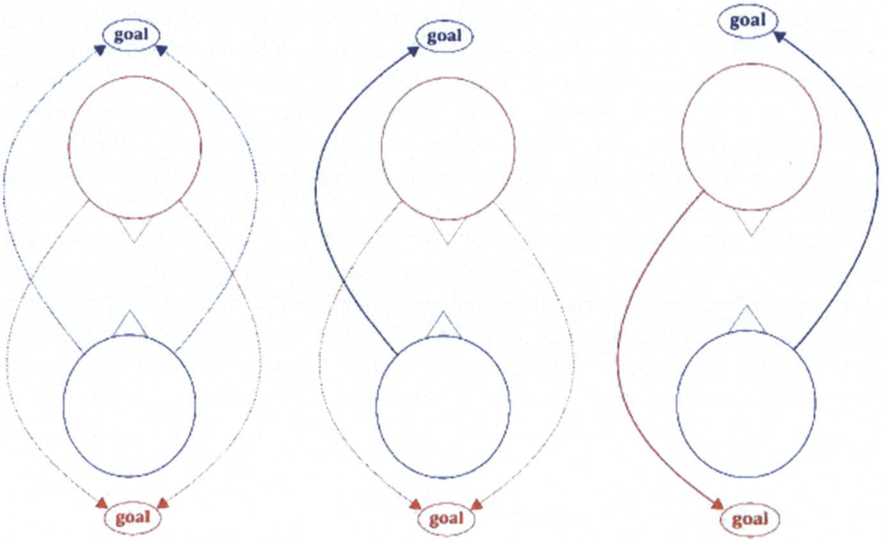

(a) Mutual uncer tainty

(b) Blue makes and communicates a de cision

(c) Red can plan on the basis of this de-cision

Fig. 3 Two people trying to pass, where mutual uncertainty in the other's plan results in a standoff. When blue makes a decision and communicates it, red can plan. While this scenario is not collaborative (two agents with independent goals), it illustrates potential downsides of unpredictable behavior in interaction

these objectives, and can be evaluated in terms of modelled or measured quantities, will be reviewed in this section.

Most HRI objectives depend on the coupled human–robot behavior, but to-date there is limited work which fully and directly considers coupled team behavior. This is no accident, there are fundamental challenges to modelling interacting agents [44], and designing an agent for coupled performance [45]. Established applications of interactive robots typically use application knowledge to simplify the problem. Some examples are reviewed in [46], with notable use-cases of gearbox assembly [47], collaborative manipulation [48], assembly of prefabricated building components [49], and automotive assembly [50].

3.1 I Work Alone

In some cases, the manner in which the robot should respond to a given human state or input is obvious. When the robot is detecting the fatigue of its collaborator and changing position [27], the two systems are still coupled, but the impact of the robot's change on the human behavior can be well-predicted.

Similarly, in collision avoidance, the robot infers the position (and sometimes velocity or projected future occupancy) of the human collaborator and adjusts the trajectory of the robot to maintain distance if necessary [51]. The robot's response is done without consideration on how the change in trajectory will affect the human motion. This approach works, in part because the impact of the robot's deviation on human trajectories is relatively small, especially considered alongside the existing limitations in human modelling.

Most industry-adjacent use-cases of HRI (e.g. those seen in Fig. 2) use such 'engineered responses', where the robot's response is defined by the design engineer on the basis of intuition, domain knowledge, and knowledge of the task/application. These examples can be considered in the coupled system framework as breaking the feedback loop between robot and human in Fig. 1—the robot's behavior causes either minor or well-predicted changes to the human behavior.

3.2 Collaborative Objectives

So when must the human behavior or state be considered in the robot design objectives? These are objectives which depend on the coupled system trajectories, or human-centered objectives (e.g. operator preference satisfaction).

3.2.1 Coupled Trajectories

In some cases, the objective is stated directly over the coupled trajectories in co-manipulation. In [52, 53], the objective of minimizing jerk (time derivative of acceleration) during interaction is taken. In [54], more general motion properties are considered (acceleration, velocity). In other work, the robot's dynamics are adjusted to to minimizing task position error [55]. The robot dynamics can also be adjusted to improve task performance—e.g., the time taken on a point-to-point positioning task [56].

These objectives are tractable—they can be readily measured experimentally, and used to optimize robot behavior, but there is no necessary relation with ergonomy, usability, or task-related performance.

3.2.2 Human-Centered Objectives

In other HRI applications, the objective depends on a human state (e.g. the operator's intent or preferences). Several frameworks are proposed to accommodate this, namely game theory [57, 58] and partially observed Markov decision problems (POMDPs) [59, 60]. These frameworks are sequential decision-making problems which capture the dynamic aspect of these tasks. The POMDP formulation assumes that the robot does not directly observe the complete state of the environment—i.e. the state of the human is not completely known to the robot. The robot's behavior then aims to maximize the expected total reward (which may depend on the hidden state). In game theoretic approaches, the human is explicitly modelled as taking actions, and assumptions are made on the principles which govern these actions, typically that the human is rationally choosing actions which maximize their expected total reward [45]. Game theoretic and POMDP approaches are typically further removed from industrial applications [61], but allow the expression of objectives which depend on human objectives or behavior. Other frameworks for collaboration are proposed, such as commitments [62] or a pure communication viewpoint [63], but they are not yet operationalized with engineering methods.

Directly optimizing a less-quantified operator state such as 'satisfaction' is typically difficult. There are not many models which allow this to be estimated from sensor values, and using self-reported values, e.g. in questionnaires, presents challenges in both data quality (accuracy of self-reported values) and the quantity (model-free robot behavior design typically requires substantial data).

Uncertainty can also enter in the objective. In [64], an objective over the joint robot/human state is posed in the risk-sensitive optimal control framework, where the uncertainty in the estimation of the human state is explicitly considered. Uncertainty is also considered in POMDP formulations, with continuous [59] and discrete unobserved states [21]. However, the lack of tractable POMDP solutions for continuous states is a barrier to the widespread adoption of these methods.

Collaboration is also sometimes posed as shared autonomy or arbitration [6, 65, 66], under the viewpoint that either the robot or human determines the coupled trajectory. In the coupled systems framework, authority can be viewed as the physical ease by which an agent brings the coupled trajectory to their desired trajectory. Much of the work in mixed authority casts authority as a conserved quantity—one which either the robot or human have more of. Quite often, authority is associated with the stiffness of an agent [67], where a stiffer agent imposes their position on the coupled behavior.

3.3 Robots as Students

A third type of HRI objective is a meta-objective beyond a single task iteration—human instruction of the robot. HRI can also employ humans to flexibly re-program robots, such as formalized in learning from demonstration [68], or by the humans providing labeled data to the robot [69]. This can provide a means for online corrections to a robot's behavior, and allow for an iterative process where the human can correct the trajectories of the robot, with the robot moving towards autonomy as the task is correctly taught.

A robot can be considered as more than a passive student, by asking the right questions to clarify uncertainty in the task. This is formalized in collaborative inverse reinforcement learning, where a human and root collaborate to understand the human's preferences [70], which, once understood, can be used as a basis for action.

4 Conclusion

A major limitation to the design and deployment of HRI is predictability. When the human operates the robot as a tool, they need to form predictions on how the robot will react to their actions. This robot predictability is potentially compromised when robots infer states of the collaborator, which adds additional non-visible states to the robot, and can be further compromised when robots adapt online. Conversely, during the robot programming, good predictions are needed for how the human will respond to a robot behavior. Limitations in human predictability necessitate the use of application-specific designer intuition or in situ learning methods to inform the design of robot behavior. Data-driven learning methods are promising, but if the human behavior depends on the robot, changes in robot behavior may limit the effectiveness of historical data. While challenging, developing HRI in production environments—where human behavior is constrained by the factory context, thus improving its predictability—provides a feasible path towards robots which understand and flexibly collaborate with their human partners.

References

1. Tsarouchi P, Makris S, Chryssolouris G, Human–robot interaction review and challenges on task planning and programming 29(8):916–931
2. Krüger J, Lien TK, Verl A, Cooperation of human and machines in assembly lines 58(2):628–646
3. Bastidas-Cruz A, Heimann O, Haninger K, Krüger J, Information requirements and interfaces for the programming of robots in flexible manufacturing
4. Willems JC, The behavioral approach to open and interconnected systems 27(6):46–99
5. Hoffman G, Evaluating fluency in human–robot collaboration 49(3):209–218
6. Losey DP, McDonald CG, Battaglia E, O'Malley MK, A review of intent detection, arbitration, and communication aspects of shared control for physical human–robot interaction 70(1):010804
7. Ajoudani A, Zanchettin AM, Ivaldi S, Albu-Schãffer A, Kosuge K, Khatib O, Progress and prospects of the human–robot collaboration 1–19
8. Demiris Y, Prediction of intent in robotics and multi-agent systems 8(3):151–158
9. Duchaine V, Gosselin CM, General model of human-robot cooperation using a novel velocity based variable impedance control. In: EuroHaptics conference, 2007 and symposium on haptic interfaces for virtual environment and teleoperator systems. World Haptics 2007, Second 3oint, IEEE, pp 446–451
10. Paynter HM, Analysis and design of engineering systems: class notes for M.I.T. course 2,751. M.I.T. Press
11. Hogan N, Impedance control: an approach to manipulation. In: American control conference, 1984, IEEE, pp 304–313
12. Khan SG, Herrmann G, Al Grafi M, Pipe T, Melhuish C, Compliance control and human–robot interaction: part 1—survey, 11(03):1430001
13. Haddadin S, Albu-Schaffer A, Frommberger M, Rossmann J, Hirzinger G, The "dlr crash report": towards a standard crash-testing protocol for robot safety-part ii: discussions. In: IEEE international conference on robotics and automation, 2009. ICRA '09, IEEE, pp 280–287

14. Haninger K, Surdilovic D, Bounded collision force by the Sobolev norm: compliance and control for interactive robots. In: 2019 IEEE international conference on robotics and automation (ICRA), pp 8259–8535
15. ÂAström KJ, Murray RM, Feedback systems: an introduction for scientists and engineers. Princeton University Press
16. Roth E, Howell D, Beckwith C, Burden SA, Toward experimental validation of a model for human sensorimotor learning and control in teleoperation 101941X
17. Conditt MA, Mussa-Ivaldi FA, Central representation of time during motor learning 96(20):11625–11630
18. Burdet E, Tee KP, Mareels I, Milner TE, Chew CM, Franklin DW, Osu R, Kawato M, Stability and motor adaptation in human arm movements 94(1):20–32
19. Nikolaidis S, Nath S, Procaccia AD, Srinivasa S, Game-theoretic modeling of human adaptation in human-robot collaboration. In: Proceedings of the 2017 ACM/IEEE international conference on human-robot interaction, ACM, pp 323–331
20. Yanco HA, Drury JL, Scholtz J, Beyond usability evaluation: analysis of human-robot interaction at a major robotics competition 19(1–2):117–149
21. Roncone A, Mangin O, Scassellati B, Transparent role assignment and task allocation in human robot collaboration. In: 2017 IEEE international conference on robotics and automation (ICRA), IEEE, pp 1014–1021
22. Eyssel F, Kuchenbrandt D, Bobinger S, Effects of anticipated human-robot interaction and predictability of robot behavior on perceptions of anthropomorphism. In: Proceedings of the 6th international conference on human-robot interaction, pp 61–68
23. Dragan AD, Srinivasa SS, A policy-blending formalism for shared control 32(7):790–805
24. Li Y, Ge SS, Human–robot collaboration based on motion intention estimation 19(3):1007–1014
25. Wang C, Li Y, Ge SS, Lee TH, Reference adaptation for robots in physical interactions with unknown environments 47(11):3504–3515
26. Kang G, Oh HS, JKSeo HS, Kim U, Choi HR, Variable admittance control of robot manipulators based on human intention 24(3):1023–1032
27. Peternel L, Tsagarakis N, Caldwell D, Ajoudani A, Robot adaptation to human physical fatigue in human–robot co-manipulation, pp 1–11
28. Gopinathan S, Otting S, Steil J, A user study on personalized adaptive stiffness control modes for human-robot interaction. In: The 26th IEEE international symposium on robot and human interactive communication, pp 831–837
29. Khoramshahi M, Billard A, A dynamical system approach to task-adaptation in physical human–robot interaction 43(4):927–946
30. Rani P, Sarkar N, Smith CA, Kirby LD, Anxiety detecting robotic system–towards implicit human-robot collaboration 22(1):85–95
31. Sadigh D, Sastry SS, Seshia SA, Dragan A, Information gathering actions over human internal state. In: 2016 IEEE/RS3 international conference on intelligent robots and systems (IROS), IEEE, pp 66–73
32. Kanno T, Nakata K, Furuta K, A method for team intention inference 58(4):393–413
33. Peternel L, Tsagarakis N, Ajoudani A, Towards multi-modal intention interfaces for human-robot co-manipulation. In: Proceedings 2016 IEEE/RS3 international conference on intelligent robots and systems (IROS)
34. Kaneishi D, Matthew RP, Tomizuka M, A sEMG classification framework with less training data. In: 2018 40th annual international conference of the IEEE engineering in medicine and biology society (EMBC), IEEE, pp 1680–1684
35. Gomi H, Kawato M, Human arm stiffness and equilibrium-point trajectory during multi-joint movement 76(3):163–171
36. Medina JR, Endo S, Hirche S, Impedance-based Gaussian Processes for predicting human behavior during physical interaction. In: 2016 IEEE international conference on robotics and automation (ICRA), IEEE, pp 3055–3061

37. Erden MS, Billard A, End-point impedance measurements at human hand during interactive manual welding with robot. In: 2014 IEEE international conference on robotics and automation (ICRA), IEEE, pp 126–133
38. Tsumugiwa T, Yokogawa R, Hara K, Variable impedance control based on estimation of human arm stiffness for human-robot cooperative calligraphic task. In: IEEE international conference on robotics and automation, 2002. Proceedings. ICRA '02 1, IEEE, pp 644–650
39. Haninger K, Surdilovic D, Identification of human dynamics in user-led physical human robot environment interaction. In: 2018 27th international symposium on robot and human interactive communication (RO-MAN), pp 509–514
40. Wang Z, Wang B, Liu H, Kong Z, Recurrent convolutional networks based intention recognition for human-robot collaboration tasks. In: 2017 IEEE international conference on systems, man, and cybernetics (SMC), IEEE, pp 1675–1680
41. Wang Z, Mülling K, Deisenroth MP, Ben Amor H, Vogt D, Schölkopf B, Peters J, Probabilistic movement modeling for intention inference in human–robot interaction 32(7):841–858
42. Takagi A, Ganesh G, Yoshioka T, Kawato M, Burdet E, Physically interacting individuals estimate the partner's goal to enhance their movements 1(3):0054
43. Sunberg ZN, Ho CJ, Kochenderfer MJ, The value of inferring the internal state of traffic participants for autonomous freeway driving. In: 2017 American control conference (ACC), pp 3004–3010
44. Albrecht SV, Stone P, Autonomous agents modelling other agents: a comprehensive survey and open problems 258:66–95
45. Choudhury R, Swamy G, Hadfield-Menell D, Dragan A, On the utility of model learning in HRI
46. Villani V, Pini F, Leali F, Secchi C, Survey on human–robot collaboration in industrial settings: Safety, intuitive interfaces and applications 55:248–266
47. Cherubini A, Passama R, Crosnier A, Lasnier A, Fraisse P, Collaborative manufacturing with physical human–robot interaction 40:1–13
48. Medina JR, Lorenz T, Hirche S, Considering human behavior uncertainty and disagreements in human–robot cooperative manipulation. In: Wang Y, Zhang F (eds) Trends in control and decision-making for human–robot collaboration systems, pp 207–240, Springer International Publishing
49. Augustsson S, Olsson J, Christiernin LG, Bolmsjö G, How to transfer information between collaborating human operators and industrial robots in an assembly. In: Proceedings of the 8th nordic conference on human-computer interaction: fun, fast, foundational, pp 286–294
50. Michalos G, Kousi N, Karagiannis P, Gkournelos C, Dimoulas K, Koukas S, Mparis K, Papavasileiou A, Makris S, Seamless human robot collaborative assembly—an automotive case study 55:194–211
51. Fridovich-Keil D, Bajcsy A, Fisac JF, Herbert SL, Wang S, Dragan AD, Tomlin CJ, Confidence-aware motion prediction for real-time collision avoidance 1:0278364919859436
52. Ranatunga I, Cremer S, Popa DO, Lewis FL, Intent aware adaptive admittance control for physical Human-Robot Interaction. In: 2015 IEEE international conference on robotics and automation (ICRA), IEEE, pp 5635–5640
53. Dimeas F, Aspragathos N, Reinforcement learning of variable admittance control for human-robot co-manipulation. In 2015 IEEE/RSJ international conference on intelligent robots and systems (IROS), IEEE, pp 1011–1016
54. Lawitzky M, Kimmel M, Ritzer P, Hirche S, Trajectory generation under the least action principle for physical human-robot cooperation. In: 2013 IEEE international conference on robotics and automation, IEEE, pp 4285–4290
55. Ranatunga I, Lewis FL, Popa DO, Tousif SM, Adaptive admittance control for human-robot interaction using model reference design and adaptive inverse filtering, 25(1):278–285
56. Modares H, Ranatunga I, Lewis FL, Popa DO, Optimized assistive human–robot interaction using reinforcement learning 46(3):655–667
57. Dragan AD, Robot planning with mathematical models of human state and action

58. Li Y, Carboni G, Gonzalez F, Campolo D, Burdet E, Differential game theory for versatile physical human–robot interaction 1(1):36
59. Brooks C, Szafir D, Building second-order mental models for human-robot interaction
60. Munzer T, Toussaint M, Lopes M, Preference learning on the execution of collaborative human-robot tasks. In: 2017 IEEE international conference on robotics and automation (ICRA), IEEE, pp 879–885
61. Mutlu B, Roy N, Săbanović S, Cognitive human–robot interaction. In: Springer handbook of robotics, Springer, pp 1907–1934
62. Castro VF, Clodic A, Alami R, Pacherie E, Commitments in human-robot interaction
63. Unhelkar VV, Yang XJ, Shah JA, Challenges for communication decision-making in sequential human-robot collaborative tasks. In: Workshop on mathematical models, algorithms, and human-robot interaction at R:SS
64. Medina JR, Lorenz T, Lee D, Hirche S, Disagreement-aware physical assistance through risk-sensitive optimal feedback control. In: 2012 IEEE/RSJ international conference on intelligent robots and systems, IEEE, pp 3639–3645
65. Reddy S, Dragan AD, Levine S, Shared autonomy via deep reinforcement learning
66. Javdani S, Srinivasa SS, Bagnell JA, Shared autonomy via Hindsight optimization
67. Gomes W, Lizarralde F, Role adaptive admittance controller for human-robot co-manipulation
68. Rozo L, Calinon S, Caldwell DG, Jimenez P, Torras C, Learning physical collaborative robot behaviors from human demonstrations 32(3):513–527
69. Arumugam D, Lee JK, Saskin S, Littman ML, Deep reinforcement learning from policy-dependent human feedback
70. Hadfield-Menell D, Russell SJ, Abbeel P, Dragan A, Cooperative inverse reinforcement learning. In: Advances in neural information processing systems, pp 3909–3917
71. Nemec B, Likar N, Gams A, Ude A, Human robot cooperation with compliance adaptation along the motion trajectory, 42(5):1023–1035

Human–Robot Collaboration Using Visual Cues for Communication

Iveta Eimontaite

Abstract The present chapter addresses the fundamental roles played by communication and mutual awareness in human/robot interaction and co-operation at the workplace. The chapter reviews how traditional industrial robots in the manufacturing sector have been used for repetitive and strenuous tasks for which they were segregated due to their hazardous size and strength, and so are still perceived as threatening by operators in manufacturing. This means that successful introduction of new collaborative systems where robotic technology will be working alongside and directly with human operators depends on human acceptance and engagement. The chapter discusses the important reassuring role played by communication in human–robot interaction and how involving users in the design process increases not only the efficiency of communication, but provides a reassuring effect.

1 Introduction

Although industrial robots have been established in the manufacturing sector since the seventies and are mainly being used for repetitive and strenuous tasks (such as welding, stacking, etc.; [1]), robots and automated technology are still perceived as threatening by operators in manufacturing. This is particularly important as with current advances in sensor technologies, robots are being rolled out onto the shop floor without being segregated from the workforce, which raises additional opportunities, challenges, and requirements from the operators. With technology development, highly connected, intelligent systems will enable processes that are more efficient, productive, and responsive to customer demands [2], and yet, the success of robotic technology working alongside human operators depends on its acceptance and engagement. The reports show that from around 40–90% of the innovations fail due to consumers or users rejecting it [3]. Not surprisingly, the same is relevant in the manufacturing domain—technology needs to be operated by human operators,

I. Eimontaite (✉)
Industrial Psychology and Human Factors Group, Cranfield University, Cranfield, UK
e-mail: Iveta.Eimontaite@cranfield.ac.uk

M. I. Aldinhas Ferreira and S. R. Fletcher (eds.), *The 21st Century Industrial Robot: When Tools Become Collaborators*, Intelligent Systems, Control and Automation: Science and Engineering 81, https://doi.org/10.1007/978-3-030-78513-0_5

and therefore their acceptance is essential for technology to achieve its full potential. Working closely and interactively with a robot will inevitably require greater communication clarity between the operator and their robot partner.

This chapter will discuss why communication in human–robot interaction is important and needed; how to involve users in the design process to increase efficiency of created communication and what task related and psychological effects it has on the user. The final section will review the ongoing challenges that affect the development of productive and stimulating communication between manufacturing operators and robots.

2 Why Do We Need Communication Between Humans and Robots?

Interaction by definition implies a level of communication. For us humans this is the basis of our society from allowing us to express ourselves to enabling collaboration with other members of group. The majority of information is transferred through verbal communication including conversations, visual information (images, graphics, symbols) and non-verbal communication (gestures, nods, gaze). Visual communication is considered to be more effective in conveying complex issues to large audiences.

The same social interaction principles apply in the human robot interaction. Not surprisingly, the majority of the research being done is in the social robotics field. In this field a robot's main purpose is to interact and provide companionship to the user. However, the varying levels of communication become important particularly within the manufacturing environment as robots are being designed for collaboration. Audio, text and graphic types of communication provide certain benefits and drawbacks. Firstly, the main benefit of verbal interaction is that it places a reduced demand on mental workload and enables a greater focus on the task, however, it is challenging in a noisy work environment. Visual information presentation has the drawback of increased mental workload and requires users to divide their attention between the task and the information being communicated. However, it is optimal for a variety of environments and people, as it is not affected by the noise that so often accompanies manufacturing work. Afterall, 80% of people remember what they see, compared to 10% what they hear and 20% of what they read. In addition, the graphic modality has the benefits of requiring little experience and training to use [4], it can be beneficial for people with different information processing abilities such as dyslexia [5] and in most cases can use universally understood symbols, providing support to people from different countries and cultures [6]. Some research suggests that the modality of information communication (audio, text, graphic) does not affect trust and user experience [7], others have found that displaying information graphically has in fact a minimal effect on increase in workload [8].

The reason clear communication of information is critical to effective working relationships is that it not only enables collaboration, but it also provides physical and psychological assurance and has direct benefits to the user. As the reliance on the autonomy, complexity and safety of robots increase, human operators need confidence in robotic co-workers and their capacities so that true collaboration can occur. One of the essential requirements for this confidence to be built is an appropriate level of trust [9, 10]. Too low or too high levels of trust (i.e. not checking the robot partner, or checking too much) leads to errors and greater task completion time [11]. Another issue is that robots in manufacturing can still feel threatening and can be perceived to be taking control from humans. Feeling out of control results in higher stress levels [12, 13]. Understanding the requirements of unfamiliar situations, having the necessary knowledge and information, results in empowerment and sense of control [14] as well as a decrease in stress levels [13, 15, 16]. Finally, an individual's cognitive load has to be considered, as it is often already high in manufacturing [17] and there can be little capacity beyond undertaking a complex activity for monitoring co-workers' progress. This issue is exacerbated if users feel they do not have enough information or training on a task. While increased cognitive load can lead to decreased concentration on task performance and an increased number of accidents [18, 19], establishing effective communication measures, in fact, can reduce the amount of information necessary for efficient decision-making [20].

Communication between co-workers, irrespective if it is between human partners or a human–robot team, is necessary for both an efficient and positive work environment. It can not only increase productivity and safety, but empower users, make them feel in control, less stressed and result in lower cognitive load. However, this raises the question what information is essential and needed from the user to understand robots and built a successful collaboration in manufacturing.

3 What Information Needs to Be Communicated in Manufacturing?

One way of testing how the information communication affects the user is to ask for their input on what information they need while they complete a task and interact with a robot. This information can be gathered through two approaches: (i) semi structured discussions and activities, which encourage the free flow of information, and (ii) to test already pre-designed information through experiments and structured interviews. The section below will discuss these approaches and their outcomes within two use cases.

3.1 Free Flow Approach

Free flow approach allows to engage users on various creative tasks. These tasks normally are designed to have two layers of information they are gathering. The first layer asks for user input in the design or evaluation of the technology. The second layer allows to assess users' attitudes towards the technology in question. This section will discuss a delivery and results of this type of work in manufacturing industry.

Two co-creation workshops were organised with eight manufacturing operators aiming to clarify what information users need to operate robots with confidence and to discuss various ways of increasing their acceptance of robots and confidence in working with them [21]. During the workshops, participants engaged in numerous hands-on tasks from designing the signage for interaction with a robot, to designing the workcell environment. These activities allowed the research team to capture the changes in attitudes and anxieties towards robots throughout the workshops without increasing the participants' performance bias-tendency to respond on the task to confirm researchers' predictions.

The main findings of the workshop showed that to accept robots, users feel they need more information regarding safety and an instant feedback on their own performance. This was particularly evident from a signage design activity during workshop 1 where participants were asked to design signs they would need while working alongside the robot (Fig. 1b). Participants mostly concentrated on designing caution/do not touch signs. Out of the 29 signs produced, 18 were related to "do not touch", while 5

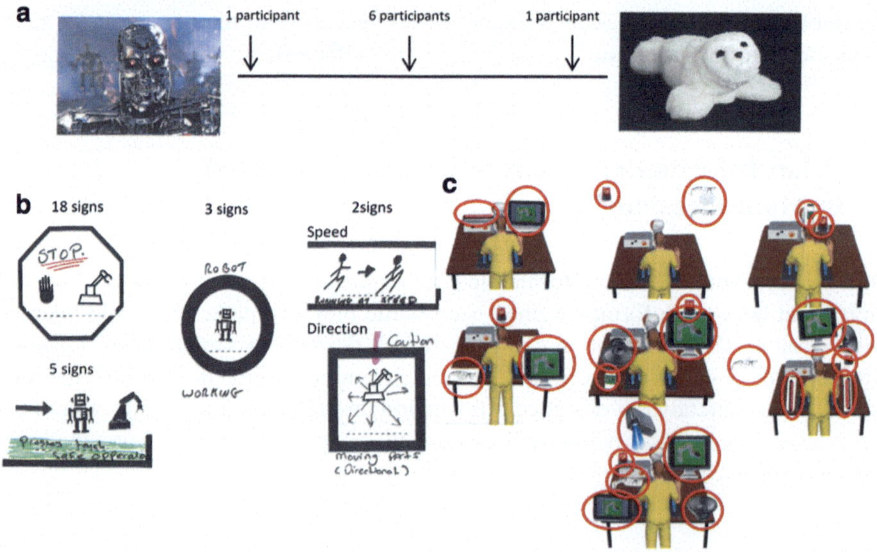

Fig. 1 Results from key activities during the design workshops; **a** Participants' attitude towards the robots during the first design workshop; **b** Important information communication via signs; **c** Information communication methods in the workplace

indicated "it is safe to touch the robot", and only 2 signs were of an instructional type. Furthermore, participants indicated that when they concentrate on the task, they get "tunnel" vision, and warning signals should attract their attention in various ways (Fig. 1c). Also, an indicator above the workplace should notify the supervisor about emerging problems.

During the second workshop participants got more comfortable operating the robot and tried pushing its limits; pushing harder, and faster until the robot "jams". This indicated participants' reinforcement learning; which states that if users don't know the limits of the robot before interaction, they test the boundaries and then try to keep within them. Furthermore, although the signs said "do not touch", the more comfortable the participants became with the robot the more likely they were to take the bolt from the robot, touching it despite being told not to, because they wanted to complete the task faster. These two examples emphasise the need to communicate deep knowledge about the robot and its operation to its operators.

Finally, the communication about the robot abilities and user needs should capture and assess changes in user attitudes and anxiety. During the first design workshop, the signage design task showed that participants were concerned about human safety (Fig. 1b; 18 out of 29 signs were related to keeping distance from a robot). The biggest change happened after a discussion about whether robots could increase safety while working with spot-welding equipment. This change was assessed by presenting participants with a scale where on one end there was an image of a Paro robot—a soft robotic baby seal used for companionship, while at the other end there was an image of the Terminator robot. A change could be seen towards the end of the first workshop when participants indicated that they view robots not as friendly as Paro, but also not as dangerous as the Terminator (Fig. 1a). This tendency remained during second workshop; the majority of participants used only 2 or 3 information communication methods, although one participant still used all of the provided communication methods (Fig. 1c). The usage of all of the communication methods in the workplace served as an indicator of the anxiety levels they experienced which later was confirmed via discussion.

3.2 *Structured Approach*

Structured approach differs from the free flow mainly in predefined hypothesis and variables which are being tested. The environment and manipulation of the variables is controlled and only certain questions allow participants free input. An example of an experimental approach discussed further was aimed to determine which visual cues are most effective in indicating robot behaviour and/or how they can be improved [22]. Participants in the study viewed six videos lasting 20 s each. The videos portrayed an operator working with a robot: together they lifted a component from a shelf and placed it on the table. In the left top corner of the video there were visual cues indicating what the robot is communicating to the operator. While the task in the video remained constant, there were six sets of the information presented in the top

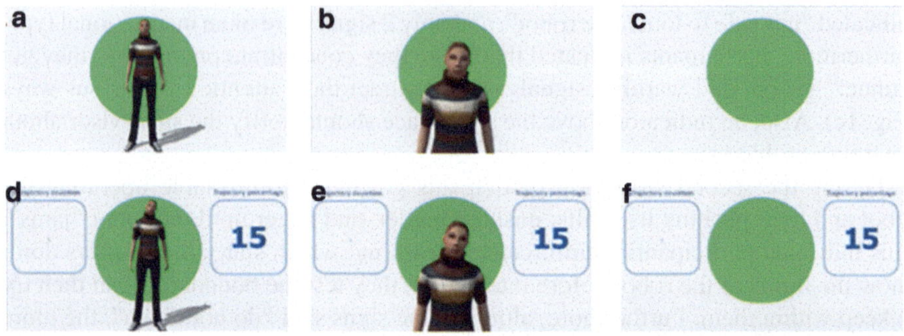

Fig. 2 Visual cues presented to participants in the study. Left to right: full-size avatar, half body avatar, no avatar. The top row indicates visual cues without graphical signage (trajectory deviation and trajectory percentage), and the bottom row presents visual cues with graphical signage

left corner (Fig. 2). The background circle which was present in all the visual cues indicated accuracy (keeping to the ideal trajectory for task completion). With the deviation, the colour changed from green to red. The bottom row of Fig. 2 presents the addition of trajectory deviation and trajectory percentage imagery. Finally, the Avatar was represented in two forms: half body (torso) and full body size. The Avatar was designed to use head nods to indicate the trajectory correction direction once participants started to deviate from the optimal trajectory.

After watching each video participants were presented with an open-ended question asking about the meaning of the visual cues. In addition to this question, there was a slider type question asking participants to indicate the clarity of the signage from 0 "Extremely unclear" to 100 "Extremely clear". Each condition/video was presented in a separate block of the questionnaire in a counterbalanced order. After these six conditions were presented to participants, follow up questions about their overall impressions of the visual cues were asked. The images of six different visual cues were displayed and participants were asked to choose one or more of the cues which they would like to use while performing a similar task with a robot. This question was followed by two open-ended questions asking participants to describe the reason for their choice and what they would change to make the information clearer.

The main question of the study was to determine which set of visual cues participants rated as clearest in terms of indicating what the robot would do next and the question relating to the clarity of the signs was investigated with inferential statistics. The results showed that the background circle with trajectory deviation and trajectory percentage (Fig. 2f) was rated significantly higher than any visual cues with the avatar (Fig. 2a–d); however, there was no significant difference comparing the background circle with trajectory deviation and trajectory percentage (Fig. 2f) to just the background circle (Fig. 2c). Yet, looking at the background circle, the difference reached significant levels only when compared this interface to interface containing only the avatar (without trajectory deviation and trajectory percentage; Fig. 3).

The qualitative results confirm the quantitative findings indicating that 75% of the participants preferred the user interface with the background circle and trajectory

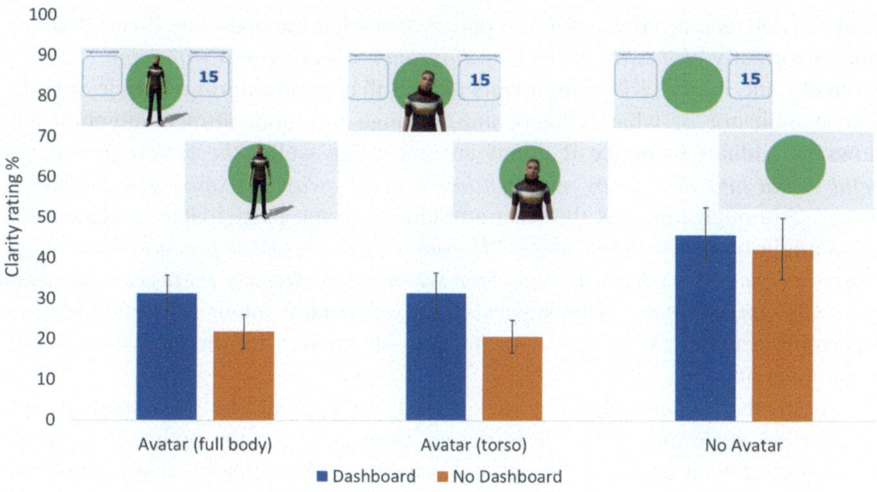

Fig. 3 Clarity ratings in percentage across conditions (±SEM)

percentage and trajectory deviation (Fig. 2f), 15% preferred the avatar torso with trajectory percentage and trajectory deviation (Fig. 2e), 5% only with the background circle (Fig. 2c), and 5% favoured the avatar torso without trajectory deviation and trajectory percentage (Fig. 2b). The full body size avatar (with or without trajectory deviation and trajectory percentage, Fig. 2a, d)) was not chosen by any participants. To investigate the reasons for such preferences, participants' answers to the open-ended questions were analysed. Comments relating to each part of the visual cues (avatar, trajectory deviation and trajectory percentage, background circle) under the investigation were separated and summarised below.

Although some participants appreciate that the avatar "*makes [it] more comfortable to interact with a robot*", other participants expressed dissatisfaction with it: "*I already have my partner that bosses me around at home, I don't need another one in the workshop*". Overall, participants suggested that the half-size avatar might be more useful; however, the movement and expressions need to be more purposeful and the size of the face should be bigger—"*if the face was bigger and maybe you could perceive the gestures better*". For example, the avatar could be showing what needs to be done, in what position an operator might stand and how to hold the part.

The trajectory percentage and trajectory deviation graphics were not intuitive to all participants, and some indicated that watching several videos helped them to understand it better. There was also confusion about what trajectory percentage meant—a few participants noted that percentage meant "*task completion*", one said that "*the numbers provide an accuracy regarding the trajectory of the object*", and other participants interpreted that it was the deviation from the right path. Some participants suggested using a 3D positioning of the arrow or "*use two arrows from the front and side point of views*". Also, trajectory percentage and trajectory deviation could be used as a reference point to understand what the operators themselves are doing incorrectly "*I'd add a trajectory diagram and compare it with mine*".

Finally, the majority of participants agreed that the background circle was the clearest indicator of what is happening. Change in colour attracts attention and allows individuals to notice if things are not going well: "*the colour provides a useful visual hint that grabs my attention straight away—to know if something is wrong*". Although this was the least ambiguous, some participants indicated that it also might be confusing at times: "*Presumably the brighter green the button the better - but exactly in what way is unclear. When it goes dimmer, there is no clue as to why it has gone dimmer*". This suggests that background colour is a good indicator of performance, but on its own does not provide enough information for accurate decision making.

Participants indicated that the background circle should be accompanied with a benchmark scale indicating the trajectory from good (on the optimal trajectory) to bad (high deviation from the optimal trajectory). Participants also commented that it might be useful to introduce commands or a feedback display to keep the communication less ambiguous. Some of the participants indicated this could be done via audio feedback to the user.

These studies investigated how users would like the communication between themselves and a robot to happen. Although there are lots of differences between the engagement of the users, they produced similar findings. First of all, participants main concern was the clarity of the visual information presented. Both studies revealed that information needs to be presented unambiguously and clearly, i.e. the size and position of the visual information is important. Secondly, for the confident completion of the task, participants indicated that instant comprehension or provided training on the visual information beforehand is needed. Furthermore, participants' qualitative comments from studies revealed what common aspects are needed from visual information communication:

- emphasizing the importance of information relating to the performance (when it is safe/ not safe, and also deviation from accuracy)
- a limited number of communication methods or symbols are required
- audio information should be used to attract attention if something unexpected happens (to distract from tunnel vision) or to indicate how to improve accuracy in a few words
- a communication method that shows the operator how to complete the task.

The results and suggestions that participants made during these two studies have been applied to further the development of information communication methods during human–robot collaboration. Although not all suggestions were possible for the follow-up studies due to technical or practical implementations, the majority of their needs were reflected in the improvements that were tested experimentally. The findings from this experiment are discussed in the following section.

4 How Can Communication Affect User Work and Wellbeing?

Clear information provision has been shown to improve safety and decrease the number of accidents in manufacturing settings [18] and on the road [19]. Navigation of unfamiliar settings has been found to take less time with clear information [23, 24] and it has enabled performance with minimal training (i.e. IKEA instructions). However, to achieve these aims the communication needs to be efficient. The previous section presented and discussed two approaches of how to involve users to define the information that needs to be communicated. In this section two follow-up studies are discussed that show how the refined information communication was affected by and delivered in static (not moving) and dynamic (adjusting to the operator movement) communication modes. It also discusses the importance of the testing environment: laboratory and general public users versus manufacturing environment and operators. The laboratory experiment settings allow for a controlled environment with measured manipulation of variables. The manufacturing field study, on the other hand, provides the validation of the laboratory studies and shows the practical application of the findings. Both approaches are complimentary and show the importance for understanding collaboration.

4.1 Laboratory Test

In this section two laboratory studies will be discussed. Both of these studies were conducted within the controlled research environment of a laboratory with participants made up of students and staff available on site. Although the studies used different experimental designs, and explored different communication modes (static vs. dynamic) the results of both studies are to a certain level consistent.

The first study tested 90 volunteer students and university staff and explored the effects of static graphical signage while co-working on an industrial type task with a robotic KUKA iiwa arm. One third of the participants were presented with task relevant signage indicating (A) robot movement speed, (B) robot movement plane (x and y axis), (C) stop and start symbols indicating passive and active states of the robot, (D) the robot is operated by touch, (E) operational area, and (F) the force needed to move the robot, (Fig. 4). The other 30 participants received information that was not relevant to the task completion consisting of robot payload, sound emitted, operational temperature, not water resistant, not to attempt to repair it, and the operational time. The final group of participants did not receive any signage.

The results of the pick and place task where participants had to collaborate with a robot to remove M5 bolts from 18 narrow tubes showed that accuracy (number of mistakes or down-time due to incorrect operation of robot) was lowest in the experimental group, while in the group with task irrelevant signage and no-signage groups accuracy did not significantly differ.

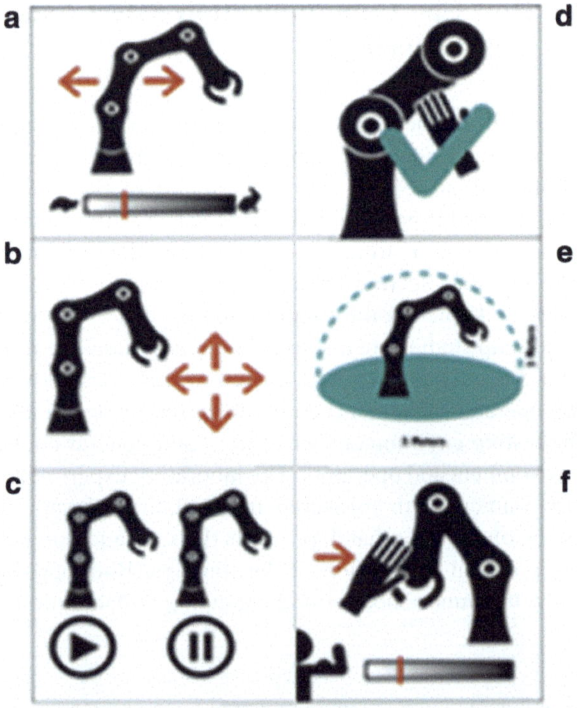

Fig. 4 Experimental condition signs

The emotional reactions of the participants were captured during the experiment using facial reading technology. Interestingly, when assessing the participants' emotional reactions during the task by analysing their facial expressions, successfully completed trials showed positive valance (emotion) in both of the signage conditions, however, the control condition which had no-signage showed negative emotional valance. Although this result was only at a trend level, it suggests that having some information about the task—irrespective of its relevance—increases satisfaction with the task. Furthermore, graphical signage affected the participants' robot anxiety. The moderated regression showed that with increasing accuracy post experimental scores on the behavioural subscale of the RAS [25] were reduced compared to pre-experimental scores on this scale, but only in the experimental group. This shows that graphical signage can reduce the robot anxiety after interacting with a robot, but only when the information is meaningful and relevant to the task.

The second example is a study conducted by Ibarguren et al. [22] investigating dynamic graphical signage within three repeated measures conditions: Avatar (torso) with trajectory deviation and trajectory percentage (the Avatar condition; Fig. 5), the No-Avatar condition trajectory deviation and trajectory percentage and the control condition of no-signage. The main aim of the study was to examine how collaboration with a dual arm robot with interactive user interface can aid the manufacturing task

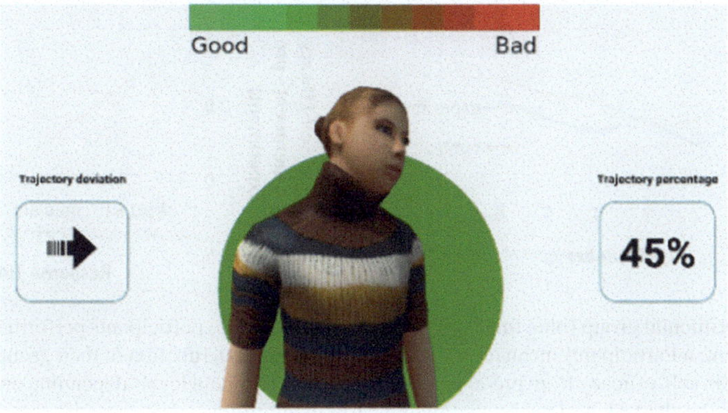

Fig. 5 Experimental condition user interface with Avatar (torso), background circle and trajectory deviation and trajectory percentage

of component removal from a shelf and placement on a desk. Twelve research centre employees with varying degrees of experience working with robots took part in this study.

The behavioural results are consistent with the previously mentioned study [26] in the sense that the user interface with an avatar showed the lowest deviation from the optimal path (highest accuracy) compared to the other two conditions, although the task completion time was not significantly affected. Additionally, the researchers explored how different interfaces affected mental workload as measured by the NASA TLX [27]. The results confirmed that the highest mental workload was observed in the no interface condition, while the lowest was found for the user interface with no avatar.

4.2 Field Study

Although it seems that both studies discussed so far in this section show that communication increases the accuracy of a task, it is important to keep in mind that these studies were not conducted with manufacturing employees. An additional study, which built on the static signage experiment described above by further refining the interface to include dynamic information, was conducted with manufacturing operators. The equivalent task of collecting M5 bolts from narrow tubes was conducted with two groups of participants: a control group with no-signage (20 participants) and an experimental group with dynamic signage (20 participants). The results showed that the experimental group of participants were significantly faster than the control group participants over the 6 trials controlling for tube position (Fig. 6a). Although accuracy on the task was similar between control and experimental groups (participants collected similar numbers of bolts over a similar number of trails), the experimental

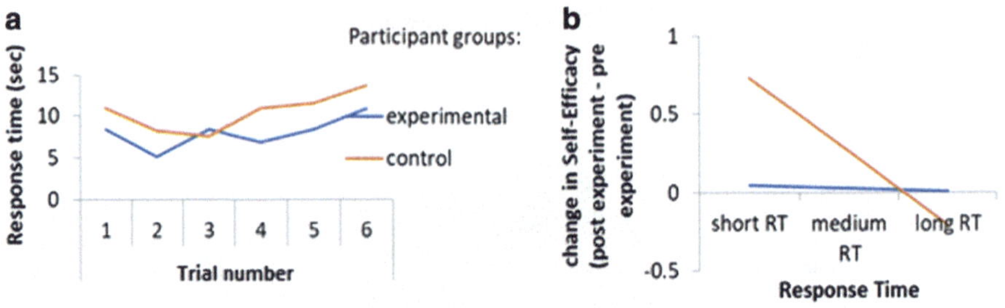

Fig. 6 Experimental group (blue line) and control group (red line) participants performance during the experiment; **a** Participants mean response time over 6 trials as a function of their group; **b** Participants change in self-efficacy from pre-experiment to post-experiment levels depending on their mean response time on the task and moderated by the participant group

group participants were on average 21% faster at completing the task over 6 trials. The main finding of the study provides evidence that graphical signage allows to complete the task quicker. The dynamic signage provided information for the participant about changes in the human robot collaborative process and therefore could help to more quickly complete each trial without needing to unnecessarily adjust robot position (adding more time for the trial completion) while the no-signage participants did not have this benefit.

Furthermore, from the psychological aspect, information communication about the robot throughout the task suggested that there were positive improvements in terms of acceptance and even willingness to engage with the new technology. Response time predicted a self-efficacy change which was moderated by group (Fig. 6b). The self-efficacy was assessed with a Compeau and Higgins Computer Self-Efficacy scale [28]. Self-efficacy relates how confident one feels while interacting with a new technology. The average response time predicted 5% in participants' self-efficacy change in the experimental group, but it was not significant. In contrast, in the control group the participants' average response time predicted 27% of self-efficacy change and this result was significant.

Dynamic and static visual information modes might suggest that the task completion results could be affected with differences, yet, the biggest difference in the results came from the testing environment and participant population. However, irrespective of this, the discussed studies present evidence that visual information communication can improve task performance (in terms of accuracy and response time) and also influence the psychological safety of the robot user.

An explanation for the results lies in participants' sense of empowerment and knowledge of the processes they are going through. Graphical signage is designed to help people understand the requirements of unfamiliar situations, and this information can lead to greater empowerment and a sense of control [14] and decreases the levels of stress experienced [13, 15, 16]. Additionally, negative attitudes towards robots decrease after having interacted with them [29], however the decrease in anxiety depends on the robot's behavioural characteristics [30]. Not surprisingly, mental workload is also an indicator how information can help to accomplish tasks without adding additional strain to the users—the user interface with and without an Avatar showed a lower overall mental workload compared to no interface condition. Furthermore, the self-efficacy findings are consistent with the attribution theory [31]. Participants with the information about how to co-work with the robot (experimental group) have a stable confidence in their ability irrespective of their performance. That means that if they are performing slowly they attribute this to the task set-up, but not their ability. On the other hand, the control group participants might think that their performance was a representation of their ability; if they perform slowly they undervalue their ability to work with robots. Stable self-efficacy results in greater resilience to stressors [32, 33].

5 What the Future Holds

Human–robot interaction is not possible without communication. The communication between a robot and its user in manufacturing is a growing field of research providing possible benefits for all stakeholders—from increased safety, productivity, to increased wellbeing and work-life balance. Yet, to achieve its full potential, certain aspects need to be considered when developing the user interfaces and considering the options for the best modality.

The findings discussed in this chapter suggest that visual communication can increase accuracy and self-efficacy, while reducing response time, anxiety, negative attitudes, and mental workload. Although graphical information presentation has its advantages, the responses from participants suggest that dual mode (visual with audio) information communication allows individual flexibility. Further studies are needed to allow more natural interaction between the user and robot to estimate the importance of different information modes (visual, audio, or text) in different environments with varying noise levels. Finally, the importance of user involvement in the design process cannot be underestimated and testing within the environment where the application is intended to be used is essential to assess its true effect.

The future of this research field depends on the technological advancement and possibility to hold a two-way efficient communication. That is, human teams have the benefit of natural language perception and comprehension of non-verbal communication, in particular body language. If the robot is to be integrated in a dynamic work environment, the user expectation might be to have the same quality of information communication as one is used to within ‘traditional’ teams. On the other hand, some participant comments on the avatar’s usefulness pose the question whether people want robots to have an anthropomorphic form and communication within manufacturing. If so, what are the limits of this replication of real-world communication?

Acknowledgements This work is supported by the projects A-GRAfIC (funded by EPSRC Centre for Innovative Manufacturing in Intelligent Automation under grant agreement EP/IO33467/1) and SHERLOCK (funded by the European Union's Horizon 2020 research and innovation programme under grant agreement no. 820689). The researchers would like to thank all project partners and participants for their support enabling this work to be completed.

References

1. Heyer C (2010) Human-robot interaction and future industrial robotics applications. In: 2010 IEEE/RSJ international conference on intelligent robots and systems (IROS), pp 4749–4754 [Online]. http://ieeexplore.ieee.org/xpls/abs_all.jsp?arnumber=5651294. Accessed 18 Mar 2016
2. Pawar VM, Law J, Maple C (2016) Manufacturing robotics. The next robotic industrial revolution (white paper). UK-RAS
3. Castellion G, Markham SK (2013) Perspective: new product failure rates: influence of argumentum ad populum and self-interest. J Prod Innov Manag 30(5):976–979. https://doi.org/10.1111/j.1540-5885.2012.01009.x
4. Tufte ER (1993) The visual display of quantitative information, vol 2. Graphics Press, Connecticut
5. Lamont D, Kenyon S, Lyons G (2013) Dyslexia and mobility-related social exclusion: the role of travel information provision. J Transp Geogr 26:147–157. https://doi.org/10.1016/j.jtrangeo.2012.08.013
6. Ben-Bassat T, Shinar D (2006) Ergonomic guidelines for traffic sign design increase sign comprehension. Human Fact J Human Fact Ergon Soc 48(1):182–195. https://doi.org/10.1518/001872006776412298
7. Sanders TL, Wixon T, Schafer KE, Chen JYC, Hancock PA (2014) The influence of modality and transparency on trust in human-robot interaction. In: 2014 IEEE international interdisciplinary conference on cognitive methods in situation awareness and decision support (CogSIMA), pp 156–159. https://doi.org/10.1109/CogSIMA.2014.6816556
8. Selkowitz AR, Lakhmani SG, Larios CN, Chen JYC (2016) Agent transparency and the autonomous squad member. Proc Human Fact Ergon Soc Ann Meet 60(1):1319–1323. https://doi.org/10.1177/1541931213601305
9. Cameron et al D (2015) Framing factors: the importance of context and the individual in understanding trust in human-robot interaction, presented at the IEEE/RSJ international conference on intelligent robots and systems [Online]. http://iros15-desrps.chrisbevan.co.uk/papers/cameron.pdf. Accessed 5 Feb 2016
10. Hancock PA, Billings DR, Schaefer KE, Chen JYC, de Visser EJ, Parasuraman R (2011) A meta-analysis of factors affecting trust in human-robot interaction. Human Fact J Human Fact Ergon Soc 53(5):517–527. https://doi.org/10.1177/0018720811417254
11. Lee JD, See KA (2004) Trust in automation: designing for appropriate reliance. Human Fact J Human Fact Ergon Soc 46(1):50–80
12. Mathews A, Mackintosh B (1998) A cognitive model of selective processing in anxiety. Cogn Ther Res 22(6):539–560
13. Ozer EM, Bandura A (1990) Mechanisms governing empowerment effects: a self-efficacy analysis. J Personal Soc Psychol 58(3):472
14. Ussher J, Kirsten L, Butow P, Sandoval M (2006) What do cancer support groups provide which other supportive relationships do not? The experience of peer support groups for people with cancer. Soc Sci Med 62(10):2565–2576. https://doi.org/10.1016/j.socscimed.2005.10.034
15. Lautizi M, Laschinger HKS, Ravazzolo S (2009) Workplace empowerment, job satisfaction and job stress among Italian mental health nurses: an exploratory study. J Nurs Manag 17(4):446–452. https://doi.org/10.1111/j.1365-2834.2009.00984.x

16. Pearson LC, Moomaw W (2005) The relationship between teacher autonomy and stress, work satisfaction, empowerment, and professionalism. Educ Res Q 29(1):37
17. Thorvald P, Lindblom J (2014) Initial development of a cognitive load assessment tool. In: The 5th AHFE international conference on applied human factors and ergonomics, 19–23 July 2014, Krakow, Poland, pp 223–232 [Online]. http://books.google.com/books?hl=en&lr=&id=6oVYBAAAQBAJ&oi=fnd&pg=PA223&dq=%22This+has+resulted+in+identification+and+classification+of+factors+suitable+for+assessment+of%22+%22assessment+of+a+task+performed+at+a+workstation.+Future+development+of+the+tool+will+include+validation%22+&ots=yEESFsIux-&sig=fwinalYk8a3b_GNvqxiyy2DJUx0. Accessed 25 Feb 2016
18. Bahar G, Masliah M, Wolff R, Park P (2007) Desktop reference for crash reduction factors
19. Laughery KR (2006) Safety communications: warnings. Appl Ergon 37(4):467–478. https://doi.org/10.1016/j.apergo.2006.04.020
20. Chen R, Wang X, Hou L (2010) Augmented reality for collaborative assembly design in manufacturing sector. In: Virtual technologies for business and industrial applications: innovative and synergistic approaches: innovative and synergistic approaches, p 105
21. Gwilt I, et al (2018) Cobotics: developing a visual language for human-robotic collaborations
22. Ibarguren A, Eimontaite I, Outón JL, Fletcher S (2020) Dual arm co-manipulation architecture with enhanced human-robot communication for large part manipulation. Sensors (Basel) 20(21). https://doi.org/10.3390/s20216151
23. Tang C-H, Wu W-T, Lin C-Y (2009) Using virtual reality to determine how emergency signs facilitate way-finding. Appl Ergon 40(4):722–730. https://doi.org/10.1016/j.apergo.2008.06.009
24. Vilar E, Rebelo F, Noriega P (2014) Indoor human wayfinding performance using vertical and horizontal signage in virtual reality: indoor human wayfinding and virtual reality. Human Fact Ergon Manufact Service Indus 24(6):601–615. https://doi.org/10.1002/hfm.20503
25. Nomura T, Kanda T, Suzuki T, Kato K (2008) Prediction of human behavior in human-robot interaction using psychological scales for anxiety and negative attitudes toward robots. IEEE Trans Rob 24(2):442–451. https://doi.org/10.1109/TRO.2007.914004
26. Eimontaite I et al (2016) Assessing graphical robot aids for interactive co-working. In: Schlick C, Trzcieliński S (eds) Advances in ergonomics of manufacturing: managing the enterprise of the future, vol 490. Springer International Publishing, Cham, pp 229–239
27. Hart SG, Staveland LE (1988) Development of NASA-TLX (Task Load Index): results of empirical and theoretical research. In: Advances in psychology, vol. 52. Elsevier, pp 139–183
28. Compeau DR, Higgins CA (1995) Computer self-efficacy: development of a measure and initial test. MIS Quarterly 189–211
29. Stafford et al RQ (2010) Improved robot attitudes and emotions at a retirement home after meeting a robot. In: RO-MAN, 2010 IEEE, pp 82–87 [Online]. http://ieeexplore.ieee.org/xpls/abs_all.jsp?arnumber=5598679. Accessed 19 Feb 2016
30. Nomura T, Shintani T, Fujii K, Hokabe K (2007) Experimental investigation of relationships between anxiety, negative attitudes, and allowable distance of robots. In: Proceedings of the 2nd IASTED international conference on human computer interaction. ACTA Press, Chamonix, France, pp 13–18 [Online]. Available: http://citeseerx.ist.psu.edu/viewdoc/download?doi=10.1.1.517.6166&rep=rep1&type=pdf. Accessed 19 Feb 2016
31. Zuckerman M (1979) Attribution of success and failure revisited, or: the motivational bias is alive and well in attribution theory. J Pers 47(2):245–287
32. Prati G, Pietrantoni L, Cicognani E (2010) Self-efficacy moderates the relationship between stress appraisal and quality of life among rescue workers. Anxiety Stress Coping 23(4):463–470. https://doi.org/10.1080/10615800903431699
33. Sim H-S, Moon W-H (2015) Relationships between self-efficacy, stress, depression and adjustment of college students. Indian J Sci Technol 8(35). https://doi.org/10.17485/ijst/2015/v8i35/86802

Trust in Industrial Human–Robot Collaboration

George Charalambous and Sarah R. Fletcher

Abstract Trust has been identified as a key element for the successful cooperation between humans and robots. However, little research has been directed at understanding trust development in industrial human-robot collaboration (HRC). With industrial robots becoming increasingly integrated into production lines as a means for enhancing productivity and quality, it will not be long before close proximity industrial HRC becomes a viable concept. Since trust is a multidimensional construct and heavily dependent on the context, it is vital to understand how trust develops when shop floor workers interact with industrial robots. This chapter provides a literature review of how trust can vary between human-robot collaboration and generic human-automation interaction whilst providing recent empirical findings on the topic of trust in industrial HRC carried out at Cranfield University.

Much has already been written about the importance of 'trust' in new technology over the years. It now seems that the role of 'trust' in ensuring the success of new systems is universally accepted. However, the concept of 'trust' and the particular role it plays in a situation is very rarely examined and understood. Trust reflects a state where one is willingly placing themselves in a vulnerable position on the expectation that the other party will perform a particular action, important to the trustor, regardless of whether their action can be controlled or monitored. This means that trust is not fixed and is likely to be affected by contextual factors. For example, a skydiver's trust in their parachute opening correctly will naturally be formed by technical factors about that situation that are completely different from, say, a parent's

George Charalambous completed a Ph.D. at Cranfield University funded by the EPSRC Centre for Innovative Manufacturing in Intelligent Automation and is currently a senior human factors consultant for SYSTRA Scott Lyster.

G. Charalambous (✉)
SYSTRA Scott Lister, London, UK
e-mail: gcharalambous@systra.com

S. R. Fletcher
Industrial Psychology and Human Factors Group, Cranfield University, Cranfield, UK

M. I. Aldinhas Ferreira and S. R. Fletcher (eds.), *The 21st Century Industrial Robot: When Tools Become Collaborators*, Intelligent Systems, Control and Automation: Science and Engineering 81, https://doi.org/10.1007/978-3-030-78513-0_6

trust in the behaviour of their children at school! Equally, the same principles are applicable to a context/setting where humans are interacting seamlessly in close physical proximity to industrial robots. In this industrial robotic setting, a worker's trust in a robot is a specific type of trust made up from very specific attributes of the system/environment. Unfortunately, very little—if any—of the literature on this subject considers the concept of trust in this specific context.

The purpose of this chapter is to:

(a) Introduce the concept of 'trust' as a key human factor to facilitate successful interactions between human operators/users and automated systems—Sect. 1.
(b) Provide a literature review of trust in generic human-automation interactions and then highlight how trust can vary in relation to human–robot interactions—Sect. 2.
(c) Introduce the concept of industrial human–robot collaboration and how the industrial context adds a layer of complexity which only recently has been explored. On this basis, trust is discussed as a fundamental element that can enhance the effectiveness of the collaboration—Sect. 3.
(d) Discuss the outputs of recent empirical works on the topic of trust in industrial HRC carried out at Cranfield University—Sect. 4.

1 Introduction to Trust

Trust represents an essential feature of daily human interactions, whether it is with other people or with interfaces and systems. In human-to-human relationships, Rotter [1] described trust in terms of relying on behaviour, verbal or written statements or promises from others. Mayer et al. [2] defined trust as the willingness of "*a party to be vulnerable to the outcomes of another party based on the expectation that the other will perform a particular action important to the trustor, irrespective of the ability to monitor or control that other party*" (p. 712). This introduces a degree of vulnerability—individuals are willing to put themselves in a vulnerable position by giving the responsibility for actions to another individual [3]. For example, people can trust someone if they are reliable, but lose trust when they experience being let down by them, any redevelopment of trust will then require more successful experiences of not being let down, over time, in order for them to trust again and reach previous trust levels. Similarly, trust in a non-human system is built upon experience of the system not failing, and is diminished when a system is found to be unreliable or unsafe. Trust in a robot depends on how well it is perceived/experienced to be reliable and safe.

Since trust can have a significant impact on performance outcomes, the development of trust between humans and machines cannot be neglected. Hence, extensive research over the years has focused on the development of trust in automated systems [3, 4]. With automated systems becoming highly utilised for a variety of tasks [5], trust has been highlighted to be a key human factor that can determine the success of the interaction [6].

2 Trust Literature Review

2.1 Generic Human-Automation Interactions

Trust is a vital component for the successful co-operation within any team, regardless of the entities that form it. In the context of human-automation teaming, trust can influence the willingness of humans to rely on the information provided by an automated system [7, 8]. If, for instance, a person enters a lift/elevator and pushes the button for the 5th floor but the lift stops at the 7th floor, then the human will, naturally, question the reliability of the lift system and their trust in it will be reduced. When experiences lead to a change in trust levels we see behavioural consequences. For example, studies have demonstrated that a lack of trust in an automated partner will eventually lead the operator/user to intervene and attempt to take control [9, 10]. However, on the other hand, over-trusting an automated system can result in the user losing situation awareness and/or developing complacency that means they may passively accept all information provided by an automated system and not intervene—which obviously impacts on performance and safety.

In the domain of human-automation interaction, a trust definition that has been widely cited is by Lee and See [3] and defines trust as "*the attitude that an agent will help achieve an individual's goals in a situation characterised by uncertainty and vulnerability*" [3, p. 54]. Subsequently, trust becomes a key component in human-automation interactions as the user/operator, must be willing to depend on the actions of another party.

Lee and See identified trust antecedents are based on three factors, namely purpose, process and performance. The purpose factor is related to the level of automation used, the process factor relates to whether the automated system employed is suitable for the specific task while the performance factor relates to the system's reliability, predictability, and capability. In addition, the degree of the system's transparency and observability available to the human partner has been found important for the development of trust in human-automation interaction [11]. Furthermore, task complexity has been suggested to have an impact on the level to which the human operator relies on the automated system [12, 13].

Research has also been directed to investigate people's perceived reliability of automated assistance versus human assistance [14] and machine-like agents versus human-like agents [15]. Dzindolet and colleagues [14] suggested that humans tend to view automation as being more reliable compared to a human aid, despite the same information being provided both by the automation and the human aid. Further on this, Lyons and Stokes [16] found that as the risk level increased, human reliance on automation support increased when compared to human support. Potentially this can lead to automation misuse or overtrust, which can be detrimental [6]. Therefore, calibrating appropriate levels of trust is vital for the success of the interaction.

2.2 Human–Robot Interactions

Robots and robotic systems, despite having a degree of automation, also have different attributes not possessed by general automated systems. The International Organisation of Standardisation (ISO) defines a robot as an "automatically controlled, reprogrammable multipurpose manipulator, programmable in three or more axes, which can be either fixed in place or mobile for use in industrial automation applications" (ISO10218-1, p2). As such, robots may include some particular attributes which introduce an additional layer of complexity. For instance, robots can be mobile, have different physical embodiments or various degrees of anthropomorphism, be of varying size depending on the application being used, may include an end-effector, and are often designed to fit a purpose. These attributes have the potential to tease out different human/user responses when interacting with robots than with any other, static, automated system. Subsequently, trust development in human–robot interactions may be affected by different factors.

Over the past 15 years, literature investigating trust development in human–robot teams and the factors that influence it, has expanded significantly. A key publication by Hancock et al. [17] involved a meta-analytic review in a bid to provide an initial framework of trust antecedents in human–robot teaming. Their review of 29 empirical studies provided quantification of the effects of the various factors influencing trust in human–robot interactions. Robot-related factors appeared to be a key element, and these were split in two sub-categories: performance-based factors (such as reliability, predictability, behaviour) and attribute-based factors (e.g. size, appearance, and movement). Perhaps unsurprisingly, these were found to be of primary importance in the development of trust.

Robotic attributes, such as the size of the robot, its degree of anthropomorphism and proximity to the human partner, have attracted attention from the research community. For instance, a highly anthropomorphic robot can generate high expectations that it may not be able to satisfy in practice thus resulting in the disappointment of human partner [18]. This is suggested to be a result of human establishing an emotional connection and trust with the robot as a result of its human-like appearance and demeanor [19].

Robot size is another type of physical attribute that has received attention in relation to human trust. Tsui et al. [20] found that robot size can influence human level of trust when a robot and a human bystander cross paths in a narrow corridor scenario. In particular their study suggested that participants appeared to trust a small mechanoid robot more than a larger one.

Similarly, robot performance characteristics (e.g. reliability, predictability) represent the second key category for trust development. When a human operator cannot predict what the robot is about to do, trust decreases [21] whilst similar decreases in human trust can be expected when a robotic system demonstrates poor reliable [22].

Although these studies enhance our understanding as to how trust may be fostered and developed in human–robot interactions, the bulk of the work involved robots in a social and healthcare settings, whilst work has also been directed to explore trust development in a military context [23]. Little research, however, had been directed in understanding how trust develops in an industrial context, where industrial robots may work seamlessly with human operators to complete a manufacturing task. In an industrial environment, where robots are seen as tools to enable the completion of a task, grasping how human trust develops when collaborating seamlessly in close proximity with an industrial robot can be key for a successful interaction.

3 Industrial Human–Robot Collaboration

A significant amount of assembly tasks in various manufacturing processes still require the flexibility and adaptability of the human operator [24]. For example, humans react very well in response to external influences and variabilities and are superior in physical dexterity and cognitive reasoning, such as engineering tolerances or process variations. At the same time, however, industrial robots can handle high payloads with greater speed, accuracy and repeatability, do not suffer from fatigue whilst the risk of operating in challenging workspaces (e.g. confined spaces or unhealthy environments) is minimised. In such processes, it is neither feasible nor cost-effective to introduce full automation. A hybrid solution where human operators collaborate seamlessly alongside industrial robots offers an optimum solution as the weaknesses of the robot can be complemented by the strengths of humans and vice versa. This type of industrial human–robot collaboration (HRC) gained momentum in manufacturing industries [2–5].

Aerospace manufacturing have also shown greater interest in the prospect of having more closer, and eventually seamless, collaboration between human operators and industrial robots. Work by OC Robotics and Airbus has seen the development of ‘snake-arm’ robots to assist the assembly tasks within aircraft wing boxes [24]. Currently, when the aircraft wing is closed-out in a box, aircraft fitters need to enter the wing box through small access panels while carrying power tools to perform a variety of tasks. The narrow access opening does not allow sufficient room for manual work to be carried out efficiently. This problem is particularly emphasised in wing sections where the wing is too small for a person to enter inside. At the same time, health and safety issues are raised when working within a confined space for a prolonged period of time. In such situations conventional off-the-shelf automation is impractical. A potential solution explored was the development of ‘snake-arm’ robots [24]. A standard single arm industrial robot is used which employs a snake-arm and represents a long slender ‘proboscis’. In this way, the slender snake-arm section can advance into the wing box or any restricted section where human operators cannot reach. The snake-arm robot can follow a route into the section under investigation either by joystick control or from a pre-determined of path. In addition, the system has been designed to allow automatic operation without the operator being present, semi-automatic where the operator is initiating a program and supervises the robot and manual tele-operation where the robot is controlled via a robot control system [24].

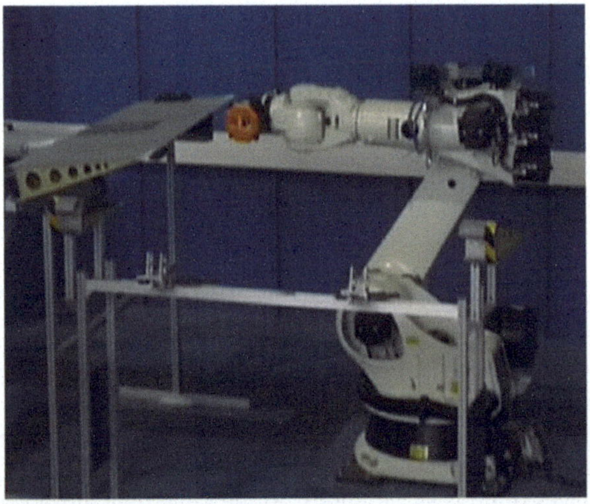

Fig. 1 A metrology assisted demonstrator cell using a high payload industrial robot

Apart from the snake-arm robot, the concept of industrial HRC has also been considered to optimise the equipping of aircraft with internal services such as attachment of aerodynamic surfaces with the use of industrial robots. To date, this area of aircraft manufacture remains exclusively manual [26]. The typically lengthy assembly methods and tight tolerances utilised in aerospace manufacturing of ±0.25 mm or less, have historically made the application of off-the-shelf automation almost impossible [27]. Walton et al. [26], however, suggested that a potential solution to overcome these challenges would be a metrology assisted human–robot collaborative system. The aim of this system is to optimise the assembly process by utilising an industrial robot to position the parts whilst the human operator performs the attachment process which requires high level of dexterity. A metrology assisted demonstrator cell was developed at Cranfield University where a typical equipping process is performed using realistic parts (Fig. 1).

This initiation demonstrated that a HRC system can be the solution into labour intensive aircraft equipping processes. The robotic partner will execute the "non-value adding" process of accurately positioning the surface while human operators will be utilised to perform the highly dexterous task of attaching the component.

In light of these technological advancement in metrology systems and industrial robotics, health and safety regulations have been updated to reflect that in some circumstances it is safe and viable for humans to work more closely with industrial robots [28]. Integrating humans and robots within the same workspace is bound to raise human factors challenges. As discussed in Sects. 2 and 3, one key human factors element that can determine the success of such a system is the degree of trust of the human operator in the robotic teammate [29]. With the concept of industrial HRC being embraced further, trust needs to be explored in depth in order to achieve successful acceptance and use of industrial robotic teammates. Furthermore, given that industrial robots come in various shapes, sizes and degrees of anthropomorphism depending on the application being used, understanding how trust develops in this setting is crucial.

4 Measuring Trust in Industrial Human–Robot Collaboration

To understand how trust develops between human workers and industrial robots, it is important to be in a position to measure it. Existing measures of trust have been heavily focused on automation, such as automated teller machines [12] and automated process control systems [30–32]. Additionally, trust measures have been developed for human interactions with military robotic systems [23], and Schaefer [33] developed a trust scale to evaluate changes in trust between an individual and a robot in general. However, the context of these previous works has been different from that of an industrial HRC setting. In military human–robot teaming, the functions and goals of both agents are obviously very different from those in an industrial scenario, and even the characteristics of industrial robots can be quite different to those possessed by military robots. Whilst the development of other measures provided a springboard to understanding trust, trust development between human operators and industrial robots will be influenced by context-specific factors that are quite different from generic or other context human-automation interactions so a trust scale that has been created from generic or alternative context scenarios is unlikely to be suitable. The difference in robotic attributes discussed in Sect. 3 mean that alternative factors can be at play influencing human trust development in the industrial environment and, therefore, an effective scale needs to measure these particular contextual attributes. After extensive searching of previous research work, to our knowledge no measure has been developed specifically to evaluate trust in industrial HRC. With significantly increasing deployment of robotics in industry, this presents a major gap in design capability so there is a need to develop a measurement tool for this purpose. The following section describes the work carried out by the authors to develop a psychometric scale for evaluating trust in industrial HRC.

4.1 Development of the Trust Scale

Exploratory study

Due to little understanding regarding the influence of trust in an industrial context, the authors started out their objective to create a new context-specific measurement tool for the industrial context by carrying out an exploratory study to gather participants' opinions in a qualitative fashion. This approach led to the generation of trust

Element	Trust-related themes	Lower level theme	Frequency
Robot	Performance	Motion of the robot	21
		Robot and gripping mechanism reliability	20
	Physical attributes	Robot's size	18
		Robot's appearance	15
Human	Safety	Personal safety	17
		Robot's safety features	11
		Safe programming of the robot	6
	Experience	Priorexperiences with robots	14
		Mental models of robots	9
External	Task	Task complexity	15

Fig. 2 Trust-related themes that emerged from the exploratory study

related themes relevant to the industrial robot context. Specification of the procedures, methods and data analyses that were employed in this initial exploratory work are available in Charalambous et al. [34]. The output results revealed a number of trust-related themes which were grouped in three high level elements, namely: robot, human and external. Figure 2 shows these themes and the frequency of occurrence in participants' responses.

In line with previous research, the robots' performance was one the most highly discussed themes among participants. Specifically, the motion of the robots was found to be a key trust-relate topic not least the way the robots moved and the speed at which they grasped the components. Participants views elaborated extensively that the robots employed a smooth and fluid motion when moving which was found to be comforting. Also, participants highlighted that the speed at which it grasped the components allowed sufficient time to react and predict what it was intending to do.

Another prevailing attitude among participants was the perceived reliability of the robots and the gripping mechanisms. Participants discussed that they could trust the robots because they completed the respective tasks accurately. Also, participants paid attention to the gripping mechanism of each of the robots. Participants felt they could trust the robot because the gripping mechanism did not drop the components during the collaboration.

Physical attributes received attention by the participants. The majority of participants elaborated that the robots' size had an influence on their perceived trust in the robot upon first encounter. Most of the participants felt intimidated by the size of the medium-scale robot prior to the interaction. The dominant view is that upon encountering the robot, participants felt intimidated by its size and appeared to be worried about interacting with it. Some participants discussed that the robots' general appearance can influence their trust. Participants found that a robot with a simple design is preferred to interact with as it is perceived less like a robotic machine which increases trust.

Safety was among the most frequently discussed themes. 17 participants mentioned that trust in the robots was influenced by their feeling of personal safety during interaction. Participants' comments suggested the main safety concern was to avoid being hit by the robot. Also, six participants elaborated they had faith that the robot had been programmed correctly by its operator. In addition, prior experiences with robots received attention by participants. 14 participants suggested that any prior experiences interacting with industrial robots would have influenced their trust. Specifically, participants elaborated that having prior exposure to similar robots would have reduced their initial anxiety. In addition, nine participants commented their trust in the robot was affected by their mental models. It appeared that participants had pre-conceived notions of robots, mainly through mainstream movies, and these had an initial influence in their level of trust as some expressed a general belief the industrial robots are fast, jerky and quite intimidating.

Task complexity was the only external related trust theme emerging through the interviews. 15 participants discussed that the complexity of the task had an influence on their trust towards the robot. Participants commented that the interactive task was not significantly challenging and this aided them to have greater trust in the robot.

As a result of these trust-related themes generated by the exploratory study, a questionnaire was developed with twenty-four items relevant to each low-level theme. Reverse-phrased items were introduced to minimise participant response bias, whilst the items were randomly placed in the questionnaire.

The next step towards the development of the trust scale was to carry out a series of industrial HRC tasks whilst using the questionnaire to gather participants' level of trust. The data would then be subjected to a factor analysis to determine the factors affecting trust in industrial HRC and to produce the trust scale. This is described in Sect. 5.

Scale development

Three human–robot trials in laboratory conditions were carried out using three different types of robots. Tasks represented potential industrial scenarios where humans and robots would collaborate. Three independent groups of participants were recruited. Upon completing the task, participants completed the survey developed from the exploratory study. Although the purpose is not to replicate the working methods already reported in detail in Charalambous et al. [34], it is deemed appropriate to provide a high-level description of the work carried out to provide the necessary context.

Trial 1 consisted of 60 participants (15 female, 45 male; M = 30.6, SD = 9) from Cranfield University. 19 participants reported having some experience with robots and automation while 41 reported having no prior experience with robots and automation. Trial 2 included 50 participants took part (13 female, 37 male; M = 30.9, SD = 9.6) from Loughborough University. 20 participants reported having some experience with robots and automation while 30 reported having no prior experience with robots and automation. Finally trial 3 consisted of 45 participants (19 female, 26 male; M = 30.7, SD = 10.3) from Cranfield University. 17 participants reported having some experience with robots and automation while 28 reported having no prior experience with robots and automation.

All three studies were an independent design at laboratory conditions. In trial 1, participants interacted with a single arm industrial robot to complete an assembly task. In trial 2, participants interacted with a twin arm industrial robot to complete an identical task to study 1. In trial 3, participants interacted with a single arm industrial robot to complete a pin insertion task.

For each trial a different industrial robot was used with varying payload capability. The tasks employed were of comparable level of complexity and represented a simplified form of typical assembly tasks found in manufacturing settings. For instance, in Trials 1 and 2 participants collaborated in close proximity with industrial robots (45 kg payload robot for trial 1 and 20 kg payload robot for trial 2) to complete a simple fitting assembly—the robot would pick up and present two sets of drain pipes (reflecting the non-value adding part of the task) whilst the user fitted the plastic fittings provided (reflecting the value-added part of the task). In trial 3, the robot (200 kg payload capability) lifted a representative aerospace sub-assembly, placed it into position and the user then applied two metallic pins to secure into the bearings. The reason for choosing different types of robots and, somewhat, different tasks (trial 1 and 2 vs. trial 3) was to gather trust data using the questionnaire developed in realistic scenarios (so far as is reasonably practicable) of industrial human–robot collaboration likely to be encountered in manufacturing settings. Figures 3, 4 and 5 show the industrial robots and materials used in these trials.

Due to employing different types of robots and having some variability in the tasks for each trial (albeit of comparable complexity), an exploratory data analysis was carried out to identify whether participant responses to the questionnaire differed significantly between the three trials. The findings suggested that whilst participants

Fig. 3 Trial 1 material and equipment used: single arm industrial robot (left); laser scanner (centre); assembly task materials (right)

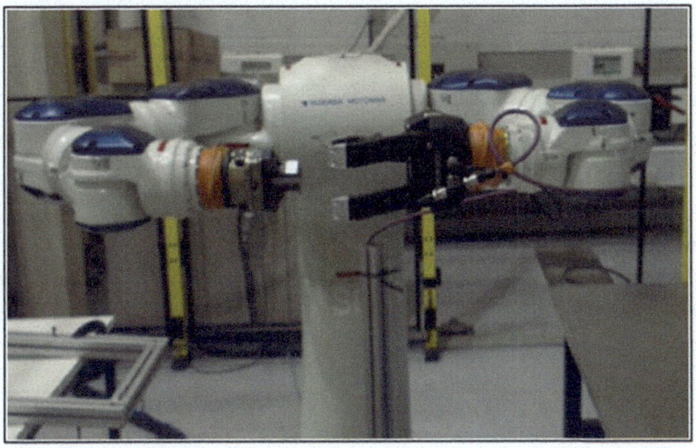

Fig. 4 Twin arm industrial robot used for Trial 2

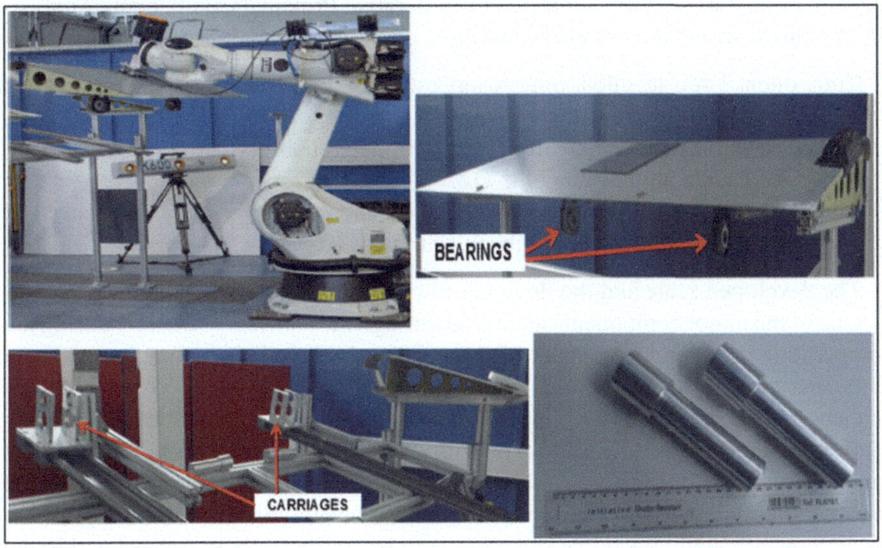

Fig. 5 Trial 3 material and equipment: industrial robot (top left); aerospace sub-assembly (top right and low left); metallic pins for securing the sub-assembly (low right)

in trial 1 experienced somewhat higher trust in the robotic teammate (M = 96.75, SE = 1.160), when compared to the participants in study 2 (M = 93.88, SE = 1.359) and study 3 (M = 95.51, SE = 1.527); this difference was not statistically significant $F(2) = 1.228$, $p > 0.05$. On this basis, the data from the three trials were merged into a single dataset, providing 155 cases. This new dataset was subjected, initially to a Preliminary Reliability Analysis, to eliminate questionnaire items which did

Scale item	Major components
The way the robot moved made me uncomfortable	Robot's motion and pick-up speed
The speed at which the gripper picked up and released the components made me uneasy	
I trusted that the robot was safe to cooperate with	Safe Co-operation
I was comfortable the robot would not hurt me	
The size of the robot did not intimidate me	
I felt safe interacting with the robot	
I knew the gripper would not drop the components	Robot and gripper reliability
The robot gripper did not look reliable	
The gripper seemed like it could be trusted	
I felt I could rely on the robot to do what it was supposed to do	

Fig. 6 The developed trust scale with the three major factors affecting trust in industrial HRC

not contribute to the overall reliability. A total of eleven items were removed and the remaining thirteen were subjected to an iterative Principal Components Analysis (PCA); this revealed the emergence of three components:

- Component 1 was labelled 'Safe co-operation', it consisted of four items and had a Cronbach's alpha of 0.802.
- Component 2 was labelled 'Robot and gripper reliability', it consisted of four items and had an alpha value of 0.712.
- Component 3 was labelled 'Robot's motion and pick-up speed', it consisted of two items and resulted in a Cronbach's alpha of 0.612.

The developed scale and the three major components are shown below (Fig. 6).

One of the major components identified through the analysis was safety during the co-operation between the human and the industrial robot. This finding is consistent with earlier work, suggesting that a positive level of perceived safety can be a key element for the successful introduction of robots in human environments [35]. This is no surprise, as interacting in close proximity with industrial robots of high payload capability can be intimidating whilst Health and Safety of the employees working in a seamless collaboration scenario is paramount. This finding is in line with previous research which highlighted that if a robot is to be successfully integrated within the human environment, it must be first perceived as safe by the human partner [36].

The performance aspects of the robotic system and specifically, the reliability of the robot and the end-effector (i.e. gripping mechanism) was the second trust related component. Robot reliability is in line with previous and more recent literature [37]. As discussed earlier the meta-analysis by Hancock and colleagues [17] suggested that robot performance-based factors (e.g. reliability) had the highest impact on trust. Furthermore, van der Brule and colleagues [37] reconfirmed that a robot's task performance influences human trust. The findings of this study highlight once again the criticality of a reliable robot system. An unreliable robot will eventually decrease operator's trust which in turn will be detrimental for accepting and using the robot.

Also, considering that humans are far more sensitive to automation errors which can lead to a significant drop in trust [31] robot reliability becomes a very important aspect.

Interestingly, the reliability of the end-effector (in this instance the gripping mechanism) appeared to have an impact on trust. To our knowledge, this context-specific element has not appeared in previous literature. This is of particular relevance to industrial HRC, since the gripping mechanism is a vital component of an industrial robot. The gripping mechanism is the mean with which the robot will manipulate components and interact with the human partner in a collaborative task. As industrial robots come in a variety of gripping mechanisms depending on the task being utilised for, findings suggest that the reliability of the gripping mechanism is an important determinant for trust development. When the reliability of the gripping mechanism decreases, human trust in the robotic partner decreases.

The third trust component was relevant to the robot's motion and the component pick-up speed. It appears that the motion of the robot is an important factor for the development of trust. This is in line with previous research indicating that robot's movement can assist the human partner to predict and anticipate robot's intentions [38]. A fluent, non-disruptive robot movement can put the human partner at ease and foster trust. This is particularly important for an industrial environment where the robot is likely to be collaborating in close proximity with a human operator. Furthermore, industrial settings can be cluttered with people, machinery, tools and other equipment therefore it is important for other operators to predict the robot's movement. Also, the final component suggested that the speed at which the gripping mechanism picks-up components has an impact on the development of trust. Similar with the previous component (robot and gripper reliability) the robot's gripping mechanism appears to have an important role in the development of trust.

Somewhat surprisingly, the analysis did not suggest that the appearance of the robot was a contributing component to trust development. Previous literature in the domain of social robotics provides contradicting results in terms of the effects of robot appearance on user preferences; some suggest robots should not be too human-like in appearance whereas others indicate that more human-like appearance can engage people more [39]. Prakash and Rogers [40] found that human perception of a robot tends to vary based on the robot's human-likeness. According to their findings, humans tend to over-generalise the capabilities of a very human-like robot. At the same time however, earlier literature stressed that anthropomorphic appearance should be treated with care in order to match the appearance of the robot with its abilities without generating unrealistic expectations to the human user [35]. This finding may further strengthen the view that people perceive industrial robots as tools used to complete a task. It is acknowledged however that our understanding on this is still growing and further work should be geared to explore the effects or industrial robot appearance on human trust.

Summary of the scale development

The outputs of this work provide an initial platform to enable the quantification of trust in a unique and largely unexplored context such as industrial-human robot collaboration. Furthermore, they highlight the key factors that determine trust development in this specific context. As discussed at the beginning of this chapter it is important to remember that trust is a state influenced by contextual factors and there is a need for such a measurement tool for the context of industrial HRC design. Having the ability to distill the factors that foster trust in an industrial human–robot collaboration context enables system designers and organisations to focus on the system characteristics that can affect users' perception of trust. Finally, the reliability levels achieved by the three sub-scales/components is encouraging. Component 1 and component 2 achieved reliability of 0.802 and 0.712 respectively which is above the generally acceptable level of 0.7 [18, 41]. The third component which achieved 0.612, although slightly below the 0.7 mark, it still reflects a good reliability figure given that there are only two items in this sub-scale.

5 Conclusion

Development of this new industrial HRC trust scale has been a worthwhile venture as there was a need for a tool that would reliably measure this type of 'trust' for the purposes of research and development to enhance system design and efficacy. To that end it has already been used in a number of subsequent research studies in the UK and Europe and found highly effective for gauging levels of user trust in relation to system characteristics. An important ongoing area of work is that the trust scale is being used to develop key behavioural rules such as the relationship between robot speed and trust, robot size and trust, robot autonomy and trust, etc.

In summary, the trust scale developed provides an initial practical tool to quantify the perceived level of human trust in an industrial human–robot collaboration setting. The tool provides practical Human Factors implications that would assist the successful deployment of collaborative industrial robots in manufacturing settings. First, the tool offers the opportunity to quantify trust specifically in industrial HRC. Second, the three major factors identified in the scale highlight to system designers and engineers understand the key system characteristics that can affect operators' perception of trust in industrial HRC. For instance, the scale identified three key design aspects fostering trust industrial HRC namely, perceived safe cooperation, perceived robot and gripping mechanism reliability and perceived robot's motion. Therefore, emphasis needs to be given on these system characteristics. Third, this scale can assist to examine the relationship of each operator and enhance awareness regarding personal tendencies. For example, a poor score on a sub-scale (e.g. robot and gripping mechanism reliability) or on the entire scale can identify those operators in need for further training.

Acknowledgements The research was funded by the EPSRC Centre for Innovative Manufacturing in Intelligent Automation.

Special thanks to the laboratory technical staff at Cranfield University, Mr. John Thrower, and Loughborough University for their assistance, support and providing technical expertise in carrying out the human-robot collaboration trials.

References

1. Rotter JB (1967) A new scale for the measurement of interpersonal trust. J Pers 35(4):651–665
2. Mayer RC, Davis JH, Schoorman FD (1995) An integrative model of organizational trust. Acad Manag Rev 20(3):709–734
3. Lee JD, See KA (2004) Trust in automation: designing for appropriate reliance. Human Fact J Human Fact Ergon Soc 46(1):50–80
4. Madhavan P, Wiegmann DA (2007) Similarities and differences between human-human and human-automation trust: an integrative review. Theor Issues Ergon Sci 8(4):277–301
5. Murphy R, Burker J (2010) The safe human–robot ratio. In: Barnes MJ, Jentsch F (eds) Human-robot interactions in future military operations. Ashgate, Surrey, UK, pp 31–51
6. Parasuraman R, Riley V (1997) Humans and automation: use, misuse, disuse, abuse. Human Fact J Human Fact Ergon Soc 39(2):230–253
7. Freedy A, de Visser E, Weltman G, Coeyman N (2007) Measurement of trust in human-robot collaboration. In: Proceedings of the 2007 international conference on collaborative technologies and systems. Orlando, FL
8. Park E, Jenkins Q, Jiang X (2008) Measuring trust of human operators in new generation rescue robots. In: Proceedings of the 7th JFPS international symposium on fluid power. Toyoma, Japan, pp 15–18
9. de Visser EJ, Parasuraman R, Freedy A, Freedy E, Weltman G (2006) A comprehensive methodology for assessing human-robot team performance for use in training and simulation. Proc Human Fact Ergon Soc Ann Meet 50(25):2639–2643
10. Steinfeld A, Fong T, Kaber D, Lewis M, Scholtz J, Schultz A, Goodrich M (2006) Common metrics for human-robot interaction. In: Proceedings of 2006 ACM conference on human-robot interaction. Salt Lake City, UT, pp 33–40
11. Verberne FM, Ham J, Midden CJ (2012) Trust in smart systems: sharing driving goals and giving information to increase trustworthiness and acceptability of smart systems in cars. Human Fact J Human Fact Ergon Soc 54(5):799–810
12. Parasuraman R, Molloy R, Singh I (1993) Performance consequences of automation-induced 'complacency.' Int J Aviat Psychol 3(1):1–23
13. Mazney D, Reichenbach J, Onnasch L (2012) Human performance consequences of automated decision aids: the impact of degree of automation and system experience. J Cognit Eng Decis Making 6(1):57–87
14. Dzindolet MT, Pierce LG, Beck HP, Dawe LA, Anderson WB (2001) Predicting misuse and disuse of combat identification systems. Mil Psychol 13(3):147–164
15. de Visser EJ, Krueger F, McKnight P, Scheid S, Smith M, Chalk S, Parasuraman R (2012) The world is not enough: trust in cognitive agents. In: Proceedings of the 56th annual HFES meeting, pp 263–267
16. Lyons J, Stokes C (2012) Human-human reliance in the context of automation. Human Fact J Human Fact Ergon Soc 54(1):112–121
17. Hancock PA, Billings DR, Oleson KE, Chen JY, De Visser E, Parasuraman R (2011) A meta-analysis of factors influencing the development of human-robot trust. Aberdeen Proving Ground, MD 21005-5425: U.S. Army Research Laboratory

18. Bartneck C, Croft E, Kulic D, Zoghbi S (2009) Measurement instruments for the anthropomorphism, animacy, likeability, perceived intelligence, and perceived safety of robots. Int J Soc Robot 1(1):71–81
19. Li D, Rau PL, Li Y (2010) A cross-cultural study: effect of robot appearance and task. Int J Soc Robot 2(2):175–186
20. Tsui KM, Desai M, Yanco HA (2010) Considering the bystander's perspective for indirect human-robot interaction. In: Proceedings of the 5th ACM/IEEE international conference on human-robot interaction. Osaka, Japan
21. Ogreten S, Lackey S, Nicholson D (2010) Recommended roles for uninhabited team members within mixed-initiative combat teams. In: The 2010 international symposium on collaborative technology systems. Chicago, IL
22. Dzindolet MT, Peterson SA, Pomranky RA, Pierce LG, Beck HP (2003) The role of trust in automation reliance. Int J Hum Comput Stud 58(6):697–718
23. Yagoda RE, Gillan DJ (2012) You want me to trust a ROBOT? The development of a human–robot interaction trust scale. Int J Soc Robot 4(3):235–248
24. Ding Z, Hon B (2013) Constraints analysis and evaluation of manual assembly. CIRP Ann Manuf Technol 62(1):1–4
25. Buckingham R, Chitrakaran V, Conkie R, Ferguson G, Graham A, Lazell A, Lichon M, Parry N, Pollard F, Kayani A, Redman M, Summers M, Green B (2007) Snake-arm robots: a new approach to aircraft assembly (No. 2007-01-3870). SAE Technical Paper
26. Walton M, Webb P, Poad M (2011) Applying a concept for robot-human cooperation to aerospace equipping processes (No. 2011-01-2655). SAE Technical Paper
27. Devlieg R (2010) Expanding the use of robotics in airframe assembly via accurate robot technology (No. 2010-01-1846). SAE Technical Paper
28. ISO 10218-2 (2011) Robots and robotic devices—safety requirements for industrial robots, Part 1: robots. International Standards Organisation, Geneva, Switzerland
29. Chen JY, Barnes MJ (2014) Human-agent teaming for multirobot control: a review of human factors issues. IEEE Trans Human-Machine Syst 44(1):13–29
30. Muir BM, Moray N (1996) Trust in automation. Part 1. Experimental studies of trust and human intervention in a process control simulation. Ergonomics, 39(3):429–460
31. Jian JY, Bisantz AM, Drury CG (2000) Foundations for an empirically determined scale of trust in automated systems. Int J Cogn Ergon 4(1):53–71
32. Master R, Gramopadhye AK, Melloy BJ, Bingham J, Jiang X (2000) A questionnaire for measuring trust in hybrid inspection systems. Paper presented at the Industrial Engineering Research Conference. Dallas, TX
33. Schaefer KE (2013) The perception and measurement of human-robot trust. Doctoral dissertation, University of Central Florida, Orlando, FL
34. Charalambous G, Fletcher S, Webb P (2016) The development of a scale to evaluate trust in industrial human-robot collaboration. Int J Soc Robot 8(2):193–209
35. Bartneck C, Kulic D, Croft E (2009) Measuring the anthropomorphism, animacy, likeability, perceived intelligence, and perceived safety of robots. Int J Soc Robot 1:71–81
36. Shiomi M, Zanlungo F, Hayashi K, Kanda T (2014) Towards a socially acceptable collision avoidance for a mobile robot navigating among pedestrians using a pedestrian model. Int J Soc Robot 6(3):443–455
37. van den Brule R, Dotsch R, Bijlstra G, Wigboldus DH, Haselager P (2014) Do robot performance and behavioral style affect human trust? Int J Soc Robot 6(4):519–531
38. Mayer MP, Kuz S, Schlick CM (2013) Using anthropomorphism to improve the human-machine interaction in industrial environments (part II). In: Duffy VG (ed) Digital human modeling and applications in health, safety, ergonomics, and risk management. Human body modelling and ergonomics. Springer Berlin Heidelberg, pp 93–100
39. Bartneck C, Kanda T, Mubin O, Al-Mahmud A (2009) Does the design of a robot influence its animacy and perceived intelligence? Int J Soc Robot 1(2):195–204

40. Prakash A, Rogers WA (2014) Why some humanoid faces are perceived more positively than others: effects of human-likeness and task. Int J Soc Robot 7(2):309–331
41. Nunnally JC (1978) Psychometric theory, 2nd edn. McGraw-Hill, New York

Adapting Autonomy and Personalisation in Collaborative Human–Robot Systems

Angelo Marguglio, Maria Francesca Cantore, and Antonio Caruso

Abstract Adaptability in manufacturing context is the ability of either the workplace topology defined by the combination of hardware and software components or any of the workplace's parts (i.e. hardware and software components and humans) to rapidly modify their characteristics and behaviours to cope with changing circumstances. Manufacturing systems need to be "Adaptive" to an ever-changing environment. This means also put together humans and automation taking advantage of each other's strengths to balance flexibility and productivity requirements in an easy and cost effective way. To achieve this challenge, companies need to re-design production systems. Enabling "Adaptive and Smart Manufacturing Systems" was indeed considered one of the challenges for Europe 2020 strategy to be addressed promoting market-oriented projects by bringing together private and public resources. This chapter aims to introduce Europe 2020 strategy for the manufacturing sector, the role of European Factories of the Future Research Association (EFFRA) as industry-driven association promoting the development of new and innovative production technologies and pre-competitive research by engaging in a public–private partnership with the European Union called 'Factories of the Future'. Some success stories in the "Adaptive and smart Manufacturing Systems" research domain will be introduced and, in particular, the project A4BLUE (Grant ID 723,828) will be illustrated in more detail for its interesting results addressing the expected scientific challenge and scope of FOF-04-2016 (Continuous adaptation of work environments with changing levels of automation in evolving production systems).

https://www.effra.eu/effra.

A. Marguglio (✉) · M. F. Cantore · A. Caruso
Engineering Ingegneria Informatica S.p.A., Rome, Italy
e-mail: angelo.marguglio@eng.it

M. I. Aldinhas Ferreira and S. R. Fletcher (eds.), *The 21st Century Industrial Robot: When Tools Become Collaborators*, Intelligent Systems, Control and Automation: Science and Engineering 81, https://doi.org/10.1007/978-3-030-78513-0_7

1 Introduction

Modern manufacturing systems need to deal with an ever-changing environment due to short term changes caused by human (e.g. worker's anthropometric, cognitive and sensorial characteristics, skills, etc.) or production related variability (e.g. work orders re-schedule, human resources re-allocation, automation failures, etc.) or long term changes caused by market's demands and company's strategy (e.g. introduction of new products that need new processes and expertise, etc.) and technology advancements that can help to allocate more tasks for automation or imply changes to the way workers perform tasks and require the introduction of new skills, or introduce new control systems, etc.

In this context of ever-changing demands, manufacturing systems (i.e. assembly system) need to put together humans and automation taking advantage of each other's strengths to balance flexibility and productivity requirements in an easy and cost effective way.

To achieve this challenge, companies need to re-design production systems and implement an adaptive automation strategy to increase adaptability, lower efforts for setting up and executing operations, compensate workers' limitations, and increase workforce satisfaction to improve levels of organisational commitment and retention.

The Europe 2020 strategy underlines the role of 'technology' for tackling the challenge of increasing Europe's economic growth and job creation. European and global societal challenges can be addressed investing in Key Enabling Technologies (KET) which can help innovative ideas to be turned into new products and services that create growth and high-skilled adding-value jobs.

In such a context, the role of manufacturing is crucial, since advanced manufacturing systems play a critical part in making KETs and new products competitive, affordable and accessible; multiplying their societal and economic benefits [1].

In 2009 the manufacturing community formed the international non-for-profit association European Factories of the Future Research Association (EFFRA), with the scope to promote pre-competitive research on production technologies within the European Research Area by engaging in a public–private partnership with the European Union called 'Factories of the Future'. The partnership aims to bring together private and public resources with the aim of launching hundreds of market-oriented cross-border projects throughout the European Union. Such projects produce demonstrators and models to be applied in a wide range of manufacturing sectors.

The Factories of the Future PPP identifies and realises the transformations by pursuing a set of research priorities along six research and innovation domains. Each of these domains embodies a particular aspect of the required transformations towards the factories of the future.

To reach this transformation, the Factories of the Future PPP, as well as Horizon 2020,[1] requires a critical mass of stakeholders and clear industrial commitment.

[1] https://ec.europa.eu/programmes/horizon2020/what-horizon-2020.

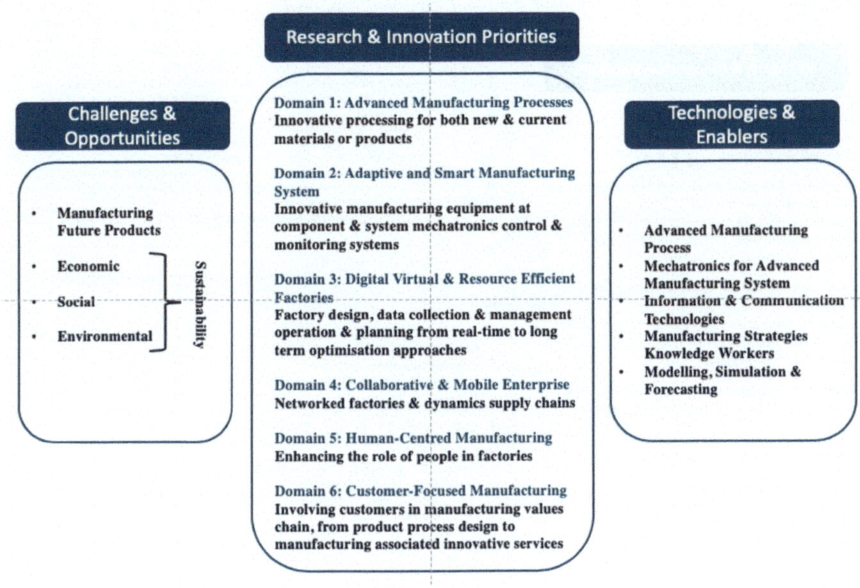

Fig. 1 The factories of the future roadmap framework

Addressing the challenges and opportunities with the right technologies and enablers along the lines of the research innovation domains constitutes the framework of the Factories of the Future roadmap (Fig. 1).

Through the EFFRA Innovation Portal, information is collected for each FoF project, in an organised structure, following a specific taxonomy. The following illustrations show how the FoF calls (2014, 2015, 2016) [2] and the projects are mapped (particularly how they relate to the roadmap's research domains) (see Fig. 2).

The graphic below illustrates the mapping of the 268 FoF projects that have successfully applied to FoF calls (2014, 2015, 2016) [2] on the six domains of the roadmap through the EFFRA Innovation Portal. This chart shows how projects have covered the six domains of the roadmap (Fig. 3).

With the advent of Industry 4.0, the modern manufacturing world has shifted its focus to further increasing automation, where advanced workplaces are replacing existing working stations. Moreover, human–machine collaboration is placing human operators in the centre of attention. Among these six research domains, the "Adaptive and smart manufacturing domain" is strictly connected to the themes that we want to explore more in this chapter.

In the following paragraphs some insights will be collected on the success stories that gave their contribution in the "Adaptive and smart manufacturing domain" and a focus will be done on the A4BLUE project, its aims, its solutions and on the reference architecture that can enable them.

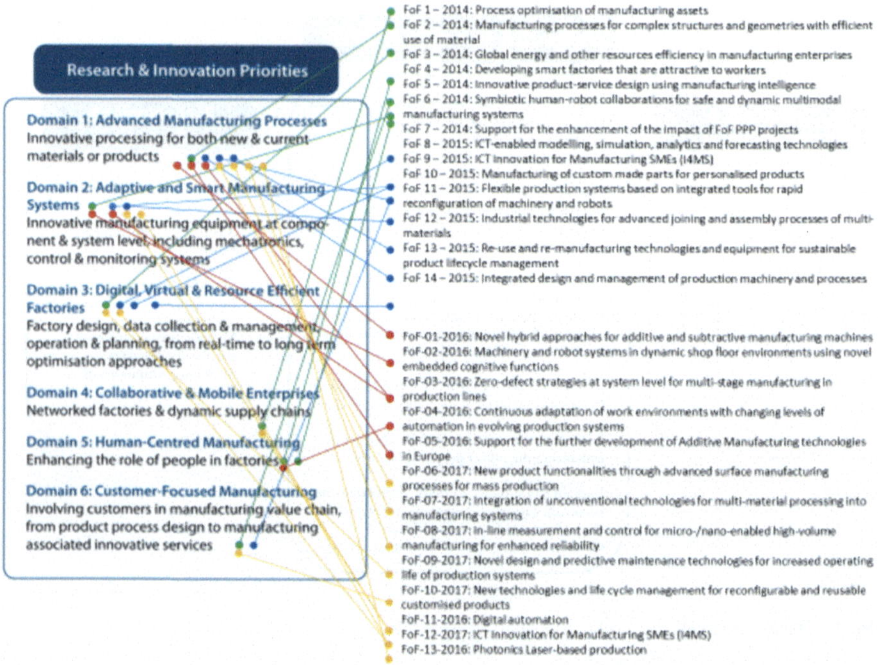

Fig. 2 Connections between EC published call topics and the research and innovation priorities of the Factories of the Future 2020 roadmap

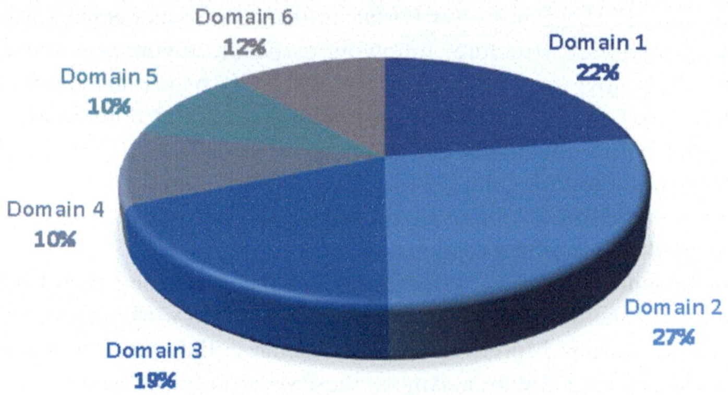

Fig. 3 Projects per roadmap domain

2 The Current State in the Adaptive and Smart Manufacturing Systems Domain

The topic of "Adaptive and Smart Manufacturing Systems" is one of the main advanced manufacturing topics and is among the research and innovation domains identified by EFFRA in the multi-annual roadmap from the contractual PPP under Horizon 2020 [1].

The set of research priorities under this domain focuses on innovative manufacturing equipment at component and system level, including mechatronics, control and monitoring systems.

The research priorities, recommended under this domain, aim at future European manufacturing systems and processes that adapt in an agile manner to varying market and factory demands thanks to intelligent robots and machines that cooperate both among themselves and with persons in a safe, autonomous and reliable manner. This domain can include two main areas:

1. Sub-Domain 2.1: Adaptive and smart manufacturing devices, components and machines includes:
    - Flexible and reconfigurable machinery and robots
    - Embedded cognitive functions for supporting the use of machinery and robot systems in changing shop floor environments
    - Symbiotic safe and productive human–robot interaction, professional service robots and multimodal human–machine–robot collaboration in manufacturing
    - Smart robotics—scaling up for flexible production and manufacturing
    - Mechatronics and new machine architectures for adaptive and evolving factories
    - Mechatronics and new machine architectures for high-performance and resource-efficient manufacturing equipment
    - Micro-precision into micro- and macro-production equipment
    - High-performance and resource-efficient manufacturing equipment by applying advanced materials
    - Multi-disciplinary engineering tools for mechatronics engineering
    - M2M cloud connectivity for future manufacturing enterprises
    - Intuitive interfaces, mobility and rich user experience at the shop floor
2. Sub-Domain 2.2: Dynamic production systems and shop floors includes:
    - Adaptive process automation and control for a sensing shop floor
    - Dynamic manufacturing execution environments for their smarter integration into dynamic and agile shop floors
    - Monitoring, perception and awareness on manufacturing
    - Intuitive interfaces, mobility and rich user experience at the shop floor
    - Mass customisation and integration of real-world resources.

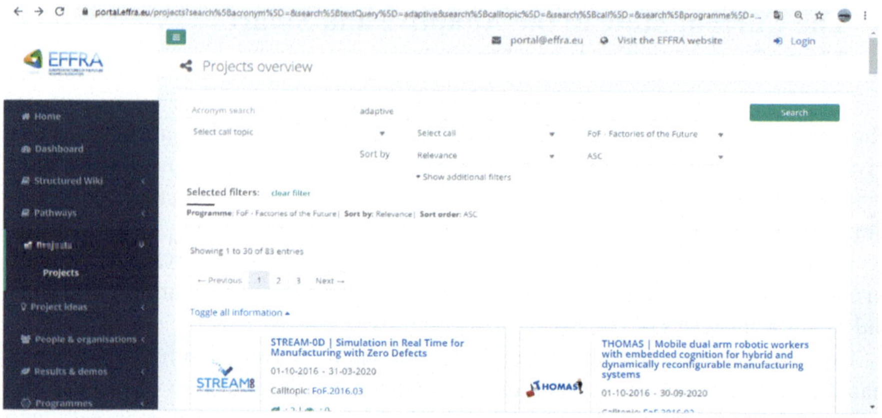

Fig. 4 EFFRA portal—projects

3 Projects Overview by EFFRA Portal

Through the EFFRA portal[2] it is possible to find F0F projects aimed to promote "Adaptive and smart manufacturing" (Fig. 4).

Using "adaptive" or "automation" as keyword it is possible to find several projects. To mention some of them among the last ones:

THOMAS[3]—Mobile dual arm robotic workers with embedded cognition for hybrid and dynamically reconfigurable manufacturing systems (01-10-2016-30-09-2020): The productivity of the serial production model is compromised by the need to perform changes in the production equipment that cannot support multiple operations in dynamic environments. Low cost labour is no longer an option for EU manufacturers due to the fast rise of wages and the increasing costs of energy and logistics. Manual tasks cannot be fully automated with a good ratio of cost vs robustness using standard robots due to: high product variability, dedicated process equipment and high cost of maintenance by expert users. The answer to this challenge lays in the creation of production concepts that base their operation on the autonomy and collaboration between production resources. The vision of THOMAS is: "to create a dynamically reconfigurable shopfloor utilizing autonomous, mobile dual arm robots that are able to perceive their environment and through reasoning, cooperate with each other and with other production resources including human operators.

CREMA[4]—Cloud-based Rapid Elastic Manufacturing—(01-01-2015-01-01-2018) concluded in early 2018. It aimed to simplify the establishment, management, adaptation, and monitoring of dynamic, cross-organisational manufacturing processes

[2] https://portal.effra.eu/projects.

[3] https://portal.effra.eu/project/1647.

[4] https://portal.effra.eu/project/1423.

following cloud-manufacturing principles. CREMA developed the means to model, configure, execute, and monitor manufacturing processes, providing end-to-end support for Cloud manufacturing by implementing real systems and testing and demonstrating them in real manufacturing environments.

MAShES[5]—Multimodal spectrAl control of laSer processing with cognitivE abilities—(01-12-2014-01-12-2017) aimed to develop a breakthrough compact imaging system for RT closed-loop control of laser processing. It will be built on a novel multispectral optics and multi-sensor arrangement in the VIS-MWIR spectrum. Absolute temperature, geometry, and speed will be imaged accurately and reliably. RT process control, and cognitive re-adjustment and process quality diagnosis will be embedded. Three key project results are being taken further with a view to commercialisation: (1) an embedded electronics and control system, with a simplified version with one camera for LMD RT control already commercialized, (2) a cognitive control system and interface, consisting of concepts and software modules for RT-control and the quality diagnosis system of LMD and laser welding; and (3) a complete MAShES system which integrates a camera system capable of multimodal monitoring and multispectral imaging in the visible-infrared range, and embedded system for RT monitoring and control of laser processing, and system control software for autonomous control configuration and readjustment capabilities and quality diagnosis.

SatisFactory[6]—A collaborative and augmented-enabled ecosystem for increasing SATISfaction and working experience in smart FACTORY environments—(01-01-2015-31-12-2017) established a collaborative and augmented-enabled ecosystem with the overall aim of increasing satisfaction and working experience in smart factory environments. The SatisFactory solution and its technology products were demonstrated and evaluated in three industrial sites in Italy and Greece. SatisFactory generated 27 demonstrators including a collaborative platform, A.R. glasses and training and a gamification platform.

AREUS[7]—Automation and robotics for European sustainable manufacturing—(01-09-2013-31-08-2016) project aimed to improve the sustainability of robotic manufacturing by providing a set of integrated innovative technologies and engineering platforms, intrinsically interdisciplinary, modular and configurable. AREUS envisioned and investigated a novel intelligent factory architecture focused on energy efficiency and sustainable robotic manufacturing.

[5] https://portal.effra.eu/project/1434.

[6] https://portal.effra.eu/project/1441.

[7] https://portal.effra.eu/project/1043.

4 ACE Factories Cluster

In this context, five projects funded under the European Union's Horizon 2020 research and innovation programme (A4BLUE,[8] Factory2Fit,[9] INCLUSIVE,[10] HUMAN[11] and MANUWORK[12]) worked in parallel and constituted the Human-Centered Factories (ACE Factories[13]) cluster. These projects developed solutions for manufacturing work environments that adapt to each individual worker. In the past, people were expected to adapt to machine requirements. Now, automation systems are being developed that can recognise the users, remember their capabilities, skills and preferences, and adapt accordingly. Adaptation can also make work organisation more flexible so that individual preferences are taken into account in task distribution. New automation approaches, with workers at the center, will complement people's capabilities and ensure higher performance, adaptability and quality [3].

A4BLUE—Adaptive Automation in Assembly For BLUE collar workers satisfaction in Evolvable context (01-10-2016-30-09-2019). The project proposes the development and evaluation of a new generation of sustainable and adaptive workplaces dealing with the evolving requirements of manufacturing processes and human variability. For this, A4BLUE is working to introduce adaptive automation mechanisms for an efficient and flexible execution of tasks, ensuring a constant and safe human–machine interaction as well as advanced and personalized worker assistance systems including VR/AR and knowledge management capabilities to support them in the assembly and training related activities.

Factory2Fit—Empowering and participatory adaptation of factory automation to fit for workers (01-10-2016-30-09-2019). The solutions developed in Factory2Fit will engage and empower future factory workers and bring increased worker motivation, satisfaction and productivity by giving the worker motivating feedback of his/her wellbeing and work performance, and adapting the work environment according to personal skills and preferences. The solutions also engage workers to share knowledge, participate in designing their work, and take responsibility of their own learning and skills development. The solutions will also allow workers to take responsibility for their own wellbeing and skills development, as well as provide tools to share their tacit knowledge.

HUMAN—HUman MANufacturing (01-10-2016-30-09-2019). HUMAN aims at developing a platform that is contextually aware of both the factory and the human operator, identifying when an intervention is required in order to support the operator in performing their tasks with the desired quality, whilst ensuring their well-being.

---

[8] http://a4blue.eu/.

[9] https://factory2fit.eu/.

[10] http://www.inclusive-project.eu/.

[11] http://www.humanmanufacturing.eu/.

[12] http://www.manuwork.eu/.

[13] http://ace-factories.eu/.

INCLUSIVE—Smart and adaptive interfaces for INCLUSIVE work environment (01-10-2016-30-09-2019). INCLUSIVE aims at covering the increasing gap between machine complexity and user capabilities by developing a smart and innovative HMI that accommodates to the workers' skills and flexibility needs, by compensating their limitations (due to age, disabilities or inexperience) and by taking full advantage of their experience. To achieve this, the developed HMI system needs to be able to measure the sustainable cognitive load of the human operator, to adapt the automation functions and support and train low skilled operators to accomplish a complex automation task properly, also by integrating a virtual environment and an industrial social network.

MANUWORK—Balancing Human and Automation Levels for the Manufacturing Workplaces of the Future (01-10-2016-31-03-2020). MANUWORK aims to focus on the development of an integrated platform for the management of manufacturing workplaces of the future that will be characterized by the complementarity between humans and automation. MANUWORK develops human-automation load balancing methods that determine the optimal trade-off between automation and human involvement at a shop-floor workplace, taking into account the needed flexibility for the process, the available skills (offered both from human and machines), a safe integration of humans and automation into the process and the overall load of the line.

4.1 Human-Centricity in ACE Factories Projects

A central component of Industry 4.0 is its human-centricity, described as development towards the Operator 4.0 concept [4]. Operator 4.0 refers to smart and skilled operators of the future, who will be assisted by automated systems providing a sustainable relief of physical and mental stress and allowing the operators to utilise and develop their creative, innovative and improvisational skills, without compromising production objectives. The factory floor solutions addressed by the five projects that make up the ACE Factories cluster reinforce the concepts introduced by the original Operator 4.0 topology [4] in the following areas:

- Off-the-job and on-the-job training and guidance
- Industrial social networking and knowledge sharing
- Collaborative robot
- The adapted workplace
- Empowering feedback on work well-being and work achievements.

All the five projects developed interesting solutions and analysis through their use cases following the EFFRA's vision: "Factories 4.0 and Beyond" for the key priority "Human–robot collaboration" in these areas:

- Robots and workers as members of the same team throughout the factory
- Safety and beyond (ergonomics, productivity, adaptability, acceptance)

- Importance of human factors (user experience, trust, comfort, feeling of safety)
- Higher levels of perception and adaptability.

5 A4BLUE Project

The main objective of this 3-year project is the development and evaluation of a new generation of sustainable and adaptive workplaces dealing with the evolving requirements of manufacturing processes and human variability.

A4BLUE introduced adaptive automation mechanisms for an efficient and flexible execution of tasks, ensuring a constant and safe human–machine interaction as well as advanced and personalised worker assistance systems including virtual/augmented reality and knowledge management capabilities to support them in the assembly and training related activities. Furthermore, A4BLUE provided methods and tools to determine the optimal degree of automation of the new assembly processes by combining and balancing social and economic criteria to maximize long term worker satisfaction and overall process performance.

Its main goals are:

- Adaptability by providing an open, secure, configurable, scalable and interoperable adaptation management and assistance system.
- Interaction by providing a set of safe, easy to use, intuitive, personalized and context aware multimodal human-automation interaction mechanisms.
- Sustainability by providing methods and tools to determine the optimal degree of automation of the new assembly processes that combine and balance social and economic criteria to maximize long term worker satisfaction and overall performance.

5.1 Reference architecture (RA)—The current 'State of the Art' (SoA) in Manufacturing workplace

Pre-requisites for digital platforms to thrive in a manufacturing environment include the need for agreements on industrial communication interfaces and protocols, common data models and the semantic interoperability of data, and thus on a larger scale, platform inter-communication and inter-operability. As it is the case for any industry-relevant innovation, standards need to be considered, including work on reference frameworks or architecture models.

A *Reference Architecture* (RA) is a synthesis of best practices having their roots in past experience. In A4BLUE project, where the business value of applying innovative adaptive patterns to the smart factory was explored, starting from an effective RA was of paramount importance. The goal was twofold: on the one hand, to leverage valuable experience from large and respected communities; on the other, to be consistent and compatible with the mainstream evolution of the smart factory.

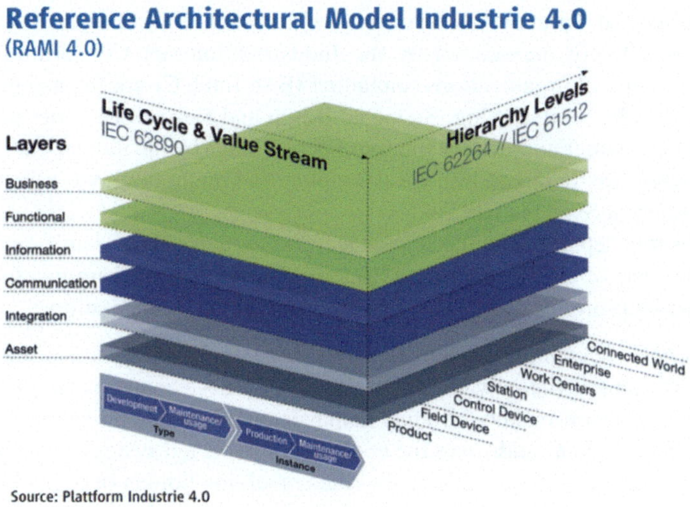

Fig. 5 RAMI 4.0 reference architecture

With the aim to design the **A4BLUE Adaptive framework**, we considered some well-known and accepted *generic RAs* as sources of inspiration: the RAMI 4.0 model defined in the Platform Industrie 4.0, the IIRA architecture defined by the Industrial Internet Consortium and the FIWARE for Industry Reference Architecture developed in several European and other national/international research projects.

One of the more interesting architectural approaches in the field of Industrie 4.0 is the **Reference Architectural Model Industrie 4.0** (abbreviated **RAMI 4.0**) [5], since it combines the crucial elements of Industrie 4.0 in a three-dimensional layer model for the first time. As its name clearly states, it is the outcome of Platform Industrie 4.0,[14] the German public–private initiative addressing the fourth industrial revolution—i.e., merging the digital, the physical and the biological worlds into *cyber-physical production systems*. This specification was firstly published in July 2015 and provides a first draft of the reference architecture for the Industrie 4.0 initiative trying to group different aspects in a common model and to assure the end-to-end consistency of "*… technical, administrative and commercial data created in the ambit of a means of production or of the workpiece*" across the entire value stream and their accessibility at all times. The RAMI consists of a three-dimensional coordinate system that describes all crucial aspects of Industrie 4.0. In this way, complex interrelations can be broken down into smaller and simpler clusters (see Fig. 5).

For the A4BLUE RA some of the RAMI 4.0 conceptual framework as its own was adopted, simplifying communication with the external communities of developers and users.

[14] http://www.plattform-i40.de/I40/Navigation/EN/Home/home.html.

The **Industrial Internet Reference Architecture (IIRA)**[15] has been developed and is actively maintained by the Industrial Internet Consortium (IIC), a global community of organizations (including IBM, Intel, Cisco, Samsung, Huawei, Microsoft, Oracle, SAP, Boeing, Siemens, Bosch and General Electric) committed to the wider and better adoption of the Internet of Things by the industry at large. The IIRA, is a standards-based architectural template and methodology for the design of Industrial Internet Systems (IIS).

IIRA has four separate but interrelated *viewpoints*, defined by identifying the relevant stakeholders of IIoT use cases and determining the proper framing of concerns. These viewpoints are: business, usage, functional and implementation.

- The *business viewpoint* attends to the concerns of the identification of stakeholders and their business vision, values and objectives. These concerns are of particular interest to decision-makers, product managers and system engineers.
- The *usage viewpoint* addresses the concerns of expected system usage. It is typically represented as sequences of activities involving human or logical users that deliver its intended functionality in ultimately achieving its fundamental system capabilities.
- The *functional viewpoint* focuses on the functional components in a system, their interrelation and structure, the interfaces and interactions between them, and the relation and interactions of the system with external elements in the environment.
- The *implementation viewpoint* deals with the technologies needed to implement functional components, their communication schemes and their lifecycle procedures.

Overall, the functional viewpoint tells us that control, management and data flow in IIS are three separate concerns having very different non-functional requirements, so that implementation choices may also differ substantially. On the other hand, the implementation viewpoint describes some well-established architectural patterns for IIS: the Three-tier, the Gateway-mediated Edge Connectivity and Management and the Layered Databus. The *Three-tier architectural pattern* distributes concerns to separate but connected tiers: Edge, Platform and Enterprise. Each of them play a specific role with respect to control and data flows, as depicted in Fig. 6.

In A4BLUE, which deals with platforms rather than solutions, the functional and implementation viewpoints described in the IIRA are the most useful especially regarding the *control domain* and the *operations domain*, where the focus is on reading data from sensors, applying rules and logic and exercising adapting control over the physical world and IT systems. For the A4BLUE RA some conceptual organization of the components foreseen in the IIRA were exploited, by targeting a more general compliance with the layered architecture and the extensive use of databus (even if in A4BLUE project an Event-Driven approach was preferred). Moreover, it is quite interesting to see that RAMI's I4.0 Components, IIRA Entities and A4BLUE modules can serve the same purpose of creating a digital *live representation* of a real-world object (thing, machine or person) that can be integrated into applications.

[15] http://www.iiconsortium.org/IIRA.htm.

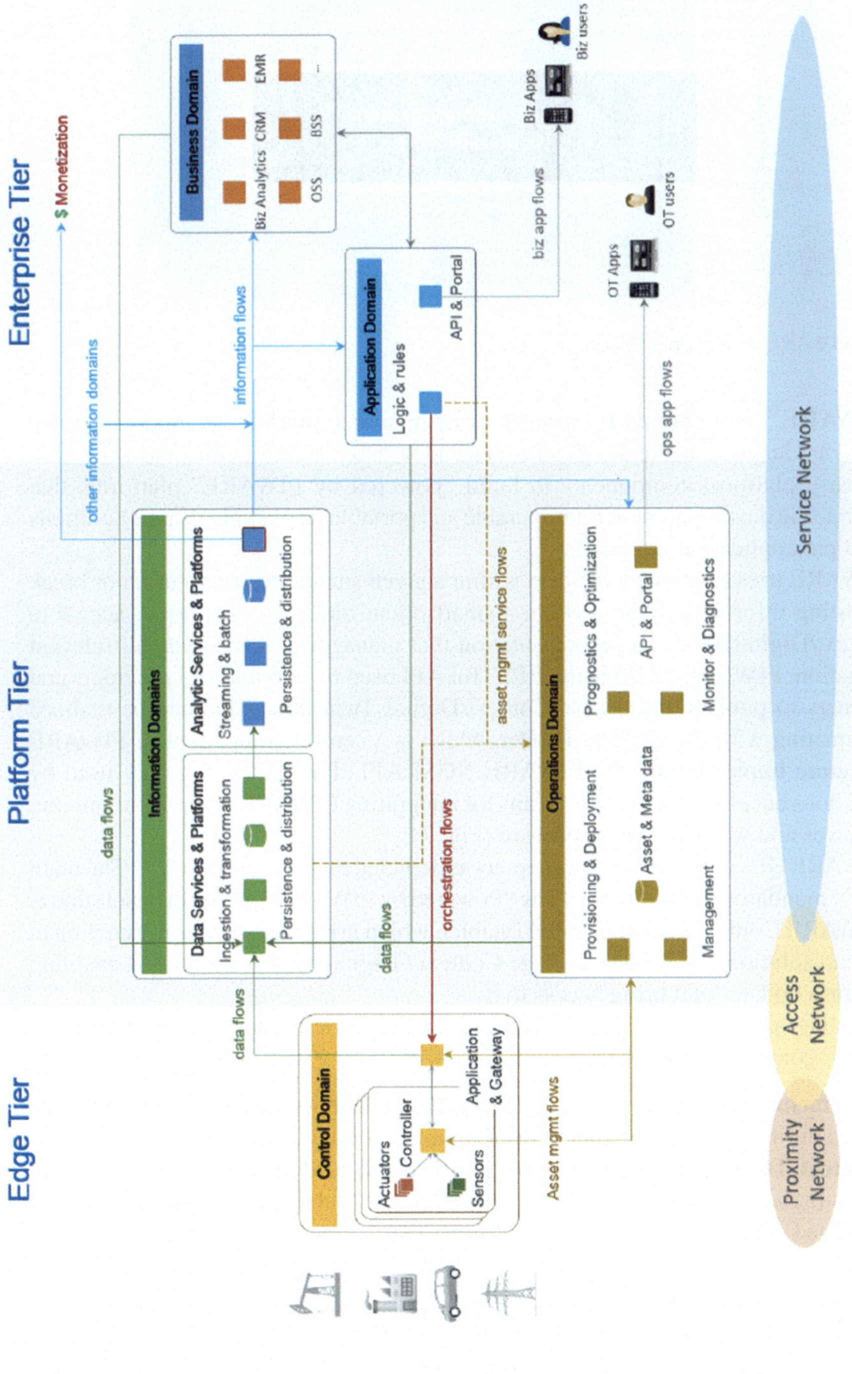

Fig. 6 IIRA three-tier architecture pattern (implementation + functional viewpoint)

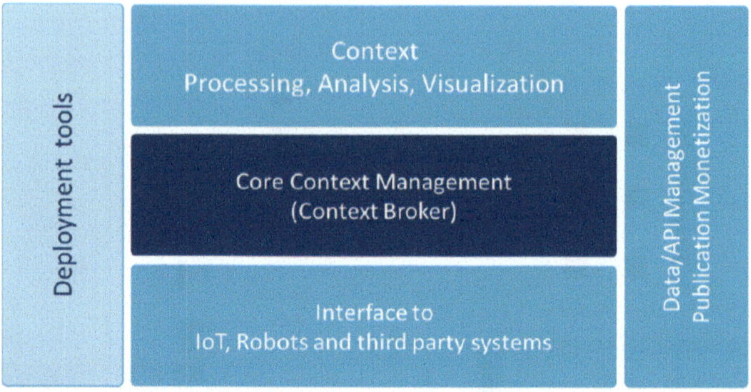

Fig. 7 FIWARE overall architecture

FIWARE[16] is a curated framework of open source platform components (also referred as Generic Enablers—GE-) which can be assembled together and with other third-party platform components to build "powered by FIWARE" platforms that accelerate the development of interoperable and portable (replicable) Smart Solutions in multiple application domains.

FIWARE tries to manage the data within a given smart vertical solution or break the existing information silos within a smart organization by supporting access to a Context/Digital Twin data representation that manages at large-scale all relevant information. FIWARE NGSI is the RESTful API used by context data providers and consumers to publish and access Context/Digital Twin data. This can be realized by interacting with the Context Broker, which is a central component of FIWARE architecture implementing the FIWARE NGSI API. The API is not only used by applications but also provides the means for integrating FIWARE components among themselves and with 3rd party software (Fig. 7).

FIWARE GEs are organized in chapters as depicted in the figure above.[17] The main and only mandatory component of any "Powered by FIWARE" platform or solution is the FIWARE Context Broker Generic Enabler, which brings a cornerstone function in any smart solution: the need to manage Context / Digital Twin information, enabling to perform updates and bring access to it.

Building around the FIWARE Context Broker, a rich suite of complementary FIWARE components are available, dealing with:

- **Interfacing with the Internet of Things (IoT), Robots and 3rd Party systems**, for capturing updates on context information and translating required actuations.
- **Context Data/API management, publication, and monetization**, bringing support to usage control and the opportunity to publish and monetize part of managed context data.

---

[16] FIWARE developers catalogue. (n.d.). Retrieved from https://www.fiware.org/developers/catalogue/.

[17] FIWARE developers page. (n.d.). Retrieved from https://www.fiware.org/developers/.

- **Processing, analysis, and visualization of context information** implementing the expected smart behavior of applications and/or assisting end users in making smart decisions.

The catalogue contains a rich library of components with reference implementations that allow developers to put into effect functionalities such as the connection to the Internet of Things or Big Data analysis, making programming much easier. FIWARE core platform model facilitating IaaS and SaaS required of application domains, on this basis GEs applications achieve already defined standards, provide APIs for interoperability, represent application domains or design granularities. We can think of a GE as macroscopes were the highest level interface is a simple controller providing a wide and in scope view of operations (attributes control functions of a system); GEs from different domains are Macroscopes on the domain: implement abstract Macroscopes concretely and provide API access via REST HTTP to trigger GE behavior. Modelling a GE is identified within UML use cases. GE specification have some properties:

- Addressing: IP address and port numbers
- Recognition: control syntax, parser, interpreter and semantic rules
- Multimodal: APIs, protocols, drivers
- Structured Data: XML, JSON, IC
- Formal Operation: state machine, dispatcher, DOM nodes
- Ad hoc Network Communication: HTTP/s, request methods, asynchronous, client/server, URIs
- Modular Design: object-oriented architecture, methods and functions, listeners, callbacks
- Behavioral: multithreaded, parallel, imperative, result combining, verifying, transformation, bidirectional communications
- Security: channel encryption, message encryption, authentication, authorization
- HCI: GUI, hardware interaction, multimodal UI, accessibility, human actors
- Interoperability: networked API server, configuration parameters, legacy system integrators, RPC, REST.

In 2016, the European Commission published The 2016 Rolling Plan for ICT Standardisation[18] in which ETSI was requested to create an ISG (Industry Specification Group) aimed at definition of a standard Context Information Management (CIM) API with FIWARE NGSIv2[19] (current API specification implemented by the FIWARE Orion Context Broker) as basis. In the beginning of 2017, ETSI created the CIM ISG[20] which produced a first version, in January 2019, of the so named ETSI NGSI-LD API specifications.[21]

---

[18] https://ec.europa.eu/growth/content/2016-rolling-plan-ict-standardisation-released_en.

[19] http://fiware.github.io/specifications/ngsiv2/stable/.

[20] https://www.etsi.org/committee/cim.

[21] https://www.etsi.org/deliver/etsi_gs/CIM/001_099/009/01.02.02_60/gs_CIM009v010202p.pdf.

This ETSI NGSI-LD API specifications are compatible with the FIWARE NGSIv2 API specifications, adding new features bringing support to Linked Data. It is planned that the FIWARE Orion Context Broker will evolve in line with the future ETSI NGSI-LD specifications, integrating the developments carried out in Orion-LD.[22]

Several implementations of the NGSI-LD API are emerging, most of which have been incorporated in the FIWARE Catalogue: Scorpio[23] or Stellio.[24] The only exception to this rule at the moment is Djane[25] despite conversations are taking place to also include it.

In 2018, the European Commission (EC) formally adopted FIWARE Context Broker technology as CEF Building Block[26] within their Digital CEF (Connecting Europe Facility) Building Blocks program. This means that the EC officially recommends Public Administration and private companies of the European Union (EU) to adopt this technology in order to foster development of digital services which can be replicated (ported) across the EU.

During 2019, the FIWARE Foundation with other relevant organizations has launched an initiative towards definition of Smart Data Models.[27] The goal is to provide a common set of data models, with their corresponding mappings into JSON/JSON-LD which, in combination with NGSIv2/NGSI-LD ensures portability and interoperability of smart applications. The initiative is experiencing a growing momentum, involving multiple organizations and projects contributing data models in multiple domains: Smart Cities, Smart Agrifood, Smart Manufacturing, Smart Energy, Smart Water, Smart Destinations, etc.

FIWARE for Industry (F4I) is a multi-project initiative aiming at developing an ecosystem of FIWARE-enabled software components, suitable to meet the challenges of Manufacturing Industry business scenarios, as indicated by Industry 4.0 vision. F4I originates at the end of 2015 as the exploitation booster of the FITMAN FP7 FI PPP project (https://www.fiware4industry) which developed Open Source reference implementations of Smart-Digital-Virtual Factory scenarios by integrating 14 FIWARE Generic Enablers with 15 original Manufacturing Industry Specific Enablers.

In the figure below, the FIWARE architecture on Smart Manufacturing is presented, which is based on a platform powered by FIWARE (Fig. 8).

[22] https://github.com/Fiware/context.Orion-LD.

[23] https://github.com/ScorpioBroker/ScorpioBroker.

[24] https://github.com/stellio-hub/stellio-context-broker.

[25] https://github.com/sensinov/djane/.

[26] https://ec.europa.eu/cefdigital/wiki/display/CEFDIGITAL/Context+Broker.

[27] https://github.com/smart-data-models.

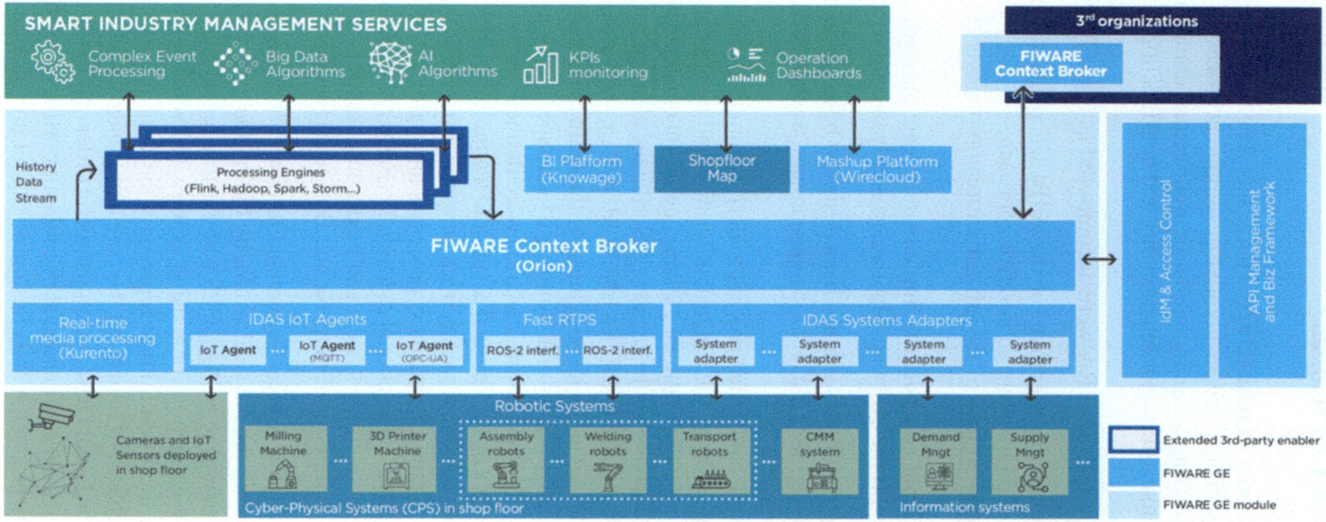

Fig. 8 Reference Architecture for smart industry management system powered By FIWARE

5.2 Development of Architecture—How Was the A4BLUE-RA Developed?

The **A4BLUE Reference Architecture** (RA) has been designed using an architecture-centered, scenario- driven, iterative development process. The A4BLUE RA is the result of assembling a certain number of architectural components in some well-chosen forms to satisfy the major functionality and non-functional requirements of the system.

The objectives of the A4BLUE RA were decomposed as in the following:

- To develop a **reference architecture** for the A4BLUE platform for the implementation of its solution by using an iterative approach;
- To define the **logical structure** of the infrastructure components in the A4BLUE stack;
- To define the **functional components** implementing each infrastructure component in order to support the evolving adaptive assembly system concept.

The A4BLUE Adaptive Framework has been designed based upon the following pillars: virtualization, integration, adaptation management, worker assistance support and monitoring as shown in Fig. 9.

To further structure the envisioned process to derive from the Reference Model a FBB Specification, Fig. 9 shows the suggested iterative and incremental approach.

Starting from defining a foundational glossary, common concepts and principles (actors, features analysis) have set the scene to describe the elements of the architecture at a conceptual level (i.e. the REFERENCE MODEL); then a more functional analysis has been conducted to the identification of the main building blocks needed for the implementation stage (Functional Building Blocks Specification). Then the set of services required by the application scenarios were defined; then these functions have been detailed in terms of background assets and technologies to be used during development; this technological view has been enriched also with a decomposition in sub-components presenting the business processes among them.

According to the A4BLUE Functional & Modular (F&M) Architecture, the functionalities of an adaptive assembly system can be decomposed into three high-level *Functional Domains*—Shopfloor, Enterprise and Business (Fig. 10).

- **Shopfloor Layer**: the lower layer is intended to ease the interconnection of the A4BLUE Platform with the physical world, by hiding the complexity of dealing with shopfloor IT systems (e.g. PLC, CPS and existing legacy system) as well as dealing with human interactions (e.g. using gesture and voice commands).
- **Enterprise Layer**: the middle layer represents the core part of the A4BLUE Platform, being in charge of managing the core components needed for adaption management using an Event Driven Architecture in order to provide the assistance services. This layer will be also enhanced by tools supporting the tactical decision-making processes by producing and consuming digital information coming from the other layers.

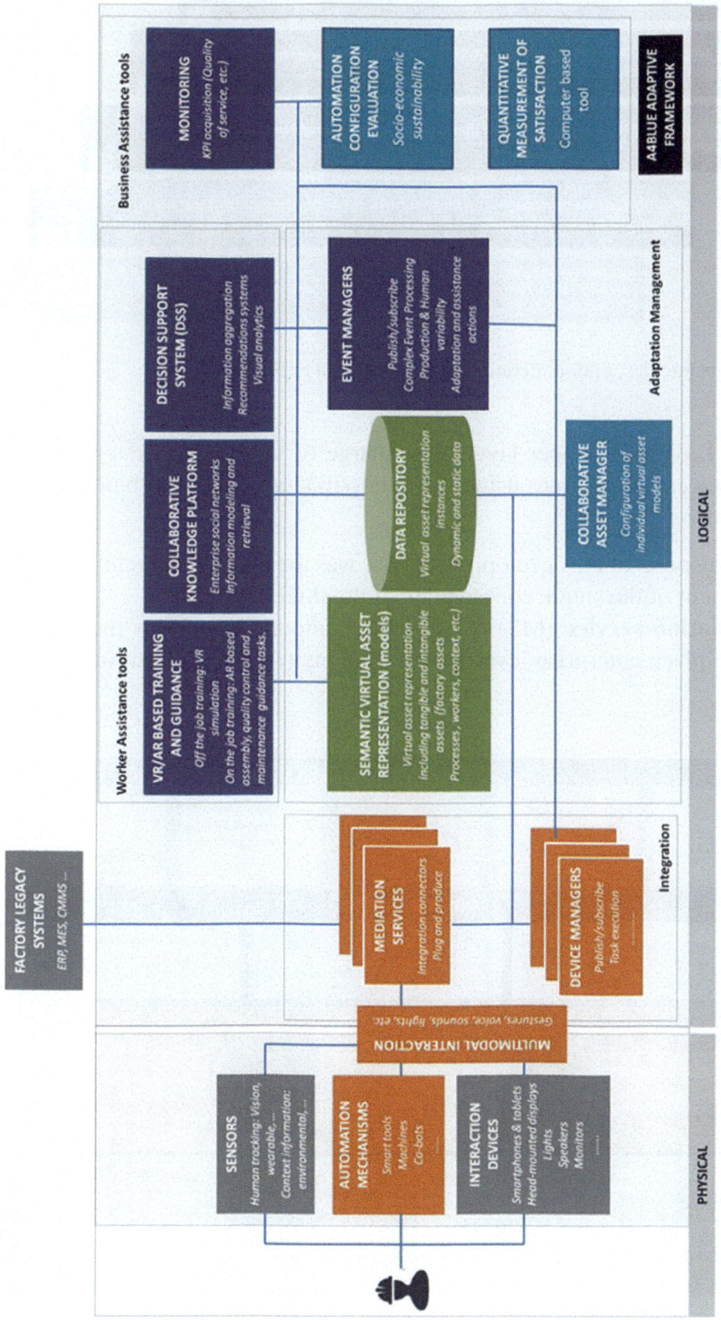

Fig. 9 A4BLUE reference implementation

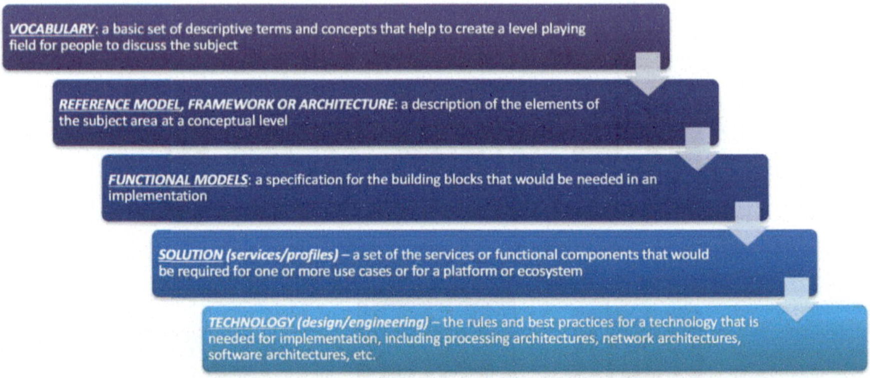

Fig. 10 Architecture description: iterative and incremental approach

- **Business Layer**: the upper layer is in charge of supporting strategic decision-making process (sometimes using off-line tools), targeting both blue- and white-collar workers (Fig. 11).

The starting point of the development phase was the F&M architecture, especially the development of the single components of the three domains.

The **Mediation service** (MS) (Shopfloor) component supports the integration of already in place enterprise level legacy systems (e.g. Manufacturing Execution Systems, etc.).

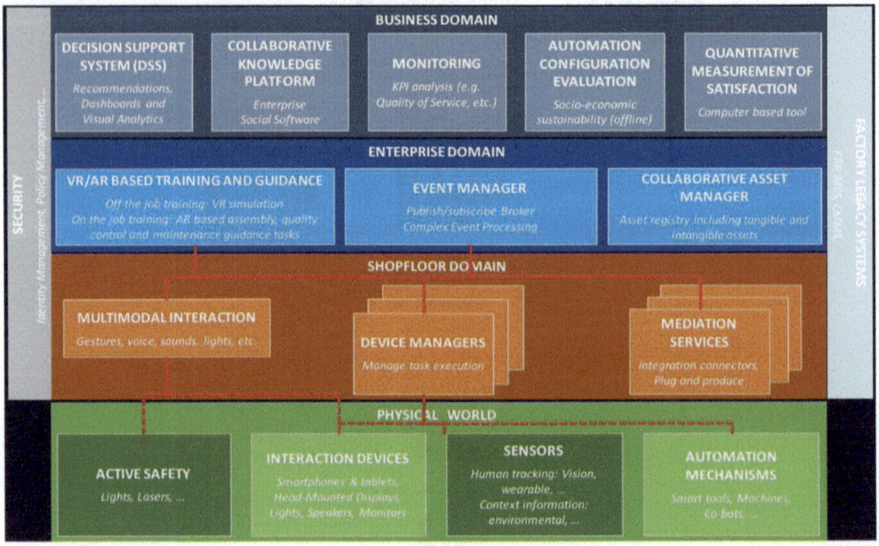

Fig. 11 A4BLUE Functional and Modular (F&M) architecture

The Legacy mediation agent is legacy system dependent and represents the connector that supports the bidirectional data exchange process between an already in place legacy system and the Mediation management services component. The MS Event Adapter supports publish and subscribe capabilities and adapts the process data collected through the Mediation management services to the event format supported by the Event Manager component. It also transforms events coming from EM component into process data to update the legacy system (e.g. information collected during the execution of the operations performed by the automation mechanisms).

The **Automaton mechanisms** (Shopfloor) involve both robots and smart tools.

The Local automation controller controls the Automation hardware (e.g. robot, smart tool) that could involve auxiliary devices (e.g. camera, force sensor, etc.) to improve the accuracy of the process in order to get a better final result, collects updated automation data values and allow remote execution of exposed methods. Furthermore, the automation mechanism can involve a graphical user interface to support user interaction (Automation GUI).

The automation mechanism includes OPC UA technology to support standard based plug and produce integration with the A4BLUE solution. This involves OPC-UA Server aiming to: (1) register automation semantic information (i.e. involving available automation data and methods); (2) register communication information (IP, port) into the OPC UA discovery server; (3) provide updates of the monitored automation data and execute automation methods. Optionally it could include a repository to persist relevant data (i.e. Automation repository).

The **Device manager** (DM) (Shopfloor) component supports plug and produce capabilities by enabling the integration of external automation mechanisms in a standardized mode (i.e. OPC UA based). It involves both the discovery and operation processes supporting the plug and produce approach.

The **EVENT MANAGER** (EM) (Enterprise) foresees the implementation of the Publish/Subscribe Context Broker that allows to manage all the whole lifecycle of context information including updates, queries, registrations and subscriptions. EM is able to register context elements and manage them through updates and queries. EM provides an API REST web service implemented in Java.

In every data intensive scenario, it is needed a component in the architecture able to mediate between entities (a physical thing or part of an application), such as data/event producers (e.g. sensors or IT systems), and consumer applications (e.g. a smartphone application or AR tools).

In A4BLUE scenario it is possible to manage different sources of context: context information may come from many sources, varying over the time; in this context, entities are the virtual representation of all kinds of physical objects ("things") in the real world.

In order to simplify the event management in a highly distributed system such as A4BLUE, it was necessary to define an **Event Taxonomy**, i.e. a controlled vocabulary which consists of concepts in a certain topic, which is structured in a hierarchical structure and whose aim is to represent all the relevant situations that it is possible to occur in A4BLUE scenario.

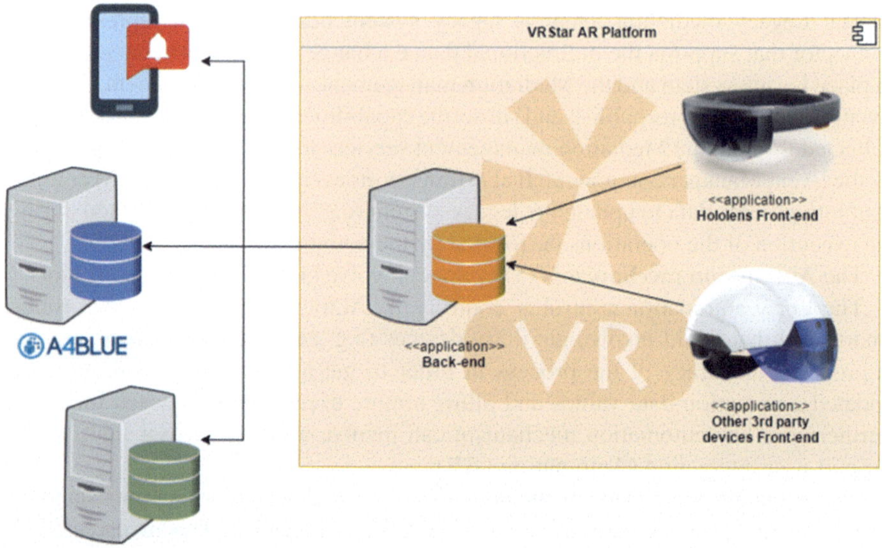

Fig. 12 The VRStar AR Platform and its modules

The A4BLUE **Collaborative Asset Manager** (CAM) (Enterprise) represents the implementation of a virtualized assets representation model (VARM) knowledge base, in charge of supporting the virtualisation and representation of Tangible Assets (TA) and Intangible Assets (IA), producing the Virtual Asset Representation (VAR) that contains the result of the adoption of the VARM model into the A4BLUE case. CAM exposes (standard) interfaces for retrieval of the assets from the knowledge base, by M2M interactions (i.e. using specific APIs) or by H2M interactions (i.e. using a dedicated GUI).

Structurally, the CAM is a web-based application written in Java, consisting of a core API able to use specific driver to handle a RDF STORE. The interface to the RDF STORE is represented by the RDF4J[28] technology, which will be discussed in detail in the paragraph on the implemented technological solution.

The A4BLUE **VR/AR based training and guidance system** (Enterprise) is composed by two separate modules: a back-end server application and a front-end Augmented Reality player (Fig. 12).

The back-end server application runs on a dedicated machine and it acts as a bridge between the existing A4BLUE framework and the AR-devices. Once set, the back-end application is able to communicate with the AR device in order to send/receive data in real-time during the training session. The EM and the CAM are used as main interfaces to access A4BLUE framework resources and data.

The front-end consists of a native application installed in the Augmented Reality HMD device that receives data and asset from the back-end and renders a 3D scene accordingly. The player is also able to capture events such as gesture and voice

[28] https://rdf4j.org/documentation/.

commands and send them back to the server application in order to update its status in real-time and propagate this information to the A4BLUE Context Broker.

Through the **Collaborative Knowledge Platform for Manufacturing** (KM) (Business), blue and white collar workers are assisted in Collaborative Decision Making (CDM) processes, adding new services to handle conflict resolutions and to support knowledge take up from the workers, in order to feed a lesson learned database built upon structured and unstructured knowledge. The KM platform has been built upon **OPENNESS** (OPEN Networked Enterprise Social Software), the results of an ongoing private research project under development at ENGINEERING R&D Department: a platform fully developed using open source technologies, and leveraging on relevant results from research fields such as Open Innovation, Collective Intelligence, Enterprise Social Software.

The **Decision Support System** (Business) component aims to support workers on relevant decisions for the assembly, maintenance, inspection operations. It aggregates relevant information produced in the domain of the A4BLUE system and provides visual analytics capabilities to support workers in the decision-making process. Furthermore, it supports the management of multichannel notifications to notify intervention requests (e.g. maintenance, assembly collaboration, inspection, etc.).

The **Monitoring** (Business) component is aimed to support the collection of key performance indicators (KPIs) produced in the domain of the A4BLUE solution to support the assessment of the impact of the introduction of the A4BLUE solution during the experimentation and evaluation phases. The A4BLUE framework will provide a user-friendly graphical interface to allow such analysis and support whit-collar operators in both strategical monitoring and decision making process.

The **Automation Configuration Evaluation** (ACE) (Business) component is aimed to help the production planner in defining the optimal level of automation for the production processes.

The ACE component is based upon three sub-components. The user will interact with the tool by using a graphical, web based, user interface. The database consists of information entered by the user, through the user interface, and persisted for future use. The ACE Management Service consists of the business logic of the software. It interacts with the Data Repository by storing and retrieving the required information.

The **Computer-based tool for Quantitative Measurement of Satisfaction** (CQMS) (Business) component enables workers to complete questionnaires, which can be used to assess levels of worker satisfaction and other aspects of wellbeing in relation to human-automation systems and wider work environment characteristics.

6 Future Trends and Transformations

To have an idea about future trends in manufacturing sector it's worth looking at EFFRA Vision for a manufacturing partnership in HORIZON EUROPE [6]. The four main components of impact are in line with the 'challenges and opportunities' described in the FoF 2020 roadmap:

- **Competitiveness**: Any European manufacturing company has a constant need to strive for excellence. This requires producing top quality goods, being highly efficient in terms of costs and resources, while being extremely responsive to market and customer needs, and using and offering creative and innovative solutions. More than ever companies can only achieve this via cooperation and a strong integration in value or knowledge networks or eco-systems.
- **Planet**: Environmental sustainability was already high on the agenda in FoF 2020. Recent reports on climate change and the impact of waste on our society keep raising the importance of energy and resource efficiency in manufacturing, including the need for circular and low-footprint and low-carbon approaches.
- **People**: Future innovations need to provide a better understanding as to how employees are creating and modifying their own jobs in their network and how new technologies and social innovations will be introduced and used by the current work force. The technological transition will require to reshape the human–machine-relation, to prepare people with the right capabilities and to provide the right tools and interfaces. Design and development of advanced technologies should consider the role of the workforce at the earliest stages and should consider the available or required additional skills of the people involved. The full benefit of new tools based on advanced technologies can only be achieved by designing new work practices and by involving the employees in the co-design. It is for instance of great importance to investigate how human knowledge and skills can complement Artificial Intelligence solutions and how smooth human-AI interaction can take place.
- **Manufacturing the products of the future**: Innovative, sustainable and affordable products are only possible when reliable and performant manufacturing technology is available which ensures the integration of key technologies, fast and smooth upscaling and conformity with societal requirements [6].

EFFRA's vision for Factories of the Future in Horizon Europe has been defined with the ManuFuture Vision 2030 document. This new EFFRA vision will build upon the successful approach of the Factories of the Future 2020 roadmap but will go further, looking at the bigger picture: the manufacturing eco-system and how co-creation takes place through the different actors involved.

Among EFFRA's priorities for the Manufacturing PPP in the next EU Framework Programme Horizon Europe it is crucial the need of excellent, responsive and smart factories where the humans are at the core of the innovation process.

7 Conclusions

The Factories of the Future Public–Private Partnership (PPP) was launched a decade ago in response to a challenging economic and social situation in Europe. Industrial companies of all sizes and industrial-minded institute answered those calls whose scope was to increase European industrial competitiveness and sustainability through research and innovation activities for the timely development of new knowledge-based production technologies, systems and activities beyond the factory floor.

ACE Factories projects and projects on EFFRA portal are examples of success stories where new automation approaches, with workers at the center, can complement people's capabilities and ensure higher performance, adaptability and quality.

Among those FoF projects, 4BLUE represents an effort (within the FOF-04-2016 cluster) to specify an adaptive framework for assembly, in-line with recently introduced reference architectures for the manufacturing industry. Indeed, the A4BLUE RA envisions automation and adaptation functionalities in close proximity to factory physical processes, including real-time operations and taking into account process/product/operator variability.

The A4BLUE Platform design has some innovative characteristics that mark a difference with respect to other similar on-going efforts: software components from the FIWARE for Industry ecosystem are used as a key enabler of distributed adaptation capabilities in several assembly scenarios. Moreover, the A4BLUE Platform enables a wide range of business scenario, including their mixing with socio-economic evaluation and analysis for decision support.

It is possible to conclude considering A4BLUE and the others FoF projects likely to change and affect industrial workplaces, workforces, teams, individuals and still relevant in consideration of the future trends. All their results can become assets in the scope of adapting workplaces, where it may guide the design of ad-hoc solutions having worker satisfaction as their main driver.

References

1. FACTORIES OF THE FUTURE. Multi annual roadmap for the contractual PPP under Horizon 2020. https://www.effra.eu/sites/default/files/factories_of_the_future_2020_roadmap.pdf
2. Factories of the Future Public-Private Partnership, Progress Monitoring Report for 2017. https://www.effra.eu/sites/default/files/fof_cppp_progress_monitoring_report_for_2017_online.pdf
3. ACE Factories White Paper (2019). http://ace-factories.eu/wp-content/uploads/ACE-Factories-White-Paper.pdf
4. Romero D, Bernus P, Noran O, Stahre J (2016) The operator 4.0: human cyber-physical systems & adaptive automation towards human-automation symbiosis work systems
5. VDI/VDE GMA, Reference Architecture Model Industrie 4.0 (RAMI4.0) (2015)
6. EFFRA VISION FOR A MANUFACTURING PARTNERSHIP IN HORIZON EUROPE 2021–2027 (2019). https://www.effra.eu/sites/default/files/190312_effra_roadmapmanufacturingppp_eversion.pdf

Designing Robot Assistance to Optimize Operator Acceptance

María del Mar Otero and Teegan L. Johnson

Abstract The implementation of an automated or robotic solution can sometimes be a difficult process, especially when manual work has always been the usual way of doing things in a company. First experiences with automation can be hard and sometimes even seen as a threat for the workers and their jobs. Will I be replaced by the robot? Is it safe to work collaboratively with an automated mechanism? At the same time workers are asking these questions, there are usually economic concerns in the company such as the initially high investment and the expected benefits of such an innovative solution, where the risks can be higher than those traditional solutions which have been widely implemented and tested over the years. This chapter will show a real example of successful implementation of a robotic solution in a Spanish aeronautical company. This is one of the first contacts with a robotic device in a production area where tasks are traditionally performed manually by human workers, including the assembly of complex aeronautical equipment and its auxiliary operations. As part of the A4BLUE project funded by the European Commission, a collaborative robot cell for the deburring of Titanium parts has been developed and implemented on a serial production line at the Compañía Española de Sistemas Aeronáuticos (CESA) facilities in Madrid. The main characteristic of this development has been the involvement of all the workers in the task from the beginning of the project to the final implementation on the shop floor. Operators, shop floor managers and production engineers have all worked together to obtain the desired result. Another special characteristic of this implementation is that the worker satisfaction has been considered and measured to make sure that it has been increased. Levels of usability, mental workload and trust in the robot have been measured through surveys, questionnaires and interviews with operators, engineers and managers involved in the project to ensure satisfactory working conditions are maintained or improved. CESA's use case is a good example of fruitful collaboration

M. del Mar Otero (✉)
Héroux-Devtek, Madrid, Spain

T. L. Johnson
Industrial Psychology and Human Factors Group, Cranfield University, Cranfield, UK

M. I. Aldinhas Ferreira and S. R. Fletcher (eds.), *The 21st Century Industrial Robot: When Tools Become Collaborators*, Intelligent Systems, Control and Automation: Science and Engineering 81, https://doi.org/10.1007/978-3-030-78513-0_8

between humans and robots, taking advantage of each other strengths to obtain flexible, productive and efficient production processes where robots relieve humans from physically demanding tasks. It also shows the benefits of involving the workforce in the implementation of a new technology which will change their work methods.

1 Short Introduction to CESA

CESA belongs to a very particular group within the manufacturing industry, which are the aeronautical companies. All of these companies present some very special characteristics that clearly define the way they produce and organize their job, which is different from other industries. Automation in the aeronautical industry still focuses mainly on the machining centres, while the assembly procedures are still mostly performed manually. Small batches and high variability of products have historically been the reasons for this. However, the search for higher market competitiveness and sustainability have made it necessary to explore new solutions involving recently developed automation technologies. A collaborative robotic system seemed to be the best solution, because it allows the company to benefit from the best skills of both operators and the automation mechanism. To better understand the reasons why CESA decided to implement a collaborative robot, and the benefits of it, it is necessary to first get to know the company, its history and expertise.

CESA is the European leader in fluid mechanics and electro-mechanics, growing from the base of a 20-year experience inherited from Construcciones Aeronáuticas, S.A. (C.A.S.A.) and more than 25 years of its own experience. As an aeronautical company, CESA is an ISO 9001 & AS9100 certified aerospace company strongly committed to technological innovation in aircraft command and control systems design and manufacturing. The company has more than 40 years extensive experience in the design, development, manufacturing (including assembly and testing) and product support in key high technology state-of-the-art projects for different aircraft systems; with a special presence in electromechanical and hydraulic actuation systems and components, among others. In October 2018, CESA became part of the global group Héroux-Devtek, which is the world´s third largest producer of landing gear and has 16 facilities distributed in Spain, UK, Canada and USA. With this purchase, Héroux-Devtek became the unique shareholder by owning the 100% of CESA. Within Héroux-Devtek, CESA will continue being a reference centre in actuation systems and will have the chance to expand its customer portfolio.

CESA's facilities are located in Madrid and Seville. The industrial complex covers a total surface area of 37,000 m^2 (13,800 m^2 of built installations) in which industrial safety and environmental protection are key factors. The engineering capacity of its testing installations enables CESA to develop and certify the most technologically advanced aeronautical equipment in record time. CESA is currently an established supplier to European aircraft industries such as AIRBUS, AIRBUS Military, BA Systems, Cassidian, Alenia, etc., as well as other worldwide manufacturers such as SAFRAN Landing Systems, SAFRAN Electrical and Power, Sikorsky (USA), IAI

(Israel) or TAI (Turkey) for product lines including Landing Gear components, Flight Control, Hydraulic systems, Pneumatic/Fuel/Engine accessories or Electromechanical actuators (EMAs). For some years, CESA has also been expanding through Asia with new programs being developed with AVIC (Aviation Industry Corporation of China) and KAI (Korea Aerospace Industries).

Regarding products, CESA has a significant portfolio in both civil and military fixed wing platforms as well as rotary wing platforms. In relation to cargo door actuation, CESA has in-service components, and has additionally within the framework of European Research programs developed linear electromechanical actuators. In addition, the company has a significant involvement in R&D programs and over the past years has participated in the development of electromechanical actuators for different applications such as cargo doors, landing gear and flight controls actuation. This experience with electro-mechanic equipment includes equipment with electronic on board qualified for flight and software. Some of the latest developments are the Tail Boom system, the A400M Weigh on Wheel system, the EELT (European Extremely Large Telescope) M1 Position Actuators, and a Cargo Door electromechanical actuator with the power electronics associated for a future replacement of Hydraulic Cargo Door Actuators.

Development of such systems is possible because CESA covers the whole lifecycle of a system and/or equipment development according to customer specification. Among others, CESA is capable of performing the following activities (Fig. 1):

Highly qualified personnel work at CESA. The workforce consists of 336 people, 46% of whom hold university degrees or are specialized technicians, and 48% are highly skilled operators (Fig. 2). This expert workforce provides know-how in precision machining and heat and surface treatments of all types of advanced aeronautical materials, e.g. highly resistant aluminium, titanium and steel alloys. Currently 80% of all the parts that are manufactured by CESA, are subcontracted. However, usually the most complex parts are manufactured inside CESA. CESA has the means to manufacture, both machining and special processes, parts internally in the facilities located in Madrid.

2 Characteristics of an Aeronautical Company Production

One of the most typical characteristics of aeronautical companies is the extensive amount of manual work needed to produce each of the components, especially in the assembly area where highly experienced operators perform almost all of the tasks manually. Two of the reasons why most of the work is performed manually are the small production batches and the high variability of products. These are the main differences between the aerospace sector and the other manufacturing industries like automotive companies. As a result, it is not usually cost-effective to invest in big automation processes with such a small production of each type of equipment, particularly for the assembly process and its auxiliary operations (Fig. 3).

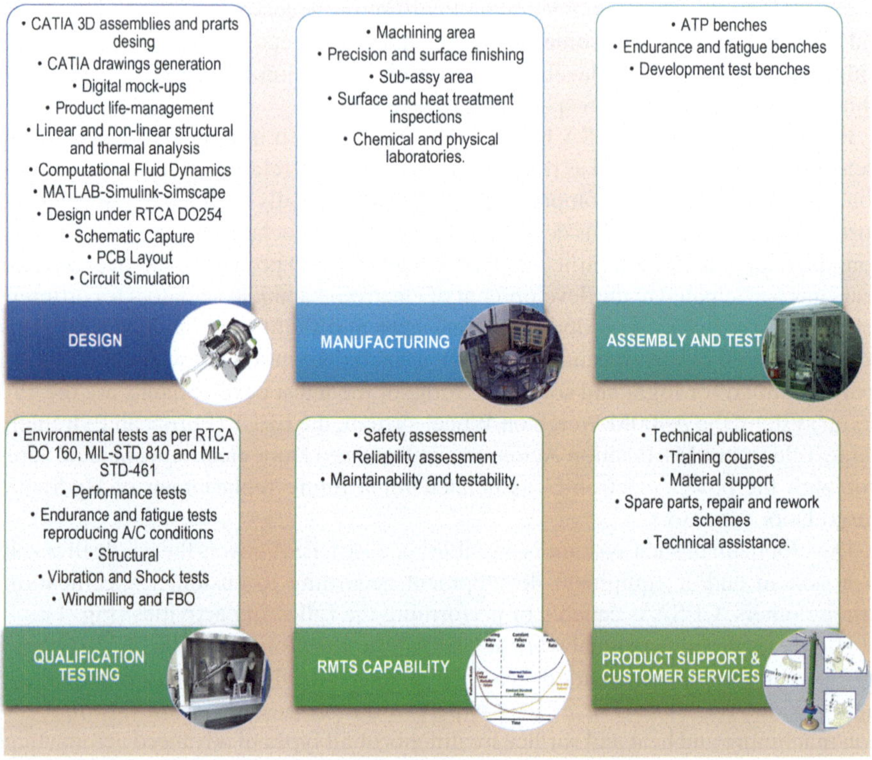

Fig. 1 CESA product lifecycle capabilities overview

Fig. 2 CESA workforce

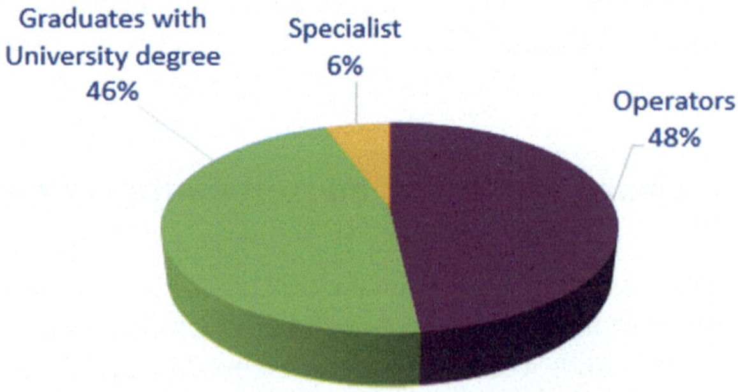

In addition, it is crucial for companies like CESA to be very flexible in order to be able to adapt to changes in demand due to market oscillations. Flexibility is also needed in the manufacturing and assembly areas to be able to change from one product to another to meet the customer´s demand. In this sense, human operators are already very flexible because they are capable of adapting to changes, learning from

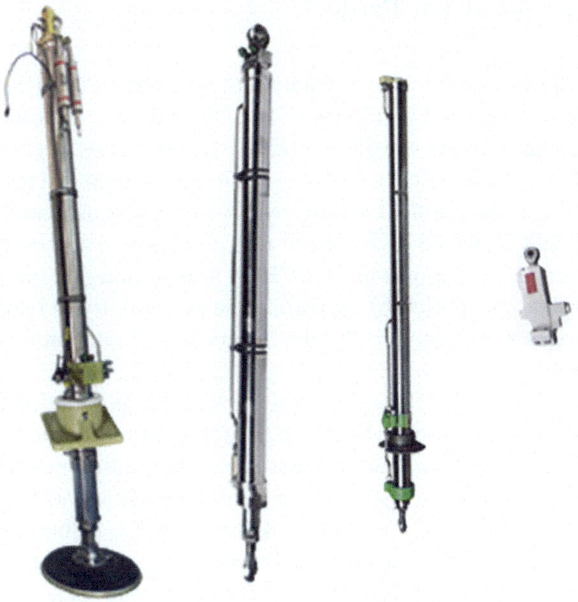

Fig. 3 Different hydraulic actuators assembled in the same module

experience, evaluating problems and finding quick solutions. However, training to gain the certifications required to be allowed to assemble each specific product can take a long time. Despite this, once they are trained there is no automation mechanism more flexible than a human worker. This training could be less necessary with an on-the-job guidance mechanism that can adapt to the different profiles of operators with varying levels of experience.

At the same time, as with any other industry, some manufacturing process can be hard, exhausting and require the operators to wear personal protection equipment to avoid risks to their health. In these cases, the implementation of automation can be a great help for improving the health and safety of workers in addition to their satisfaction.

As a summary, aeronautical companies like CESA need automation to solve their particular issues, while still retaining the same high levels of flexibility provided by humans workers. A strong collaboration between the human operators and robots can bring very important benefits for aeronautical production, by combining the flexibility of human workers and the capabilities of automation to increase productivity and efficiency and releasing operators from physically demanding tasks.

3 A4BLUE Project

The successful implementation of automation in CESA detailed within this chapter, was developed within the A4BLUE (Adaptive Automation in Assembly for BLUE collar workers satisfaction in Evolvable context) research and development project (https://a4blue.eu/). CESA participated as final end user of the A4BLUE technology in the aeronautical sector, by providing the use case scenario where technologies could be tested to solve the real problems of the industry. A4BLUE represented a very good opportunity for CESA to help mitigate the risks associated with this first robot installation by providing the opportunity to benefit from a consortium with a lot of experience in the development and integration of industrial automation and robotics.

The A4BLUE project was a 3 year project funded by the European Commission under the European H2020 Research and Innovation Program and finished in October 2019. CESA was part of a first-class international consortium led by IK4-TEKNIKER (Spain) and involving prestigious universities such as RWTH Aachen University (Germany) and Cranfield University (UK), technology providers companies such as Engineering Ingegneria Informatica S.P.A (Italy), Illogic Societa'a Respponsabilita'Limitata (Italy), CIAOTECH Srl (Italy) and Ingeniería de Automatización y Robótica KOMAT S.L. (Spain) and end user companies like AIRBUS Operations S.A.S (France).

A4BLUE provided solutions for a new generation of assembly processes, putting together workers and automation by taking advantage of each other´s strengths to enhance flexibility, productivity and worker satisfaction. A4BLUE solutions were based on adaptability to the process requirements and physical characteristics of the workers, safe and efficient human–robot interaction and sustainability. These solutions included, among others, worker personalized assistance systems based on virtual and augmented reality capable of providing off-the-job training and on-the-job guidance. Interaction between humans and robots occurring using gestures and voice commands. A4BLUE also provide methods and tools to determine the optimal balance between automation and workers presence in assembly environments to maximize long term worker satisfaction and the overall process performances.

Finally, exhausting and repetitive tasks are automated with collaborative robots, improving the physical and cognitive ergonomic conditions of the workers. Additionally, a usability methodology was established that was tailored to each of the use cases to make sure all steps were taken to ensure worker satisfaction. This was the situation for the CESA use case, with the details of this process provided in the following sections.

4 Identifying Users Requirements and Needs

The first thing to do, to make sure that A4BLUE was providing a solution to real problems and bottlenecks in the industry, was to obtain an initial understanding of the likely requirements of the potential users of the A4BLUE solutions. Opinions from individuals with first-hand knowledge of industrial processes about what should be prioritised in the design of the project's solutions were gathered via a small-scale survey of representative users from partners' organisations. Surveys were completed online by twenty-two representatives from all of the main departments of the organisations participating in the A4BLUE consortium, their customers and suppliers. The participants were presented with a set of items (questions) under each of the following headings:

- Organisational requirements
- Automation and robotics
- Communication and Interaction Mechanisms
- Work system feedback, training, and assistance
- System Security and Data Management

The participants were asked to indicate whether the survey items were: "Essential", "Desirable", or "Unnecessary" with regard to "Assembly work systems of the future". This simple method for gathering expert opinions on what would be vital, advantageous but not vital, or of no use at all in future production systems, was used to illuminate design priorities for the project.

Regarding *"Organizational Level Requirements"* the greatest "Essential" responses were seen for:

- The ability to easily reconfigure the workplace when introducing a new automated system or robotics (e.g. plug & produce capabilities)
- Continuous data collection for analysis of system performance and optimisation needs
- On-the-job work instructions that guide the worker through assembly or support processes (i.e. inspection, routine maintenance) to reduce the need for organised off-the-job training and supervision

Both the first and third of these "Essential" items show the importance of flexibility within the assembly work systems of the future, to account not only for a range of products but also for the people who may work with these systems. Therefore, this indicated that they should be a high priority in the use cases developed within the A4BLUE project. Although none of the items presented had an "Unnecessary" response greater than either the "Essential" or "Desirable" scores, some scores were greater than 10%. The highest "Unnecessary" responses were for the items bulleted below:

- The ability to self-adjust to compensate for lower training and experience levels
- The ability to self-adjust to compensate for reduced technical capabilities (older computer programs)

- Direct connection to organisational systems for post-production product service and support

Participants were also asked to provide additional comments about organizational requirements of future assembly systems features. Two participants provided similar comments concerning "Focus on recentering the work of everyone on tasks with high added value" which indicated that a large amount of time is lost with operators working on lower value tasks and the importance of increasing time on higher value tasks for the organisation.

Regarding "Automation and Robotics", the items that participants considered most "Essential" were:

- Robots have safety capabilities that move the robot away from the worker in the event of an accidental collision.
- Safety capabilities that adapt the speed of the robot according to the distance or speed of the operator.
- Robots have safety capabilities that immediately stop the robot in the event of an accidental collision.
- Safety mechanisms that make operators comfortable when collaborating with automation/robots during assembly.

Three of these four items with high "Essential" responses reflect the criticality of safety with regard to robotics and automation and, therefore, safety had to be prominent within the list of user requirements for the A4BLUE use cases. The items with the greatest number of "Desirable" responses were:

- Automated/robotic functions that will adapt to suit operator's preferred working methods
- Automation/robotics that can change themselves safely to meet varying production demands.
- Automation/robotics that can change safely on their own to meet different experience capabilities of the involved operators.

Where the "Essential" response profile reflected the need for safety, the "Desirable" responses showed the importance of meeting operator and production variations. These results showed the need for flexibility in the new manufacturing workplaces.

A greater number of "Unnecessary" responses were provided in this section than for organisational level requirements. Items with the greatest percentage of "Unnecessary" responses were:

- Robots that do not work with or in close proximity to humans.
- Automation/robotics that run at a constant rate or on a constant programme and do not change.
- Automation/robotics that can only be adapted by management.
- Robots should work safely alongside or near to an operator but on separate tasks.

The higher frequency of "Unnecessary" responses was expected for these items, as they were negatively weighted responses and used to identify whether participants are automatically responding to questions or reading the items and then responding. They were additionally included within this section to assess participant's reaction to these types of systems and working set up. The higher frequency of "Unnecessary" responses to these items provided further evidence for the need to ensure flexibility in the new systems and to introduce human robot collaboration to the list of user requirements for the new system.

The "Communication and Interaction Mechanism" items were all identified "Desirable" rather than "Essential" features of future assembly systems:

- The automation/robot/system has visual capabilities (e.g. computer systems, lights, projected messages, etc.) to display relevant feedback and notifications to operators.

Seven of these items identified by respondents as "Desirable" covered interaction mechanisms:

- The automation/robot/system has both visual and auditory capabilities to present relevant feedback and notifications.
- Automation/robot/systems that can be controlled with a computer system on a mobile devise (e.g. tablet, smartphone).

This indicated that of the four items regarding feedback, visual feedback was considered most "Desirable" and, therefore, would be most preferred. However, including both visual and auditory capabilities were also considered "Desirable" option but for fewer respondents. The item that presents only auditory feedback had the highest percentage of "Unnecessary" responses for the feedback items, reinforcing that both visual and auditory or just visual feedback should be a user requirement for the A4BLUE case studies.

The interaction mechanisms identified as "Essential" were for the two following items:

- A workstation PC with an interactive computer system that allows the operator to interact and control the automation/robot/system.
- The automation/robot/system has feedback abilities to show that it has understood a command.

These items reflect current practices in interacting with manufacturing systems so the high level of "Essential" scoring shows that participants see that these methods should be continued, or perhaps cannot see a future where they are not crucial to production processes. The "Desirable" responses on the other hand received the greatest percentage of responses which reflected what participants wanted to see in future systems. These included:

- Automation/robot/systems that operators interact with using natural speaking (i.e. non-predefined commands).

- Automation/robot/systems that can be controlled with a computer system on a mobile devise (e.g. tablet, smartphone).
- Automation/robot/systems that operators interact with using pre-defined voice commands.

The results of this survey provided a reference point from which to ensure that solution implemented on to CESAs shop floor, as a result of the A4BLUE project, would meet the expectations and aspirations of stakeholders and industrialists. A more extensive and detailed summary of these results can be found in Fletcher et al. [6]. The following sections provide more in-depth detail of the original manual process and the automated solution that replaced it, as developed by the project.

5 Use Case: Manual Deburring of a Titanium Part

CESA's selected use case was one of the auxiliary assembly tasks: the deburring of parts prior to assembly. "Deburring" is the process of removing all of the metal chips and marks that are left in the part as a result of the machining process. It is a very common process completed at CESA manually to all different parts and materials, from Titanium to Aluminium and Steel. It is an exhausting task during which the worker has to wear a mask to avoid breathing in the metal chips expelled from the part during the process. The particle suction system always runs during this process for extra protection from expelled metal chips and particles. The operator also has to rotate the part several times to reach all the areas of the part.

The process takes 130 min to complete, during which time the experienced worker uses a number of different tools which are required for each phase of the long deburring process. The deburring process consists of several phases, the longest and most exhausting part of the task is the smoothing and homogenization of the surface and removing the machining marks. Other phases such as deburring drills and sharp edges require less time (Fig. 4).

The most promising phase to automate was identified as the longest and more physically demanding task in this case: the smoothing and homogenizing of the surface and removing the machining marks (90 min). Other simple tasks like deburring the inner drill can still be easily performed by the operator manually, and in the

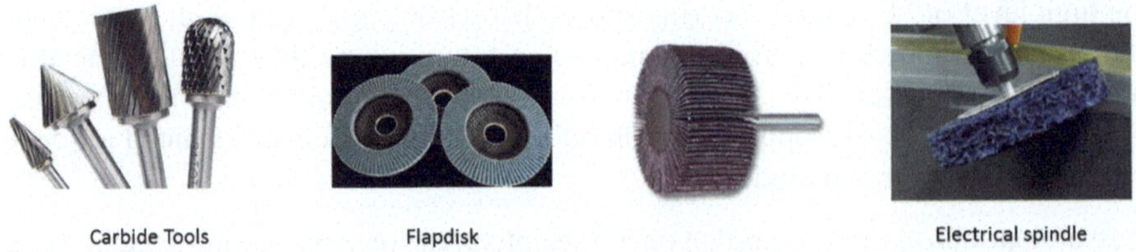

Fig. 4 Different tools used for deburring purposes

A4BLUE prototype the work is shared between the operator and the robot, leaving the more time consuming and physically demanding work for the robot. However, the automation of these shorter timed tasks can be considered in the future as an extra improvement in the process.

The level of worker experience is also crucial in terms of quality checks of the surface of the part; knowledge of how the part should look when a satisfactory finish is achieved, is gained over time and experience. A human factors study of the manual deburring process, (described in more detail later in this chapter), revealed that each operator had their own way of deburring the same part. Each operator had their preferred tools and followed a slightly different order of steps to achieve the same level of finish on the product, and these differences were learned both through experience and training.

For the purpose of the project CESA focused on a single part that was most representative of hydraulic equipment, the retraction actuator of the main landing gear of a well-known single aisle commercial aircraft. Specifically, the deburring process of its Titanium Earth End was selected as it is one of the most complex parts to deburr due to the high hardness of the material. Taking all of this into account, the deburring operation of a Titanium Earth end seemed to be the perfect candidate for new automation to improve ergonomics, reduce physically demanding tasks and allow the worker to use their time on more value-added tasks while improving productivity and maintaining the good quality levels.

6 Solution: A Collaborative Robotic Cell

From the beginning it was clear that the deburring of a Titanium part using a robot would be a challenge due to the hardness of the material and lack of available solutions on the market. So it was crucial to select the right robot that would have enough force to effectively complete the task. This was achieved by selecting the FANUC robot M-20iA/35 M with a 35 kg payload.

At the same time, CESA required the robot cell to be open and collaborative to allow and easy and safe interaction between the human and robot and easy integration on to the already crowded shop floor. Flexibility to move the robot cell from one place to another was also very important to be able to adapt to the periodical layout improvements in the manufacturing and assembly area. In order to achieve these goals no big static fences could be built, so a collaborative robotic cell was the solution.

Required safety levels in the cell were guaranteed by physical fencing on three of sides and a safety laser monitoring system was used for the open entrance to the cell to detect the presence of an operator. Two safety areas were established depending on the distance between the worker and the robot. In the less dangerous area the robot reduces its speed if a human presence is detected, however in the most dangerous and closest area the robot completely stops all work to avoid any accidents (Figs. 5 and 6).

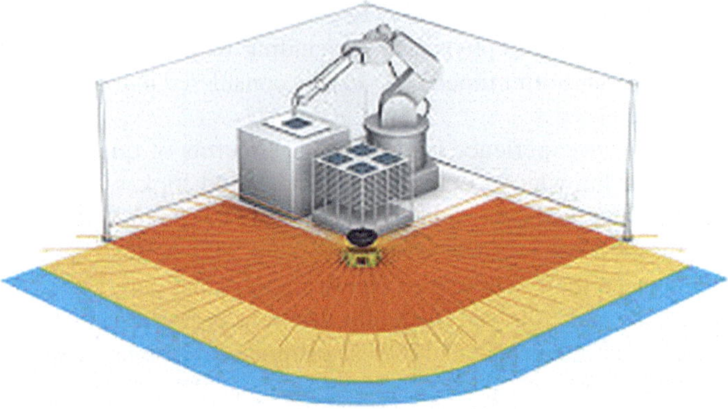

Fig. 5 Scheme of the safety laser scan and the danger areas

Fig. 6 Collaborative robotic cell for the deburring of parts

The robot also had to be easily operated by workers with no deep programming skills. Easy re-programming was important to be able to include different parts in the robot deburring process beyond this first prototype. The robot and its program had to be flexible.

Another challenge was how to achieve the desirable quality of the surface using the robot and how to check it, so that performance and quality was as effective as the original manual process. In this case the collaboration of the human workers was very important to do the final quality check of the part and reprogram the robot to rework some areas if needed. Effective collaboration between the human and robot was, therefore, crucial in this process.

7 Implementation Process

The automation implementation process was the result of the collaborative and fruitful work of all of the members of the A4BLUE consortium. In particular the Spanish company KOMAT S.L., who are experts in automation solutions, were in charge of developing the automated solution and cell for this application. They took care of the conceptual and detail design, supported CESA in the purchasing phase, built the first prototype in their facilities and implemented the final solution on CESA's shop floor.

At the same time, from the start of the process, CESA's operators and manufacturing and assembly engineers were involved in every decision. As explained before, representatives of the main departments in the company were interviewed through surveys and face to face meetings to understand their needs, the problems they had to face every day and what kind of solutions they would propose.

Then, during the development of the first prototype at KOMAT facilities in the Basque Country Spain, CESA operators and manufacturing engineers travelled to help in the refinement of the robot work. They gave suggestions on the position of the tools, and how to use each of them to achieve the best results and extend the life of the tool as much as possible. Other crucial parameters for which they provided recommended adjustments on were the force to be applied, and the duration of each pass of the tool to avoid sparks between the tools and the part which can damage the part.

In addition, KOMAT was invited to CESA to see and understand in detail the manual deburring process with experienced operators. Each phase was explained and parameters were studied, tips and best practices were shared with KOMAT to be applied in the robot program. With this information and through an iterative process between CESA and KOMAT, a first prototype of the robot program the robot was ready. Once it was ready, it was taken from KOMAT to CESA's facilities in Madrid for installation at the final location on the shop floor. This final implementation took several months of collaboration between CESA and KOMAT as the robot program continued to be refined and the robot cell was adapted to fit CESA's need and available space.

Other new features that were implemented included allowing workers remote access to the robot status while they are performing other tasks in the area. Maintenance alerts are sent to the operator's mobile phone to inform about the progress of the robot work and possible failures. This allows the worker to complete other tasks

Fig. 7 Human and robot working collaboratively at the deburring cell

far from the robot, while always having accurate information about the state of the robot and intervene only if needed.

Finally, as this was the first experience with robots for most of the workers, technical training was considered crucial and was therefore given to operators and engineers in the area on how to use and work with the robot and reprogram it if it was needed to deburr other different parts (Fig. 7).

8 Workers' Opinions on the Final Robotic Solution

One of the main goals of the ABLUE project was that its automated solutions should be designed to ensure worker satisfaction. To achieve this, a human factors assessment was carried out prior to, and after implementation of the robot to establish levels of operators' satisfaction and usability of the collaborative robot system. This provided guidance on design during development of the solution as well as a measure of impacts / benefits for the workforce that were achieved by its implementation. All human factors research was conducted by Cranfield University, with full ethical approval and in accordance with appropriate codes of practice and regulations for human research.

The first step in assessing the usability of the collaborative robot system was to identify potential issues prior to implementation of the robot and to identify the most suitable usability measures to apply once the robot system had been installed. This required an assessment of the original process via task analysis, specifically using hierarchical task analysis (HTA), task decomposition (TD) and a comparative HTA technique. Data was gathered using interviews and observations of the task, completed by two male operators with between four and ten years' experience, over two days.

HTA is a type of task analysis and involves the systematic deconstruction of the manual task activity so that it is broken down into its constituent parts to show the step-by-step sequence of individual activities and operations that are necessary to complete the task. This in-depth breakdown of individual task steps forms a hierarchy of 'goals', 'sub goals', 'operations' and 'plans' that are needed to meet the overall objective. The highest level is typically the 'overall goal' and subsequent levels are developed until sufficient to fully represent the task. By breaking the task down in this systematic way, the HTA revealed the distinction steps to the task. The TD technique was then applied to the HTA data in order to examine cognitive activities and task steps. In the TD procedure, the task elements were explored by grouping data in a series of more detailed statements related to specific aspects of the task under subheadings that were selected to answer key questions. Thus, in this case, the subheadings were chosen so that the information extracted was specific to usability of the A4BLUE assembly work systems and technologies. Usability of the currently manual process was assessed in addition to traditional factors to help to isolate and prevent any usability issues being carried forward to the new system early in the process. Frequently, usability assessments investigate objective and subjective factors and focus on effectiveness, efficiency, ease of use and accessibility [12]. This includes the identification of error and difficulties, work rate, and goal achievement.

The standard TD subheadings prescribed by Kirwan and Ainsworth [10] and used within this analysis included:

- *Purpose*: The reason for a particular operation.
- *Cues*: The cues and feedback that operators respond to (consciously or unconsciously) in order to complete the task.
- *Decision*: The decisions made by the operator in response to cues to perform the task successfully.
- *Likely errors and error correction*: The types of error and remedial actions that operators make. This subheading was directly linked to "ease of use" which is one of the formative definitions of usability [2, 12].

Additional subheadings that were used to address the requirements of this project included:

- *Variations between operators*: this information was gathered to identify if differences in workers' methods should be considered and addressed prior to implementation of the solution. Not attending to impacts of these differences could impact both usability and satisfaction with the system, which could in turn hamper engagement and uptake.

- *Identification of easy and difficult operations*: this information was collected by asking participants to indicate the degree to which they experienced "ease of use" when using the system from the options of "Easy" or "Difficult". Ease of use is a key component of usability [2, 12] and, therefore, vital to operational success.
- *Physical discomfort*: this information was gathered to provide insight into the physical fit between an operator and the work system, based on the degree of subjective comfort they reported. The responses were used to provide recommendations on where adjustable ergonomic features could be included.

Once the complexities of the task had been understood from the HTA and TD, a comparative HTA was conducted. This involved providing a breakdown of the current manual activity and the new activity resulting from introducing the collaborative robot. The allocation of tasks attributed to the human operator and the robot could then be seen alongside one another and the original manual activity and reviewed. This provided a clear understanding of how the activity would change and potential factors that could affect usability. Due to the commercially sensitive nature of the results, examples of the HTA, TD and comparative HTA for this use case cannot be provided here. However, they led to the identification of the need for an overall usability assessment, an assessment that mental workload was not over the effective performance threshold as a result of introducing the new robot, and an assessment of workers' trust in the robot. The mental workload and trust assessments were recommended as they both likely affect perceived usability and satisfaction. They are two known factors that have historically been found to negatively affect successful implementation of robotics [15]. The assessment of mental workload is concerned with whether the act of interacting with, or being in proximity to, a collaborative robot increases mental workload levels for the individual. With the introduction of a robot it was possible that operators would experience an increase in monitoring, and anticipation and planning on how to interact and respond to the collaborative robot. The consequence of this was a likely increase in cognitive demand for the operator. If implementation of the robot lead to greater workload, the acute (increased error rates and task time, etc.) and chronic effects (physiological long-term health conditions) of excessive workload could have been observed [5, 9]. Understanding the effect of introducing a collaborative robot on mental workload early was vital for building a safe and effective system [16].

Trust in automation and robotics is inherently tied to reliability [16]. A system's ability to perform reliably, particularly where an individual works in close contact with it, is critical to ensuring an effective working relationship between the robot and the person. Other factors have also been found to affect trust in robotics including the size, shape, sounds and movement of the robot [3]. A consequence of a lack of trust is the system not being used [11], which undermines the benefits anticipated from implementation. Therefore, it was essential to understand how the new human robot collaboration affected trust.

To assess these factors an open-ended survey was developed and suitable pre-established psychometric measures were identified. These measures were presented to the operators, engineers and shop floor managers involved in the process to establish their perceptions of the deburring robot cell. The following sections provide a detailed description of the procedure and the results.

9 Assessment of the Robot

9.1 Participants

Four participants took part in this study, all were involved with the development process of the robotic cell. They were all male, aged between 34 and 59 years old. Two participants were deburring operators, one participant was an engineer and responsible for lean manufacturing in the company, and the fourth was the shop floor manager who has been technically involved from the beginning of the automation implementation process.

9.2 Materials and Analysis

Data was collected using a paper-based survey which comprised of three forms. The first form contained quantitative measures of usability and workload, the second form measured trust in human–robot collaboration, and the third provided the open-ended usability survey. The separate measures of usability, trust and workload are described in more detail below.

Usability was measured using a scale and open-ended questions. The scale assessed usability using four-point semantic differentials across the four usability dimensions. These differentials included:

- Ease of use (Easy—Difficult)
- Enjoyment (Fun—Boring)
- Clarity (Clear—Ambiguous)
- Comfort (Comfortable—Uncomfortable)

To answer the survey the participants indicated the most appropriate point along each of the scales. The data was analysed by scoring the points within the scale from 1 to 4, and identifying the means and standard for each of the differentials.

The open ended survey questions were intended to provide information that could be used to contextualise the results of the semantic differentials. These questions included:

- Did you find the task easy/difficult to complete, and why?
- Did you find the instructions easy/difficult to follow, and why?

- Did you find it easy/difficult to interact with the Robot?
- Were you comfortable/uncomfortable while you completed the task?
- Is there a part of the task that you found boring or tedious, and why?
- Did you find anything annoying or frustrating about the task?
- What did you enjoy about completing the task?
- What did you particularly like/dislike about the task?
- Were there any aspects of interacting with the robot that were confusing or unclear?
- If you could change anything about the task what would you change?

Mental Workload was captured using the NASA-Task Load index (NASA-TLX) developed by Hart and Staveland (7) which measures six key dimensions of workload. The NASA-TLX is a widely used subjective workload rating tool that can be used in either its traditional or raw form. The traditional form involves weighting the Mental Workload variables bulleted below before rating the scales, whereas in the raw version the weighting process is not completed. Due to the time requirements of the experiment, the raw form was used. The use of the raw form has been validated as effective as the traditional form as a measure of workload [1, 4, 8, 13, 14]. Each participant was asked to mark the most suitable point on each of the scales. There are 20 intervals, with 21 vertical tick marks dividing the scale from 0 to 100 in increments of 5. Overall Mental Workload is calculated by averaging the scores from the six Mental Workload variables:

- Mental Demand (Very low—Very high)
- Physical Demand (Very low—Very high)
- Temporal Demand (Very low—Very high)
- Performance (Perfect—Failure)
- Effort (Very low—Very high)
- Frustration (Very low—Very high)

The scales were scored using the methodology provided by Hart and Staveland (7), and averages and standard deviations were calculated across all dimensions.

The Trust in Industrial Human–Robot Collaboration Scale [3] was used to measure participants' trust in the deburring robot. The participants were asked to respond to ten statements using a five-point Likert-like scale that ranged from strongly disagree to strongly agree. The statements are grouped under three main dimensions and included:

- Robot motion & pick-up speed

  o The way the robot moved made me uncomfortable
  p The speed at which the gripper picked up and released the components made me uneasy

- Safe co-operation

  o I trusted that the robot was safe to cooperate with
  p I was comfortable the robot would not hurt me
  q The size of the robot did not intimidate me

r I felt safe interacting with the robot

- Robot and gripper reliability

o I knew the gripper would not drop the components
p The robot gripper did not look reliable
q The gripper seemed like it could be trusted
r I felt I could rely on the robot to do what it was supposed to do

Trust data was analysed using the analysis guidelines provided by Charalambous et al. [3] and the average and standard deviation for the Overall Trust score was identified.

9.3 Procedure

As stated this research was approved and conducted in accordance with appropriate protocols for ethical human research. To comply with this, written informed consent was obtained from participants after they had been made aware of the nature of the study. The consent form was provided to the participants along with a briefing sheet describing the nature of the study, objectives, procedure and timing. The right to withdraw, including the timeframe for withdrawal, was included within both the briefing and consent forms.

The procedure detailed below was followed prior to the participants completing the surveys to have them informed about the purpose and methodology:

1. Briefing was made by the A4BLUE project manager and engineers responsible of the project to inform the participants about the content of the surveys, the implication of signing the consent and answer their questions and doubts.
2. Deburring operators were already used to working every day with the robot so they were familiar with the technology. In this case no extra demonstration was required for them to be able to complete the survey.
3. Finally the surveys were provided to each of the operators so that they could answer them and sign the consent form.

9.4 Results

Figure 8 shows that the results revealed fairly positive reactions to working with the newly installed collaborative robot. The semantic differential usability scores leaned towards the positive ends of the scales across the four differentials, suggesting a positive experience with the system.

The qualitative responses provided further insight into the qualitative results and show that all the participants found an improvement in their work with the introduction of the deburring robot. These responses indicate that a good level of usability

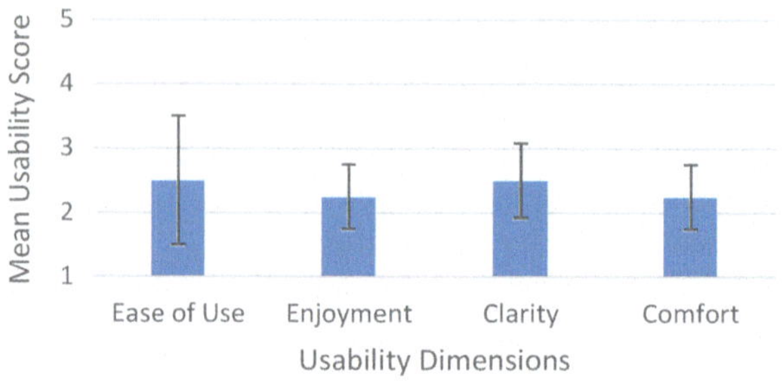

Fig. 8 Usability Semantic Differential Results (Means and SDs)

is experienced by the operators when interacting with the robot, and its introduction has meant they are able to complete different work.

> "I found the task easy to complete and it is easy to interact with the robot
>
> I did not found tedious parts. While the robot was performing the deburring I was performing other tasks.
>
> "I did not found tedious parts. I keep on working in parallel.
>
> I enjoyed completing the task"
>
> Participant 2

> "What I enjoyed the most was time saving, therefore let time to do other tasks. I particularly liked the interaction with the control screen."
>
> Participant 3

> "I liked the speed and seeing how the robot was able to perform the task."
>
> Participant 3

However, the qualitative responses reveal that more training and a more accessible user manual are required to further improve the experience of operators.

> "User manual are a little bit complex."
>
> Participant 4

The NASA-TLX results which can be seen in Fig. 9 revealed that mental workload was neither overloaded nor under-loaded, which indicates that operators are able to maintain effectiveness in their work. It is important to remark that these surveys were made once the robot was fully implemented and the workers had already experience in the robot management. It is likely that mental workload scores may have been higher at the beginning of the implementation and training.

Finally, trust levels were good, with the *Total Trust Score* found to be 39.75. As a result, the score fell within an acceptable range of trust between a lack of trust (<25) and overreliance (>50) as identified by Charalambous et al. [3]. The workers trusted that the gripper will not drop the components or damage the part and that the robot would do what it is supposed to do. The workers also felt comfortable with the speed of the robot and the gripper, this is likely because the operators decide the speed.

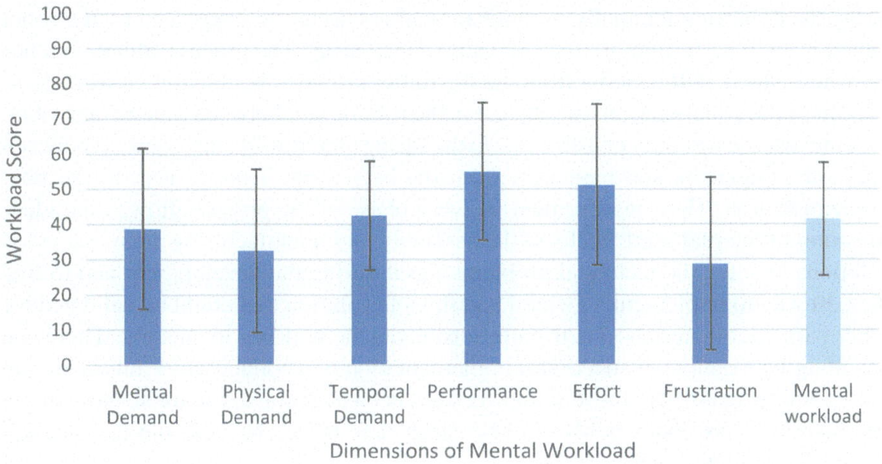

Fig. 9 NASA-TLX results (Means and SDs)

The results revealed that the operators felt relatively safe working in collaboration with the robot and interacting with it. Some answers showed that there is still margin to improve their levels of trust. This is probably because it is the first collaborative robot cell to be implemented on the CESA shop floor. It is likely that with more experience trust scores may increase.

10 Conclusion

The implementation of a robot cell at CESA as part of the assembly process can be considered a success. Several months after the end of the A4BLUE project the robot is still fully operational and integrated in the serial production of Titanium Earth Ends. Work is also been done to program more parts to be automatically deburred and to implement some hardware and software improvements in the cell.

This automated solution has increased productivity because now the operators have more free time to dedicate to other value-added tasks, like the programming and control of machining centres or preparing new parts to be deburred. At the same time, quality has been maintain at the same good levels that were achieved when the parts were manually deburred. The operator is in charge of performing the final quality check on the part and decide if it is OK to start the assembly or whether some areas need rework.

According to the results of the surveys and interviews with the operators, it is possible to say that the workers satisfaction has been improved, by releasing them from a repetitive and physically demanding task. Operators are in charge of programming the robot and collaborate with it by installing new parts, checking the quality of the work and solving maintenance problems which are more fulfilling task than doing the deburring manually. So human workers do not feel replaced by the robot but allowed to perform more fun and value-added tasks. The process, although it has been automated, still benefits from the flexibility provided by the human workers as a result of the collaboration established in the robotic cell between human and robot.

This success was only possible due to the strong and fruitful collaboration between operators, engineers and managers from the beginning of the project to the final implementation. Their involvement in every phase of the project allowed the identification of the real bottlenecks in the assembly and manufacturing areas, provided solutions to solve the technical problems faced during the development and to find the best way to implement a general and flexible solution that could be validated for other parts and materials. Their enthusiasm was the real key to success. This is an encouraging example of successful implementation of a collaborative automation in the assembly area where most of the work has been traditionally manual. Now future automation of processes is likely to be easier than before because the first attempt has been completed successfully.

Collaborative automation like this can be the key to start introducing these kind of technologies in the aeronautical industry, by taking advantage of both human flexibility and robot productivity and efficiency. These are sustainable solutions with an optimal balance between automation and human presence. Human workers remain at the centre of production but are supported by automation mechanisms that help them during their job and with physically demanding tasks.

Acknowledgements This work was supported by the A4BLUE project which received funding from the European Commission's Horizon 2020 research and innovation programme under grant agreement no. 723828. The researchers would like to thank all project partners for their support during this project and specifically Luis Salazar, Iban Azurmendi and the rest of KOMAT team for their work during the installation of the robot.

References

1. Bittner AV, Byers JC, Hill SG, Zaklad AL, Christ RE (1989) Generic workload ratings of a mobile air defense system (LOS-FH). In: Proceedings of the 33rd annual meeting of the human factors and ergonomics society. HFES, pp 1476–1480
2. Chapanis A (1981) Evaluating ease of use. In: Proceedings of IBM software and information usability symposium. IBM, pp 105–120
3. Charalambous G, Fletcher S, Webb P (2016) The development of a scale to evaluate trust in industrial human-robot collaboration. Int J Soc Robot 8(2):193–209
4. DiDomenico A, Nussbaum MA (2008) Interactive effects of physical and mental workload on subjective workload assessment. Int J Ind Ergon 38(11–12):977–983
5. DiDomenico A, Nussbaum MA (2011) Effects of different physical workload parameters on mental workload and performance. Int J Ind Ergon 41(3):255–260
6. Fletcher SR, Johnson T, Adlon T, Larreina J, Casla P, Parigot L, Alfaro PJ, del Mar Otero M (2020) Adaptive automation assembly: identifying system requirements for technical efficiency and worker satisfaction. Comput Ind Eng 139:105772

7. Hart SG, Staveland LE (1988) Development of NASA-TLX (Task Load Index): results of empirical and theoretical research. In: Advances in psychology, vol. 52, North-Holland, pp 139–183
8. Hendy KC, Hamilton KM, Landry LN (1993) Measuring subjective workload: when is one scale better than many? Hum Factors 35(4):579–601
9. Hockey GRJ (1997) Compensatory control in the regulation of human performance under stress and high workload: a cognitive-energetical framework. Biol Psychol 45(1–3):73–93
10. Kirwan B, Ainsworth LK (eds) (1992) A guide to task analysis: the task analysis working group. CRC Press
11. Lee J, Moray N (1992) Trust, control strategies and allocation of function in human-machine systems. Ergonomics 35(10):1243–1270
12. Lewis JR (2012) Usability testing. In: Salvendy G (ed) Handbook of human factors and ergonomics. John Wiley & Sons Inc., pp 1267–1312
13. Moroney WF, Biers DW, Eggemeier FT, Mitchell JA (1992) A comparison of two scoring procedures with the NASA task load index in a simulated flight task. In: Proceedings of the IEEE 1992 national aerospace and electronics conference. IEEE, pp 734–740
14. Nygren TE (1991) Psychometric properties of subjective workload measurement techniques: implications for their use in the assessment of perceived mental workload. Hum Factors 33(1):17–33
15. Parasuraman R, Sheridan TB, Wickens CD (2008) Situation awareness, mental workload, and trust in automation: viable, empirically supported cognitive engineering constructs. J Cogn Eng Decis Mak 2(2):140–160
16. Vidulich MA, Tsang PS (2012) Mental workload and situation awareness. In: Salvendy G (ed) Handbook of human factors and ergonomics, 4th edn. John Wiley & Sons, Inc., pp 243–268

The Role of Standards in Human–Robot Integration Safety

Carole Franklin

Abstract This chapter will describe the role of voluntary industry consensus standards in guiding the safe application of human–robot collaboration (HRC) in industrial settings. It will also describe the nature of voluntary industry consensus standards, providing an overview of how such standards are developed, thoughts on how standards impact innovation in the marketplace and how standards provide safety requirements for industrial HRC. The history of industrial robot safety standards will be discussed, along with areas of potential future work. Challenges and limitations of voluntary industry consensus standards will be identified, both in general and specifically in the area of closer human–robot collaboration in industry.

1 Introduction

In this chapter, we will use the term "standards" to mean "voluntary industry consensus standards." Voluntary industry consensus standards comprise a complex ecosystem of interlocking documents that describe a specific set of requirements in a given field. As their name implies, compliance with such standards is usually voluntary, meaning that it is not a requirement of law; rather, individual entities in that field choose to comply. Note that in some regions, countries, or localities compliance could be required by local law, and frequently those developing such standards make efforts to stay abreast of important regulations in nations where the developers intend the standard to be observed. Such standards are usually developed by committees of volunteers from industry, academia, and governments. The committee's goal is to identify a consensus regarding best practice which all of these disparate interest groups can support. The establishment of a robust consensus is important for voluntary industry consensus standards, since by their nature they are not legally required. For a voluntary standard to succeed—that is, for a standard to be widely adopted—all major interest categories must be satisfied that their needs have been taken into

C. Franklin (✉)
Association for Advancing Automation (A3), Ann Arbor, USA
e-mail: cfranklin@automate.org

M. I. Aldinhas Ferreira and S. R. Fletcher (eds.), *The 21st Century Industrial Robot: When Tools Become Collaborators*, Intelligent Systems, Control and Automation: Science and Engineering 81, https://doi.org/10.1007/978-3-030-78513-0_9

account in the development of the standard; if they are not so satisfied, they have no reason to comply.

The safe application of human–robot collaboration (HRC) in industrial settings has its own set of standards, developed within the framework established by the International Organization for Standardisation (ISO). The consensus standard development work is conducted by the ISO Technical Committee (TC) 299, Robotics. Industrial HRC standards are designed to fit into the larger ecosystem of industrial machinery safety standards. This makes it more practicable to integrate industrial robots, including those designed for HRC, into larger industrial systems, since the standards governing the robots are intended to fit with, and not to contradict, the standards governing the larger world of industrial machinery.

2 Current State

2.1 What Are Industry Standards?

When we say "industry standards" in this chapter, or more simply still, "standards," we mean "voluntary industry consensus standards." Let us consider what we mean by each of these terms.

Voluntary: Compliance with the standard is generally voluntary; that is, compliance is not required by law. Rather, members of the industry addressed in the standard choose to comply with its requirements. Standards can be incorporated by reference into local or national laws, but this is frequently not the case. In fact, in some contexts it can be considered beneficial for the content of a standard to avoid being incorporated into law, since technology changes continually and therefore standards must be updated every few years according to a new expert consensus; whereas the updating of laws is generally seen to be a much less frequent occurrence and a more challenging one. For example, in the U.S., one law concerning industrial machinery safety was originally passed in 1989 [14]; by comparison the voluntary industry consensus standard concerning industrial robot safety is currently undergoing its fourth update since then (in 2020). If we consider the amount of technological change during this time, we can see that standards are updated when the marketplace sees a need for updates, whereas laws can only be updated after a mustering of considerable political will. Because of this, it can be the case that the voluntary industry consensus standards are more up-to-date with the current state of the art than are legal requirements for that same industry.

Industry: Industry standards are developed through voluntary cooperation by industry experts, as well as academic and government experts. Furthermore, the work is intended to support the success of industry through establishing predictable and best-practice requirements for that industry's product or service. One benefit of such common requirements is that they enhance interoperability among the equipment produced by different manufacturers.

To take one of the best-known examples, an entire industry is made up of universal serial bus (USB) devices. These USB devices interoperate successfully around the world—and only because they comply with international standards regarding USBs. For this interoperable technology to spread globally, no national government had to decree this by law, and even if a government had done so, its decrees would stop at its borders; and yet consumers and industries worldwide have benefited.

Consensus: Consensus means that the contents of the standard have satisfied the needs of a substantial majority of committee members in order to be approved [1, 3, 6] . The exact threshold for establishing consensus varies depending on the particular standards development framework being observed, but it is typically a supermajority, or more than two-thirds of voting members. Depending on the topic, the support for a given consensus decision is usually closer to 90%, and occasionally reaches 100%. Note that consensus does not mean unanimity; for a consensus to stand, or for a consensus document to be approved, it does not require 100% approval.

Because such standards frameworks usually have some requirement for balance [1, 3, 6] among various interest categories within the voting membership of a committee, one can be confident that such a consensus represents a broad agreement among those interest categories. For example, the needs of *producers* of a certain type of product are balanced with the needs of that same product's *users*, and usually incorporate the views of academic and government experts as well. This helps ensure that the needs of one particular interest category do not dominate over the needs of the others; if it were so, the other interest categories would vote against it. Because of such broad consensus agreements among the interested parties, consensus standards can be broadly adopted. For a standard to succeed, broad adoption is crucial.

The more broadly a standard spreads, the easier it is for technology based on that standard to spread as well; see the USB example described above. On the other hand, anyone who has had to purchase electrical adapters in order to travel abroad can see the result of separate national or regional consensuses that were developed in advance of a robust international consensus.

Standard: Usually, a document containing requirements representing consensus best practices in a given field, developed according to a framework defined by a national or international body. Such a framework typically requires transparency, fairness, and balance among those with an interest in that field [1]; for example, a balance between the needs of producers and the needs of users.

In summary, a *voluntary industry consensus standard* provides a voluntary means to achieve a desired outcome to the benefit of industry, and by extension to the benefit of its shareholders, customers, suppliers, and employees, and of the society in which it operates.

2.2 Important Concepts for Understanding Standards

There are different types of standards documents produced under different standards frameworks [1, 5, 6]. In the ISO system, for example, there are three types of documents: The international standard (IS); the technical specification (TS); and the technical report (TR).

The IS and TS may contain what is known as "normative" language; that is, requirements, typically denoted by the use of the word "shall," and occasionally "may" in the sense of giving permission. Compliance with the IS or TS means that one has met all the requirements stated within the document. While one can choose whether or not to comply with a given IS or TS, once one has chosen to comply with it, one cannot pick and choose among the requirements stated within the document. The IS and TS may also contain "informative" language; that is, useful information that is not a requirement, typically denoted by use of the word "should," and occasionally "can" in the sense of stating that something is possible. The TR, the third type of ISO document, may not contain normative language, but rather, informative language only.

If both the IS and TS can contain normative and informative language, what is the difference between them? Why do we need both types of document? Indeed, in some other standards frameworks (e.g., ANSI in the U.S.) there is no "technical specification" level of document, only "standard" and "technical report." In the ISO system, however, a Technical Specification (TS) is distinct from the full-fledged International Standard (IS) [6]. While the TS can contain normative text (that is, requirements), and in that regard is the equal of an IS, a TS is the preferred type of document in the case of a new technology or new business practice in which the content is viewed as important for industry, but not as mature and stable as the content of an IS. In a rapidly-changing area of technology, the consensus best practices can also change rapidly. The TS is used, therefore, in cases where such change is anticipated. The TS therefore undergoes a systematic review in a shorter time period than an IS and is likely to be updated in a shorter time period as well. There is an expectation that the content of a TS is substantial enough to mature into a full IS in the future. Following its systematic review, the TS might be updated and its future edition published as a full IS in its own right; or, its contents might be incorporated into an updated edition of an established IS; or, it might be updated and published as the new edition of a TS.

In the Introduction to this chapter, we mentioned that standards for a given industry usually comprise a complex ecosystem of documents that work together. One level of this interaction is the practice of "nationally adopting" an international standard. This means that the IS prepared by, for example, an ISO technical committee, can be accepted by the members of that industry in a given nation, and "adopted" as a national standard. For example, the ISO 10218–1,2:2011 [7, 8] was nationally adopted in the U.S. as R15.06–2012 [15] and in Canada as Z434-2014. This means that the content of these national standards based on the 10218 series is essentially the same as the content of the ISO standards.

There are two types of national adoption [2]: first, a "direct" adoption in which no changes are made to the technical requirements of the IS; and second, an adoption "with national deviations" in which specific changes are made to the technical requirements of the IS, and these differences must be identified. When used, national deviations are typically made in order to comply with local practice and/or law. In order to maintain the "sameness" of the standard content worldwide, it is desirable to minimise national deviations that are applied to a national adoption; with changes that are of too great an extent, or are less strict than the original, there could come a point at which the national edition of the IS ceases to be the "same" as the source document. The U.S. R15.06–2012 is a "direct" adoption in which the only changes are to use American spellings instead of British ones. The Canadian Z434-2014, on the other hand, does contain national deviations; these are considered minor and in general are more conservative than the source document. In other words, compliance with the Canadian document is still compliance with the source IS, but Canada also has some requirements that are more strict than in the original.

Another concept frequently used in standards work is that of "harmonisation." To say that a document is "harmonised" is to say that compliance with this document will not cause a breach of compliance with another document on a similar, or perhaps even the same, topic. This other document could be a law, or it could be another standard in an adjacent industry, or it could be other similar national standards in nearby nations. In the European Union, for example, it is common to say that a machinery safety standard is "harmonised" with the Machinery Directive (MD), which is a law. Once a standard has been harmonised with the MD, this means that compliance with that standard can be used to "presume conformance" with the MD. Clearly, a machinery safety standard that can be used to demonstrate compliance with a machinery safety law is a more valuable document than one that cannot be so used; that is, one that is not so harmonised with the law. But to say that a standard is "harmonised in Europe" can also mean that the German version of that standard is consistent with the French and Spanish versions, and so on, which is important for cross-border commerce in the region. And though the concept is not cited as frequently in North America, it would be correct to say that the U.S. standard R15.06–2012 and the Canadian Z434-2014 are "harmonised" in the same cross-border sense, since they are both national adoptions of 10218–1,2:2011 with no, or minimal, national deviations.

A final important concept in the machinery safety domain is that of the type of standard. There are three types of standards concerning machinery safety: Type A, Type B, and Type C Standards [10, 11]. These standard types are described in ISO 12100:2010, a standard on the topic of safety of machinery. Type A standards are general standards providing basic principles and general requirements for safety of machinery; ISO 12100 is itself a Type A standard. Type B standards provide safety requirements that are applicable across a wide range of machinery, such as requirements regarding safety distances or noise (Type B1 for safety aspects), or safety-related devices, such as two-hand controls (Type B2 for protective devices). ISO 13849–1:2015 is an example of a Type B standard.

Type C standards provide requirements for the safety of a particular machine or type of machine, such as pneumatic presses, or robots. ISO 10218–1 is an example of a Type C standard. These types of safety standards refer to each other, and the committees responsible for drafting them make efforts to avoid contradicting one another. The result is that the ISO 10218 series of standards (Type C) generally abide by the principles espoused in the Type B and Type A standards for safety of machinery. Thus, by simply referring to the relevant Type A or B standard(s) (e.g., 12100 or 13849–1), the Type C standard can avoid reinventing or restating those principles, and can concentrate on covering its specific type of machinery. However, it is possible that in some cases or on certain topics, the machine-specific safety requirements contained in a Type C standard could differ from the content of a Type B or Type A standard, due to the particular characteristics or hazards of that particular type of machinery. In such cases, the Type C standard takes precedence for that particular type of machinery.

2.3 History of Standards Development for Industrial Robots

The first standard containing safety requirements for industrial robots and robot systems was the U.S. standard ANSI/RIA R15.06–1986 ("R15.06"). Following updates to R15.06 that were published in 1992 and 1999, the organization which developed R15.06, the Robotic Industries Association (RIA), approached ISO to see if there was interest in using R15.06 as the foundation for an international standard for industrial robot safety. In response, the *ISO TC 184, Automation systems and integration*, through its then-*subcommittee (SC) 2, Robots and robotic devices*, formed a working group (WG), *WG 3, Industrial safety*. The goal of this working group was to take the 1999 R15.06, update its technical content, and modify it to meet ISO guidelines and the needs of the national members of the TC 184/ SC 2/ WG 3. As a result of this work, in 2007 the TC 184 published the International Standard (IS) 10218 Part 1, Edition 1. This document, 10218 Part 1, contains safety requirements for industrial robots; however, such robots do not perform their tasks in isolation, but rather as part of a system. As a result, the working group went on to develop 10218 Part 2, containing safety requirements for industrial robot systems. In 2011, an update to 10218 Part 1 (Ed. 2) was published, along with the new 10218 Part 2 (Ed. 1). Both 10218 Part 1 and Part 2 [7, 8], taken together, are necessary to establish safety requirements for the completed robot system in the industrial environment.

Thus, in the 2011 10218 series of standards, we had the first ISO equivalent of the 1999 R15.06, containing updated technical content for both robot manufacturers and integrators, and establishing a broad international consensus. The 2011 editions of 10218 Parts 1 and 2 have been adopted in major industrial nations around the world, including the U.S., Canada, the European Union, Japan, Korea, and more. As of this writing (2020), the 10218 Parts 1 and 2 are currently undergoing their next update. When these updates are published, they will represent Edition 3 for Part 1, and Edition 2 for Part 2. Until then, the 2011 editions of the 10218 series are still in effect.

In the 2011 10218 series, collaborative robotics—enabling HRC—was a field still in its infancy. Industry experts needed to learn more about this discipline in order to establish best practices for it. The 2011 10218 permitted collaborative robotics but did not go into details about how to achieve it safely. Thus, following the publication of the 10218–1,2:2011, the working group began work on a new Technical Specification (TS) regarding safety requirements for collaborative robotics; see above for discussion of the distinction between an IS and a TS.

In 2016, the ISO TC 184/SC 2 was spun off as its own separate ISO Technical Committee, *TC 299, Robotics*. The first publication of ISO TC 299 was ISO/Technical Specification (TS) 15066:2016, Collaborative robot safety [9]. This document established the first requirements for the safe operation of industrial HRC technology. As a Technical Specification, TS 15066 is intended to work with the ISO 10218 series of International Standards (IS). For its correct application in industry, TS 15066 assumes that the robot system intended for collaborative operation meets the requirements of ISO 10218–1,2:2011. The TS 15066 contains additional requirements for the safe operation of a system intended for HRC. It is a supplementary document to the 10218 series, not a replacement.

Thus, the ISO/TS 15066–2016 represents the best knowledge available at the time about safety requirements in a rapidly-changing area of technology, e.g., human–robot collaboration (HRC). TS 15066 does contain normative requirements; however, these requirements must be reexamined within a few years of publication. It is possible that the contents of TS 15066 might be incorporated into a future update of an established IS; the most likely standard for this case would be the 10218 Part 2.

Since the publication of TS 15066 in 2016, additional consensus documents have been developed in the area of safe HRC. For example, in the U.S., the industrial robot safety community which provides U.S. experts to the ISO work also develops U.S. consensus documents under the framework of the American National Standards Institute (ANSI). This community has developed a U.S. Technical Report (TR) on testing methods for verifying and validating that a given robot system complies with the limits described for one type of HRC, "power-and-force-limited" (PFL) collaboration [19]. Another U.S. TR is in development for testing methods for a second type of HRC, "speed and separation monitoring" (SSM); its publication is expected in 2021. Although these TRs have been developed for use in the U.S., since they are meant to work with the U.S. national adoptions of 10218 and TS 15066, they can also be used with other national standards based on 10218 and TS 15066.

2.4 Philosophies of Standards Development

Judging by their effects, there are at least two major philosophies of standards development.

The first approach is to *lead* the development of technology, in order to shape it and create a market for it. The USB is perhaps an example of this. To succeed in this philosophy of standards development, it is important for the standards committee to stay ahead of technology currently in widespread use.

The second philosophy is that experts cannot establish a consensus of best practice until a technology has been observed in use for some time. To succeed in this philosophy, the content of a standard must necessarily *lag* the state of technology currently in widespread use.

Machinery safety standards, including industrial robot safety standards, are of the latter type. When human life and limb are at stake, it is important that identified best practices are of the "tried and true" variety. In this field—the protection of humans in the workplace—the risk of prematurely adopting a technology before it proves reliable is too great to permit rushing the development of consensus. In safety, if we know something works, we don't want to jump on the next shiny bandwagon until the new solution has proven to work at least as well as the current-state solution. There must be clear evidence that a solution is better than the current one before the consensus among experts will swing to embrace the new solution.

It is also important to realize that in the world of machinery safety standards, including industrial robot safety standards, the standards are addressing the use of a large installed base of capital equipment worldwide. Changing requirements for such equipment cannot be implemented overnight; and such change involves not only new capital investment for the equipment itself, but also the updating of established training materials, procedures, placards, the knowledge base of the workers who will interact with the equipment, and doubtless many other considerations as well. Thus a new solution that would drive such changes must be shown to be not only a marginal advancement over the current state but a clearly superior advancement; and it must be recognized that such changes, even when clearly justified, will take time to be implemented in the marketplace. Thus, machinery safety standards are inherently conservative documents, and this is not a shortcoming but rather, a purposeful outcome.

2.5 *Impact of Industry Standards on the Marketplace*

Because standards (remember, "voluntary industry consensus standards") are voluntary in both development and application, are market-driven, and have the intent of supporting industry, these factors govern what topics are considered important for standards-development activities to cover. The goal of a machinery safety standard is not to prescribe a particular technological solution, but rather to describe a desired outcome, leaving the decision of how to achieve that outcome to the creativity of the engineers and businesspeople of that industry. As a result, depending on their goal, the standards-developing committee might purposefully choose for their standard to "be silent" on a given topic, leaving the standard reader the freedom to discover the best solution for him or her on that particular topic. On the other hand, an interoperability standard does intend to specify a particular technology or solution in order to ensure that the machines that comply with it will successfully interoperate.

To take one example from a machinery safety standard, the ISO 10218 Part 2 requires that a conventional (non-HRC) robot system protect workers by keeping people outside of the "safeguarded space." This can be achieved by a system of physical, permanently-mounted fencing ("guards"); doors that cause a shutdown in the robot system when opened ("interlocked movable guards"); electrosensitive protective equipment ("ESPE") that cause a shutdown when the presence of a person is detected; or a combination of the above as the user determines is best for his or her budget and other considerations. The standard does not require that, for example, only physical fences be used, or only the most expensive safety scanner; requiring either of these would be "too prescriptive" of a particular solution. Either could achieve the goal effectively, but which one is "best" is left to the judgment of the standard reader in his or her particular situation. The standard describes the desired outcome ("keep people out of the safeguarded space") and is silent on which among multiple possible technologies might be used to achieve that outcome.

Do standards restrict or encourage technology and innovation? The answer is, "It depends."

A good standard provides a structure of consensus best practices within which a nearly infinite range of solutions can be developed. In that sense the standard is like the structural requirements for a sonnet or a haiku: For one's poem to be "in compliance" with the structural requirements, one must remain within the specified structure. For a haiku, this is three lines of text with the first line consisting of five syllables; the second line consisting of seven syllables; and the third line consisting of five syllables again. This is a very restrictive structure; one could say that it restricts poetic innovation. But if one wishes to say one has written a haiku, one must observe the five-seven-five structure. And to date, possible poems in the haiku form have proven, so far, to be infinite.

So it is with a machinery safety standard. To say one has developed an industrial robot system in compliance with 10218 Part 2, one must observe the requirements of that standard. To say that one has further developed an industrial robot system capable of safe HRC, one must additionally observe the requirements of TS 15066. Within those structural requirements, the possibilities are nearly infinite, limited only by human creativity.

On the other hand, a poorly-written standard might restrict innovation by prescribing a particular solution that only one company could provide; but committees are recommended to avoid this outcome. It is up to the committees of volunteer experts who are preparing the standard to self-police against including overly-prescriptive requirements, and this is why it is so important to maintain a balance of interest categories on a standards committee. Transparency in decision-making is also important, since having a balance of interest categories on a committee is not helpful if the members of one category do not feel their views will be taken into account. Committee members must feel free to speak up if they perceive an overly-prescriptive requirement being stated; and they must be listened to respectfully by others on the committee.

Once the draft standard has been developed to the satisfaction of its drafting committee, it is then put to a vote among the voting members of the standards organization. In the ISO framework, for example, national standards bodies (NSBs) are the members of ISO TCs and their SCs and WGs. These national bodies can participate in the ISO committees through multiple paths. Two common means of participating are, first, by designating expert volunteers of the NSB's country to contribute to the work of a given working group and its standards document projects; and, second, by casting a vote on behalf of the NSB's national industry as to whether a given standards document is ready to be adopted. Once a working group of experts is satisfied that its draft is ready, the draft is reviewed and voted on by the voting members of the TC. Thus, to be adopted, a standards document must satisfy a supermajority of working group members as to the content they agree to include; and then, separately, satisfy a supermajority of member countries that the resulting document will meet their needs.

Through these multiple levels of approval and consensus, the standards-development process guards against the development and adoption of standards whose requirements would be overly restrictive of technological development.

2.6 Standards Organizations

Standards organizations exist at the national, regional, and global levels. Some that started as national bodies have grown into organizations with regional or global reach. Each of these organizations generally prepares a set of procedures which govern how standards are to be developed within its framework. Thus, to prepare a document that will be published as an "ISO" standard, the document must be developed in accordance with the requirements of the ISO framework. Here are some examples of standards organizations that are currently active (not an exhaustive list):

International:

ISO - International Organization for Standardization
IEC - International Electrotechnical Commission

National/Regional:

ANSI - American National Standards Institute
BIS - Bureau of Indian Standards
BSI - British Standards Institute
CEN - European Committee for Standardisation
CENELEC - European Committee for Electrotechnical Standardisation
CSA - Canadian Standards Association
DIN - Deutsches Institut fur Normung

JISC - Japanese Industrial Standards Committee
KSA - Korean Standards Association
SA - Standards Australia

Professional/ Industry/ Other:

ASTM - American Society for Testing and Materials
IEEE - Institute of Electrical and Electronics Engineers
NFPA - National Fire Protection Association
UL - Underwriters Laboratories

As mentioned earlier, it is common for the members of international standards organizations to be the national organizations for their countries. For example, ANSI, BSI, DIN, JISC, and other national standards bodies are the official members of ISO Technical Committees and their subcommittees and working groups. In addition to coordinating their nations' participation in the international work, such bodies also typically develop national standards for their countries, including the national adoption of international standards.

2.7 Current Industry Standards Governing Human–Robot Collaboration

Currently, the industry standards setting forth the requirements for HRC in the workplace are ISO 10218–1:2011 [7], ISO 10218–2: 2011 [8], ISO/TS 15066:2016 [9, 18], and their national adoptions [15]; and in the U.S., the RIA TR R15.806–2018 [19]. The 10218 series permits HRC and the TS 15066 describes how to achieve HRC safely. Finally, the TR 806 describes test methods for verifying that forces exerted by the collaborative robot system remain within the limits described in TS 15066.

At the time of this writing (2020), the TS 15066 is a key document describing safe HRC. It describes four different types of collaborative operation [9, 4]:

1. Safety-rated monitored stop (SMS)
2. Hand guiding (HG)
3. Speed and separation monitoring (SSM)
4. Power and Force Limiting (PFL)

Let's discuss what each of these types of collaborative operation entail [9, 4].

Safety-rated monitored stop (SMS). Simply put, either the worker or the robot may move within the collaborative workspace, but not both simultaneously. With SMS, when the collaborative robot system detects a worker in the collaborative workspace, the system stops all motion. The system then monitors that the stop condition is held until it no longer detects the worker in the collaborative workspace. When the worker has left the space, the robot system may start up again automatically and at full process speed. This represents an improvement for efficiency with no reduction in safety, compared to a conventional (non-HRC) industrial robot system, since with the HRC system there is no need for the worker to restart the system manually upon leaving the safeguarded space.

Hand guiding (HG). With HG collaboration, the robot may move while the worker is present, but only if the robot motion is under the direct control of the worker. This is achieved through the use of a guiding device, activated manually by the worker, near the robot end-effector. This form of collaboration can be used to enable the robot to take the strain of lifting a heavy workpiece, while the human worker uses hand-guiding to make sensitive adjustments to the positioning of the robot, end-effector, and/or workpiece. For example, when installing seats into an automobile, the robot could bear the weight of the seat, and the human worker can maneuver it carefully into position within the autobody.

Speed and separation monitoring (SSM). With SSM, robot motion is permitted only when a certain separation distance to a human worker is maintained. The appropriate separation distance is determined uniquely for each robot system, because it depends on many factors including the relative location, speed, and motions of both the robot system and the worker. If the separation distance is not maintained (for example, if the worker approaches the robot system too closely), the speed of the robot system is reduced to a level that will mitigate risk of injury if the worker should be contacted by the robot or part of the robot system. Typically, if the worker continues to approach, the robot might be programmed to come to a stop. This is an example of "zoning," in which the robot system's operation changes depending where the human worker is detected within a series of zones.

Power and force limiting (PFL). With PFL, physical contact between the robot system and the worker is permitted, so long as the robot is limited to a pre-determined amount of force or pressure it can exert. The limits are set at levels where such contact would not cause injury to the worker [9, 13]. These limits are specified in an annex of TS 15066, and they differ according to the body part likely to be contacted. The PFL robot is usually what is meant when people use the term "collaborative robot," although, strictly speaking, it is the entire robot *system* (not the robot by itself) that is collaborative, or not. Naturally, it is important to be confident that the forces and pressures actually exerted by the PFL robot system remain within the limits described in TS 15066. TR 806 [19] describes test methods for this purpose.

2.8 Industry Standards in Development Concerning Industrial HRC

In 2020, the industrial robot safety community in the U.S. is developing a new TR providing testing methods for SSM. It is intended to achieve the same thing for SSM that TR 806 did for PFL: help people test whether the HRC robot system is performing as intended.

3 Future Trends and Transformations

3.1 Anticipating the Market's Future Needs for Industry Standards Concerning Industrial HRC

In the first few years of collaborative robot systems being installed [12], enabling initial forays into HRC, much of the focus was on PFL technologies. However, because of the limitations of the PFL robots—their very name indicates they are limited in the pressures and forces they can exert [13]—many industrial robot applications require a larger, more powerful robot for successful implementation. As a result, collaborative methods other than PFL, such as SSM and HG, have been growing in usage and will likely continue to grow. This is because SSM and HG can be implemented using conventional (non-PFL) industrial robots integrated with added sensors, control logic, and/or other equipment such as hand-guiding devices.

Therefore, in addition to the continued need for research into refining the pressure and force limits shown in TS 15066, the standards community will also need research, data, and documents that support collaboration methods other than PFL.

3.2 Technological Trends Likely to Impact Industrial HRC in the Next 5 years

There are a number of technological trends likely to impact industrial HRC in the near future. Leading examples include (not an exhaustive list): Continued advances in non-contact sensor technologies, including machine vision, laser scanners, and pressure-sensitive "skins" for industrial robots; artificial intelligence and machine learning; and growth in the capabilities and availability of autonomous mobile machines. Let's briefly explore these.

Advancing sensor technologies. With respect to sensor systems that gather data about the environment without physical contact, machine vision, laser scanners, lidar, infrared sensors and more are continuing to advance. These sensors are important for HRC because they hold the promise of enabling the robot system to sense the presence of a human worker before coming into physical contact with him or her. In particular for larger, more powerful (non-PFL) industrial robots, preventing physical contact with workers remains crucial to worker safety. With advances in sensor technology, it is becoming more feasible to avoid physical contact without the use of physical barriers; this contributes to the success of SSM, one form of HRC.

In other cases, physical contact could be permitted even though the robot arm itself is large and powerful, in HG systems as one example; but in such cases, worker safety can be enhanced through the use of pressure-sensitive surfaces or "skins" applied to the robot. Such pressure-sensing technologies could, for example, detect if a worker is in an unexpected location, or if a second worker were to approach while hand-guiding is being conducted. In such a case, although the robot is moving slowly and is under the control of a human worker performing hand-guiding, robot movement can be stopped if pressure on the robot is detected in an unexpected location. Slow and controlled motion of a powerful robot is might not be immediately harmful, since the speed might not be high enough to result in injury from initial contact; but it could become harmful, if, for example, the robot's continued motion were pushing a worker into an obstacle. Pressure-sensitive "skins" are an example of an emerging technology that can mitigate such a risk, thereby enabling the HRC market to continue to grow.

Artificial intelligence and machine learning. Other technological trends that are also likely to help HRC capabilities become more powerful and more intuitive for the human workers who operate HRC robots and robot systems include artificial intelligence and machine learning are. For example, advances in such systems and their increasingly widespread adoption have the potential to enable faster or more accurate interpretation of the data that advanced sensors collect. At the time of this writing (2020) it is still surprisingly difficult for an autonomous machine to tell whether an object it "sees" is a person or a waste bin or a structural beam or any number of other things that it might find in its operating environment. Artificial intelligence and machine learning are likely to drive advances in this area.

Autonomous mobility. Autonomous mobile machines are being adopted at ever-increasing rates. Their capabilities in autonomy exceed those of previous mobile machines, such as teleoperated devices and guided devices. Teleoperated devices are often referred to as "bots" or "robots" in everyday conversation, but they are not "robots" as the term is used in the standards community; teleoperated devices lack autonomy, since a human worker is fully controlling their actions. Demolition or construction "bots", even unmanned aerial devices ("drones"), have typically been teleoperated in this fashion. Commonly, cameras on the devices transmit images back to the human operator, and the interpretation of the visual data and planning of the next motion by the machine takes place within the worker's brain—not within the machine's control logic. Guided devices are another relatively familiar type of mobile machine; they are often referred to as "driverless industrial trucks" and in everyday conversation are often called "AGVs" ("automated guided vehicles"). While their movements are conducted without direct human control, such guided machines rely on a pre-programmed route called a guidepath; in that sense the control over their movement is programmed in advance. Autonomous mobile robots (AMRs), however, rely on no such pre-programmed paths. An AMR can plan its own navigational route to a destination, and can change routes if it detects an obstacle blocking the route it planned initially.

With respect to HRC, the robot standards community agrees that one cannot necessarily consider an AMR "collaborative" even if it has a PFL robot arm mounted to it; one must also take into account the forces exerted by the mobile platform, and the combination of forces from the mobile platform and the moving robot arm could result in the AMR exerting forces in excess of what is permitted by TS 15066 in a

PFL robot arm. However, in a larger sense, most, perhaps all, AMRs could be said to engage in "human–robot collaboration," because in the current industrial environment, AMRs tend to operate in areas where human workers are also present. This represents a paradigm shift in the safety of workers around robotics: the industrial robot arm, even a so-called "collaborative" one, is bolted into place, and therefore it *can* be separated from human workers if need be for safety by physical fences or other safeguards. A worker would need to approach the area in which the fixed-in-place robot might present hazards in order to be at risk from those hazards. Furthermore, since the worker would likely need to open, remove, shut down, or otherwise defeat the safeguards in order to approach the hazards of the fixed-in-place industrial robot system, it can be reasonably presumed that the worker has awareness that he or she is approaching or entering an area containing robot hazards. On the other hand, with the advent and increasing adoption of AMRs, the robot can approach the worker without the worker intending to enter the robot's workspace, or even being aware of having done so. As a result, in the sense of "human and robot working in the same space at the same time," AMRs can often, perhaps even usually, be said to offer "human–robot collaboration." The market's need for a safety standard covering this new paradigm has led to the recent publication of the ANSI/RIA R15.08–1-2020 [16], which describes safety requirements for industrial mobile robots (IMRs), and further work has already been begun.

3.3 How Might These Trends Affect Industrial Workplaces, Workforces, Teams, and Individuals?

Going forward, as collaborative robot systems enable partial automation of tasks, HRC will continue to spread into workplaces and industries that might not have considered automation in the past. Rather than always being "caged" away from workers, more and more robot systems will be accessible by workers, yet human safety must still be protected. As workers become more familiar with HRC robots and robot systems, including autonomous mobile robots, comfort level with such machines will increase.

As well, continued investments in training and development of workforces, teams, and individuals will be needed so that skills and knowledge keep pace with technological change.

3.4 How Might These Trends Affect Industrial HRC Standards?

With respect to industry standards for industrial HRC, there will likely be continued need for more-frequent reviews and updates of these documents, due to the pace of change in the enabling technologies. Further, as the presence of HRC grows in the workplace, there will be more opportunities to study systems of HRC, generating data and identifying best practices. These new data and insights can then be incorporated into new editions of industry standards regarding industrial HRC.

4 Conclusion

Human–robot collaboration is supported by voluntary industry consensus standards describing the safety requirements for HRC robot systems [7–9, 15, 17–19]. Such standards benefit industry by promoting consistent requirements across borders and around the world. The goal of such standards is to describe a desired outcome, while avoiding prescribing a particular solution, and when this is achieved, standards encourage technological innovation rather than restrict it. And, because of the pace of that technological innovation in the area of HRC, standards guiding the safe practice of HRC will need to be reviewed and updated relatively frequently.

References

1. American National Standards Institute (2020) ANSI essential requirements: due process requirements for American National Standards. New York
2. American National Standards Institute (2007) ANSI procedures for the national adoption of ISO and IEC standards as American national standards. New York
3. American National Standards Institute (2019) ANSI procedures for U.S. participation in the international standards activities of ISO. New York
4. Franklin C, Dominguez EG, Fryman JD, Lewandowski ML (2020) Collaborative robotics: new era of human-robot cooperation in the workplace. J Safety Res 74(2020):153–160
5. International Organization for Standardization, International Electrotechnical Commission (2020) ISO/IEC directives, part 1: consolidated ISO supplement—procedures specific to ISO. Switzerland, Geneva
6. International Organization for Standardization, International Electrotechnical Commission (2018) ISO/IEC Directives, Part 2: principles and rules for the structure and drafting of ISO and IEC documents. Switzerland, Geneva
7. International Organization for Standardization (2011) ISO 10218–1:2011, Robots and robotic devices – safety requirements for industrial robots – part 1: robots. Switzerland, Geneva
8. International Organization for Standardization (2011) ISO 10218–2:2011, Robots and robotic devices – safety requirements for industrial robots – part 2: robot systems and integration. Switzerland, Geneva
9. International Organization for Standardization (2016) ISO/TS 15066:2016, Robots and robotic devices – collaborative robots. Switzerland, Geneva
10. International Organization for Standardization (2010) ISO 12100:2010, Safety of machinery – general principles for design – risk assessment and risk reduction. Switzerland, Geneva
11. Kelechava B (2017) "ISO Type A-B-C structure for machinery standards," blog published October 27, 2017. https://blog.ansi.org/2017/10/iso-type-abc-structure-machinery-standards-ansi-b11/#gref
12. Müller C, Kutzbach N (2019) World robotics 2019 – industrial robots. Frankfurt am main, Germany: IFR Statistical Department, VDMA Services GmbH

13. Muttray A, Melia M, Geißler B, König J, Letzel S (2014) Research project No. FP-0317: collaborative robots – determination of pain sensibility at the man-machine-interface. Mainz, Germany: institut für Arbeits-, Sozial- und Umweltmedizin (Institute for Occupational, Social and Environmental Medicine), Universitätsmedizin der Johannes Gutenberg-Universität Mainz (Johannes Gutenberg University of Mainz)
14. Occupational Safety and Health Administration, Labor Department, United States (1989) 29 CFR § 1910.147, the control of hazardous energy (lockout/tagout)
15. Robotic Industries Association (2012) ANSI/RIA R15.06–2012, American national standard for industrial robots and robot systems – safety requirements. Ann Arbor, MI, USA
16. Robotic Industries Association (2020) ANSI/RIA R15.08–1–2020, American national standard for industrial mobile robots – safety requirements – part 1: requirements for the industrial mobile robot. Ann Arbor, MI, USA
17. Robotic Industries Association (2016) RIA TR R15.306–2016, technical report – industrial robots and robot systems – safety requirements – task-based risk assessment. Ann Arbor, MI, USA
18. Robotic Industries Association. (2016) RIA TR R15.606–2016, technical report – industrial robots and robot systems – safety requirements – collaborative robots. Ann Arbor, MI, USA
19. Robotic Industries Association. (2018) RIA TR R15.806–2018, technical report – industrial robots and robot systems – safety requirements – testing methods for power & force limited collaborative applications. Ann Arbor, MI, USA

Engineering a Safe Collaborative Application

Elena Dominguez

Abstract Human–robot collaboration (HRC) has become a popular topic in the manufacturing industry, as a possible solution to improve inefficiencies of 'traditional' robots, i.e. the large payload robots that have been too hazardous to be positioned anywhere near the workforce. Collaborative robots, typically of a much smaller payload and force limitation, are attractive as they offer a range of benefits and allow more appropriate allocation of human and robot capabilities. However, uptake of collaborative systems is often hindered by companies' lack of confidence and knowledge of how to ensure safety. This chapter describes the process of engineering safety in the installation and assessment of a collaborative human–robot system.

1 Collaborative Robots

In many manufacturing processes the segregation of 'traditional' robots obstructs production flow and does not allow the skills and attributes of human operators and robots to be utilized most effectively. Repetitive tasks are better allocated to a robot while operators are more skilled at flexibility and decision making. The potential benefits of using collaborative robots over traditional robot systems include the following.

- Provides a high degree of interaction with operators to utilize human skills
- Makes it possible to automate only portions of a process while leaving the more challenging tasks to the operator
- Saves costs of traditional perimeter safeguarding
- Allows the robot to continue to run while personnel approach the cell

However, to accomplish these benefits, the robot system must be safe to protect the operators who will interact with it. This creates the need for a new approach in

E. Dominguez (✉)
Pilz Automation Safety, Michigan, USA
e-mail: E.Dominguez@pilzusa.com

M. I. Aldinhas Ferreira and S. R. Fletcher (eds.), *The 21st Century Industrial Robot: When Tools Become Collaborators*, Intelligent Systems, Control and Automation: Science and Engineering 81, https://doi.org/10.1007/978-3-030-78513-0_10

safeguarding robots. First, the essential risk assessment must account for the high degree of potential interaction with operators. With traditional robots, the goal was to stop the robot if contact was possible. Now, the potential for contact cannot be ruled out so the cell must be safe even if the operator performs reasonably foreseeable tasks that expose them to unintentional contact with the robot. The assessment must examine ALL of the possible ways that persons may be contacted by the robot. The following sections go through the process of risk assessment to undertake this thorough examination of potential human contact.

2 Risk Assessment Strategy

Whenever a robot operating in an automatic cycle can contact a person, the event must be analyzed to determine the risk. This means that the balance between potential injury and the probability of that injury occurring must be set to a point where the risk is acceptable. There are several strategies to achieve this risk assessment (Fig. 1).

1. Stop the robot while permitting power to the servos. The robot can move, but the safety system monitors the position to shut down the robot if it moves.
2. Put the robot into a semi-automatic mode where it only responds to the commands of the operator, i.e. hand guided mode.
3. Limit the force and contact pressure to levels below the biomechanical limits below the thresholds for minor injury and pain while allowing the robot to continue to move.
4. Monitor the robot and objects around the robot at real time to separate or stop the robot if contact is possible. Operators may share the collaborative space while the robot responds to the monitoring system. Get too close and the robot may slow down, stop with power on, or just avoid the person and maintain the minimum separated distance.

So, risk assessment must identify the appropriate strategies and identify the performance parameters. In fact, most of the time, a combination of these strategies is used.

3 Contact Events

A risk assessment of a collaborative robot must identify all potential contact events, both intentional and unintentional, in order to determine if the risk is acceptable. These contact events must be assessed in terms of the potential severity of injury and the probability of occurrence of contact. This, of course, is the definition of risk. If the risk level is not tolerable, then measures can be taken to either reduce the potential severity of injury or the probability of contact.

Fig. 1 Risk assessment area

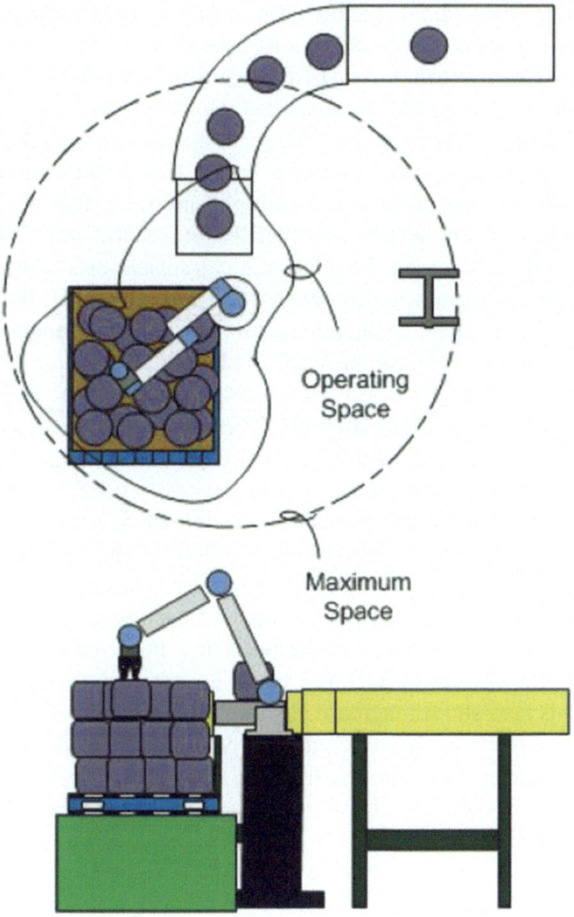

Regardless of the strategy selected, the potential contact event must be identified. A contact event is defined as a physical contact between the robot and an operator regardless if the contact is intentionally or unintentionally done by the operator. Contact events can occur in many ways. If the operator is handing a part to the robot, then the contact would be intentional. However, most of the time, potential contact events would be unintentional. For example, if the robot is moving towards a machine in order to unload a part from the fixture, and a person steps in front of the robot path, the robot could run into the operator. This would be an unintentional contact event. All contact events may be classified as a crush or an impact event.

At the base level, the difference between these two types of contact events is obvious. In a crush event, the robot would push an operator against a fixed, immovable structure. For example, the robot may trap an operator and press them against a machine that the robot was trying to move into. In an impact event, the robot contacts a person such that the operator's body would be free to fall away or get pushed aside as the robot moves through its path.

This most important difference is the duration of the event. A crush event may be short or long, that is, sustained. However, an impact event will always be short in duration. Whenever the force is sustained beyond a half second, the force is referred to as a quasi-static contact event. A transient event is one where the force exerted onto the operator has a duration of no more than one half second. Knowing this, a contact event may be classified as quasi-static, transient or a combination of the two. A crush event may be quasi-static, transient or a combination where the initial force is in the form of a short peak force and then it if followed by a sustained force.

In an HRC application, human contact with the robot should be considered inevitable. Because an HRC application is safeguarded without a perimeter guard, persons have the ability of entering the collaborative space even when strictly trained to stay out. It is human nature. Therefore, the risk assessment must account for reasonably foreseeable misuse by the operator. This includes walking into an HRC cell even if there is no reason.

An intentional contact event is defined as a task the operator must perform where the operator touches the robot, EOAT, or the part the robot is handling/processing. For example, if the operator is required to place a part into the gripper of the robot, this is considered an intentional contact event. Contact between the robot and the operator occurs at a regular basis and, thus, increases the risk of injury. While most of the current HRC applications do not design cells for intentional contact events, this area should increase as more challenging designs are created.

An unintentional contact event is defined as contact that is the possibility of occurring even though the cell is designed without this contact being necessary. This is typically classified as reasonably foreseeable misuse. However, if the cell is designed such that the operator is required to perform a task within the path of the robot when the robot is not there, the task is not misuse. Performing this task increases the risk of contact. The cell design should avoid this situation to reduce the risk of contact.

Still, a contact event may not always create an injury. Any contact event may be classified by the key physical characteristics of the event. These characteristics are as follows.

- Magnitude and duration of the force
- Magnitude and duration of the contact pressure
- Speed of the robot
- Mass of the robot
- Exposed body region of the operator, that is, where would the robot contact the body of the operator.
- Type of contact event, either crush or impact.

4 Contact Thresholds

The ISO/TS [1] 'Robots and robotic devices—Collaborative robots' is a technical specification that provides a table of biomechanical limits which establish the thresholds for minor injury and pain. Table A provides biomechanical limits for 29 parts of the human body from the head down to the lower leg. For each region, there are four biomechanical limits. The first column has contact pressure limits defined by the pain threshold and a duration of over 0.5 s. This, of course, is the quasi-static contact limit. The reasoning for using pain is that if you design a tool to create contact pressure below the pain limit, you are below the minor injury threshold. Also, the pain data was derived by conducting a university study on actual humans that determined that contact pressure was the main driver for triggering a pain sensation. The next biomechanical limit is the quasi-static force limit where the force is sustained for more than 0.5 s. Note that this limit is based on an approximation of the minor injury threshold.

The next two columns are for transient contact events where the duration of the contact is no more than 0.5 s. In this case, the limits are double the value of the quasi-static limits indicating that the human body can withstand higher magnitude without injury if the duration is short. Therefore, each contact event must be properly classified to apply the correct limits.

The robot system must be designed to keep all the contact events below these biomechanical limits. By applying these biomechanical limits, the robot system integrator can design a system that can allow contact, either intentional or unintentional, that avoids injury. Where this is not possible, the integrator can apply other risk reduction measures such as speed limits, presence sensing devices and passive measures such as padding.

When contact occurs, the robot exerts a force onto the human. Where forces are high, the potential severity can result in broken bones or worst. Thus, to operate in a force limited design that permits contact, the forces and contact pressures must be in a very lowest range of potential injury. The goal of operating a robot cell in a force limited mode is to design the system to restrict the potential severity level to a minor injury threshold. The minor injury threshold is defined on the AIS scale as an AIS class 1 where the injury is a slight laceration or deep contusion treatable with first aid.

Pain is not injury. It is a human body sensation warning that injury is possible. A study of 100 human subjects was conducted to measure pain caused by contact pressure to determine the pain threshold for 29 parts of the human body [2]. The group represented a wide range of the human population including age, gender and occupation. This group exhibited a wide range of pain tolerance. Interestingly, persons working in industry, exhibited a higher pain threshold that persons working in an office environment. The contact pressure biomechanical limits in the table are based on a 75-percentile grouping where 25 percent of the group may still feel some pain. None of the population in the university study exhibited any injury at these limits. From this observation, it was deemed that forces at this level did not cause injury.

Therefore, designing the robot system using this target for contact pressure should avoid injury.

In a risk assessment, besides identifying the potential contact events, it is also necessary to determine the body region that may be exposed. For a given contact event, the characteristics of the robot motion, robot manipulator design and tool design, will point to potential body regions that may be contacted. For example, if a person walks through the programmed path of the robot, there is the potential for an impact type contact to the torso. In addition, if the robot is moving in a path above 1200 mm, based on anthropometric data, there is the potential for head impact contact. In another example, where the robot approaches a fixed structure, there is the potential a crush type contact. Typically, the operator's hand and lower arm is exposed to contact if they reach onto the fixed structure as the robot approaches.

Each body region exhibits a degree of vulnerability to injury. On average, contact to the hand permits a higher contact pressure than the head or torso. Therefore, the design and the risk reduction measures must account for the exposed body region.

The human body is designed to take a bump or two without injury. Typically, at the low forces considered acceptable for collaborative robot contact, the body can recover from a short duration force much easier than a force of the same magnitude that is apply for a longer duration. This is mirrored in another machine, the power door on an elevator. If the door closed on someone, but withdraws in short order, the allowable force is higher than if the door continued applying the force. The same concept is applied to the biomechanical limits. Transient biomechanical limits are double the limits of quasi-static limits if the duration is no more than 0.5 s.

5 HRC Risk Assessment

For human–robot collaboration (HRC) the base process of a machinery safety risk assessment remains the same. Hazards are identified and drive the potential degree of harm. The tasks are identified and mainly drive the exposure to the hazards. The methodology chosen then combines two other factors that together with exposure predict the probability of an injury occurring. The methodology then estimates the initial risk level and evaluates this level to determine if it is acceptable or if risk reduction measures are required. Below is the ISO model for the elements of risk (Fig. 2).

The HRC risk assessment must evaluate a higher degree of exposure due to the lack of traditional perimeter guarding. There are unique risk sources created by the unimpeded exposure to the hazards inherent in operating machinery. There is no perimeter guarding to put these hazards into a safe state.

1. Robot movement
2. Robot system tooling, e.g. gripper
3. Unguarded machinery within the robot cell

The HRC risk assessment must evaluate the risk created by each of these situations.

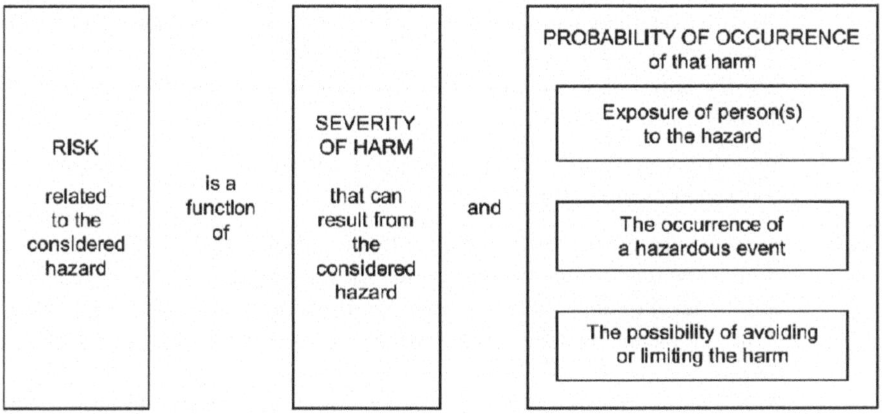

Fig. 2 Elements of risk [3]

1. Predict contact events between the robot and personnel. The type of contact event and the body region exposed are important factors driving the initial risk level.
2. Identify the hazards accessible from the tooling carried by the robot because it will be still be active.
3. Identify the accessible hazards contained within the associated machinery in the cell. This may be standalone machines such as a mill with clamps and cutter. This may also be a simple re-grip stand with clamp.

Because personnel are not impeded by safety fencing they, of course, may enter and do as they like. This must be considered. Fortunately, in an industrial environment, the activities of personnel are heavily influenced by rules and assigned jobs. We can use the following factors to predict the more credible contact events.

1. Are there tasks defined for the operator that puts them with or very near the operating space of the robot, i.e. the path of movement?

    a. Is the operator reaching into the robot cell to pick up a part?
    b. Is the operator sitting in the adjacent station such that the robot passes near their head?
    c. Is there a need for the operator to walk into the cell to empty a reject bin?
    d. Of course, where is it expected that the operator would need to enter to perform an error recovery from a part jam.
    e. Does the operator need to periodically enter the cell to replenish a part feeder?
    f. Is the cell near a high traffic aisle?

2. Are there strong motivations to enter the robot operating space to perform an undefined task?

  g. Does the operator have a habit of examining the associated machine to make a quick check of the quality of the machine process?
  h. Can the adjacent worker save time by inching over into the robot cell to start their activity early?

3. Are there strong motivations to do reasonably foreseeable misuse that puts the operator within the operating space?

  i. We must always be aware that people are curious. The good news is that after a few weeks, people will grow quickly accustom to the new robots and the level of entry falls to a low rate. It becomes just another machine.
  j. Is the cell located in a place where personnel are motivated to walk through the cell?
  k. Would an operator go directly into the associated machine to grab a part instead of waiting for the robot?

It is important to be aware, at the end of the day, that the cell must be safe regardless of what personnel do. Therefore, a collaborative robot cell must operate at an acceptable level of risk. This drives the following overall criteria.

1. If the robot and other hazards are permitted to continue operation when personnel enter the hazard zone, the potential severity must be in the reversible injury range. Only small robots are permitted to continue to move when contact occurs.
2. Robots and machinery that may cause irreversible injury including amputation, fatality or an injury that lower the long-term abilities other person, must come to a safe state prior to contact. Therefore, robots with a medium payload, for example exceeding 35 kg, should operation in a collaborative mode such as speed and separation monitoring.

Generally, it should be obvious to classify a contact event as to if it is a crush hazard or an impact hazard. Crush hazards occur when the robot moves near or comes in contact with a fixed structure such as a machine, a pallet, a conveyor, or a building column. The criteria for deciding when a crush can occur also must account for the gap that is created, whether it is zero or non-zero. The EN ISO 13,854 standard defines the minimum gap to avoid a crush for different body regions. Here is a sample of key minimum gaps.

- Torso—500 mm
- Finger—25 mm
- Hand—100 mm
- Arm—120 mm
- Upper leg—180 mm

If a potential contact event is identified where a crush is not possible, then it may be predicted to be an impact contact event.

The factors that determine the exposed body region are related to both the machinery and the posture that the human may assume at the time of the contact event. First, are there any characteristics of the machinery that place the contact high

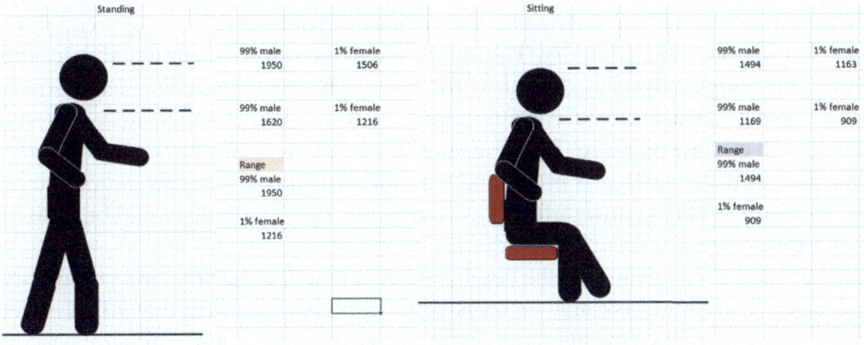

Fig. 3 Anthropometic data for head and face region

or low? Typically, a machine load station is in the proximity of 1066 mm (42 inches) above the floor. Therefore, if the posture of the operator is to be standing, contact to the head and lower limbs may be ruled out for a crush event.

One important consideration is the anthropometric data that may be used to predict the location of the different body regions. Most important is the location of the head and neck. As shown in the diagram below, which is a based on essential anthropometric data [4], the general population 99 percentile range for the for head and neck depends upon whether they are standing or sitting. If the person is standing while the robot approaches, there is a potential for a head contact if the robot path is above 1216 mm (Fig. 3).

There are four biomechanical limits for each body region.

- Quasi-static contact pressure (N/cm^2)
- Quasi-static force (N)
- Transient contact pressure (N/cm^2)
- Transient force (N)

The selection of the force limit is based on the duration of the contact. In an impact event, the duration is always brief. Therefore, the transient force limit is used. The force comes from the transfer of kinetic energy stored in the moving robot manipulator. In a crush event, the duration of the force may take one of three actions. If the robot strikes and immediately pulls away creating a very short duration. In this situation, the transient limit is used. If the robot is moving slowly and presses against the human body for a sustained period, then only the quasi-static force limit is used. Note that the force is primarily caused by the torque of the motors in the manipulator. Eventually, the robot controller will stop applying force when the force limit in the robot controller is tripped.

In some instances, the robot may impart a combination of a short duration peak force followed by a sustained duration of force until the force limit is tripped. Here the force is caused initially by the kinetic energy and followed by the torque of the motors. In this situation, both limits would be considered. The transient limit is applied to the initial peak and the quasi-static limit is applied to the sustained force.

The contact surface geometric characteristics determine if there is a potential for a contact pressure induced injury. The choice between transient, quasi-static or both follows the same analysis as described for force. The contours, edges and corners of the contact surface create concentrations of force over a small area, thereby, causing high contact pressure. The contact pressure of these contact surfaces should be measured. If the surface is broad and smooth, or if it is effectively padded, then the contact press may be insignificant. A reasonable rule of thumb would be radii of 0.5–2.5 mm.

Since risk is created through the combination of potential severity and probability of occurrence of harm, risk may be reduced by applying measures that affect either factor or both. To reduce the potential severity, the force and contact pressure need to be reduced to below the applicable biomechanical limit. There are several measures, both active and passive, may be taken. Active measures that influence the severity include the following.

- Reduce the speed limit setting in the robot controller. This will reduce the kinetic energy and, therefore, reduce the peak force.
- Reduce the force limit setting. This will reduce the sustained force that the robot may apply.
- Install a presence sensing device that will reduce the speed of the robot if persons approach.
- Program the robot to move in a defensive manner that eliminates active contact against portions of the tool or part that contain sharp edges.
- For example, modify the path program to orient the end of arm tool such that contact surface causing high contact pressure are now trailing and smoother surfaces on the tool or the robot arm are now leading.
- Program the robot to alter the path of the program to increase the gap and clearance where persons may be trapped against another fixed surface or parts of the robot arm.

Passive measures that influence the severity include the following.

- Modify the contours, corners and edges of the contact surface to increase the contact area.
- Add padding over surfaces that cannot be modified.
- Modify the end of arm tool to have compliance that will give when contact occurs.

To reduce the probability of occurrence of harm, the following measures may be considered.

Active measures that influence the probability of occurrence of harm include the following.

- Reduce the space where the robot may operate by using safety-rated space limiting settings.
- Install a presence sensing device that will stop the robot in safe standstill mode if persons approach.

Fig. 4 Collaborative robot crush hazard

- Install a pressure sensitive device on the robot manipulator or end of arm tooling to stop the robot if contact occurs.

Passive measures that influence the probability of occurrence of harm include the following (Fig. 4).

- Design the cell to minimize the interaction between the operator and the robot.
- Install awareness barriers that reduce the potential for causal contact with the robot movement in critical areas
- Install awareness markings and labels to warn personnel of the operating robot

6 Design

Now the design of the system can proceed based on the guidance of the risk assessment. However, the risk assessment only offers high level performance requirements. Putting ideas to paper in order to manufacture the robot system will require decisions to be made that meet the process requirements of the system such the throughput and qualify needed. This is now balanced with the safety requirements of the design risk assessment. For example, the throughput requirement will establish a speed for the movement of the robot. However, this speed must be compared to the allowable safe speeds identified in the risk assessment.

Since a primary safety function in a collaborative robot system the force limit, what should this limit setting be? The force limit is critical in crush events. Therefore, these force limits must be set to the values recommended in the risk assessment for the different locations where the crush event is possible. There may be an overall force limit setting based on the most vulnerable body region exposed in the risk assessment. There may be reasons to change the force limit in special circumstances such as the need for the robot to exert a level of force required for the process. This does not negate the safety requirements and may require a more elaborate programming scheme such as approaching the fixed structure at the safe force limit and only switching to the higher limit once the gap is small enough, e.g. 4 mm, to

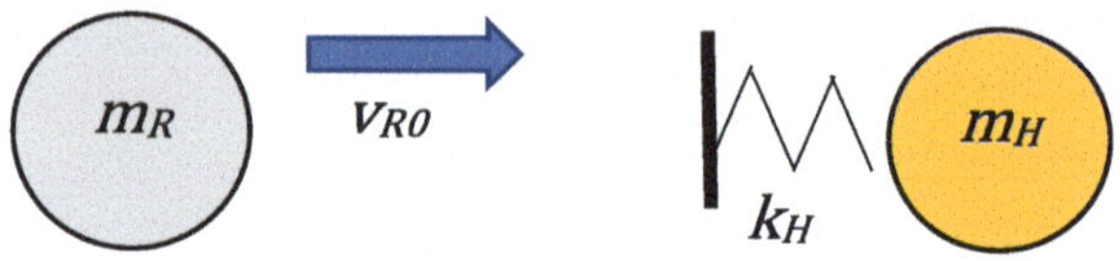

Fig. 5 Free space impact contact event

eliminate the possibility of a crush hazard. Humans behave in this matter by judging when a higher force is needed. The robot must be taught.

The speed limits are also guided by the risk assessment. There will be at least one speed limit applicable to impact contact events and there should be one for the approach speed for crush contact events.

For free-space collisions, the force produced is an impulse of short duration. Therefore, the transient force biomechanical limit for the exposed body region is used. Where personnel are permitted to walk into the cell, the chest is the appropriate limit. The Mechanics equation in the ISO TS 15,066 can then be applied to calculate the allowable speed for the robot (Fig. 5).

First, the robot effective mass, m_R, is estimated using the equation in the ISO TS 15,066. Typically, the entire arm is assumed to be moving. However, the base that is bolted to the floor does not move. Therefore, the moving mass of the arm is approximately 80% of the mass of the manipulator (Fig. 6).

$$m_R = \frac{M}{2} + m_L$$

m_R—Robot effective mass, kg

M—Total mass of moving parts of the robot, kg

m_L—Payload including tool and workpiece, kg

ω—Payload including tool and workpiece, kg

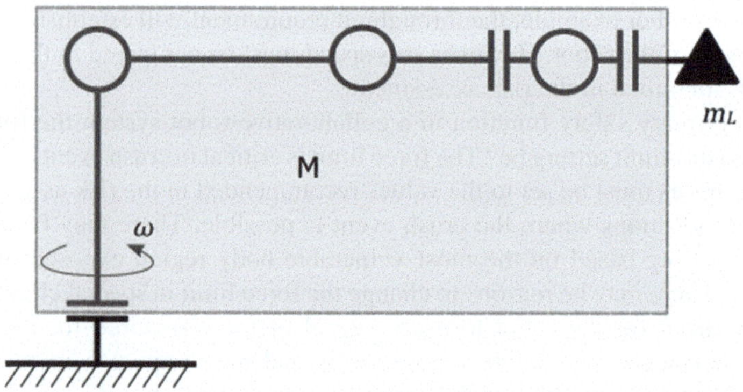

Fig. 6 Simplified mass distribution model from ISO TS 15066 [1]

With this, the allowable speed may be calculated.

$$V_{RO} = \frac{F_{maxT}}{\sqrt{\frac{m_R m_H}{(m_R + m_H)} k_H}}$$

F_{maxT}—Transient biomechanical limit, N

V_{RO}—Allowable speed, m/s

m_R—Robot effective mass, kg

m_H—human body region mass, kg

k_H—human body region stiffness, N/m

For example, if the moving mass of a robot is 18 kg and the payload is 5 kg, then the effective mass is 13 kg. The allowable speed for free-space collision is then calculated for chest contact. The chest has a mass of 40 kg and a stiffness of 25,000 N/m. The allowable speed is then 0.565 m/s or 565 mm/s in the typical units for robot TCP speed. Caution, even though this appears to be a precise number, the data for human characteristics varies and is conservative.

Most collaborative robot cell integrators quickly realize that the allowable speed limit for impact is too low to make the cycle time required. In order to achieve the move faster, the robot cell can include a laser scanner to slow down the robot when the operator approaches. Provided that the robot cell is co-existing, that is, it is designed to work without a high degree of human interaction, the robot can operate at the high speed required to meet the required cycle most of the time. On the occasion that an operator enters the cell, the robot can be automatically slowed down to the allowable speed. Once the operator leaves, the robot can then automatically return to its full programmed speed.

There is an allowable speed for approaching fixed rigid structures that would result in a force peak that is within the quasi-static biomechanical limit. This occurs in locations where the robot goes to pick up a part and where the robot goes to drop off the part. The following equation may be used to determine this allowable approach speed (Fig. 7).

$$v_{RO} = \frac{F_{maxQ}}{\sqrt{m_R k_H}}$$

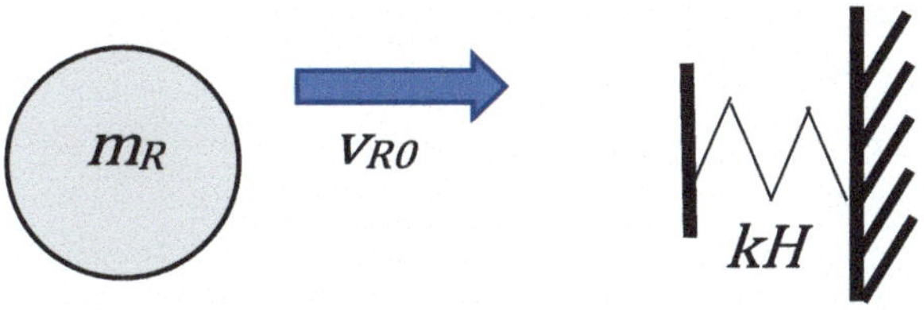

Fig. 7 Crush contact event against a fixed structure

F_{maxQ}—Quasi-static biomechanical limit, N

V_{RO}—Allowable speed, m/s

m_R—Robot effective mass, kg

k_H—human body region stiffness, N/m

For example, if the same robot from the previous example approaches the drop off point, it still has the same effective mass of 13 kg. In this example, the hand is the exposed body region. The hand has a mass of 0.6 kg and a stiffness of 75,000 N/m. The allowable approach speed should be limited to 0.284 m/s (284 mm/s). This should keep the initial force peak to stay below the transient limit. The force limit should be set to the quasi-static biomechanical limit to prevent the robot from exerting a sustained force above the Q-S limit.

The robot should be designed with smooth surfaces and a manipulator linkages that cannot create crush zones as the robot moves through its programmed path. However, all robots have possible motions that may create crush zones. The robot program should be implemented to avoid these positions. Safety-rated joint and space limits should be applied to prevent the robot from assuming these positions. Where these crush zones cannot be avoided, then other protective measures are warranted, such as padding or safety-rated presence sensing devices. Two common crush zones occur at the joint between the lower and upper arm linkages and around the wrist area.

The End-Of-Arm-Tooling must also operate safely. If there is a gripper, the accessible force must be below the biomechanical limits. This limit must also meet the required circuit performance level required by the risk assessment or the gripper must be inherently safe. To be inherently safe, it must be incapable of delivering a force above the biomechanical limits even if the robot commands a higher force.

So far, the focus of the design has been the force level. It is also necessary to keep contact pressures below the biomechanical limits for contact pressure. Where sustained pressure is exerted, then the quasi-static limit is applied. Where the pressure is a duration of 0.5 s or less, then the transient limit applies. In order to design the system to respect these limits, the potential contact surfaces must be identified. At the design level, the geometry of the mechanical elements within the cell can be modified to prevent high contact pressure areas. Typically, the corners and edges of the contact surfaces must be made smooth and wide. A rule of thumb would be to design the radius of all corners and edges to be in the range of 1 to 2.5 mm. Where surfaces cannot be modified, then other measures are required such as padding and smooth covers. This treatment should be applied to all potential surfaces including things like valves and hose fittings.

7 Validation

As with any machine, the safety system must be validated to confirm the effectiveness of the protective measures. A validation is a procedure involving testing and analysis to confirm that the safety circuits and safety devices used on a machine provide the risk reduction required by the risk assessment to achieve a safety machine. This is common to all machines. However, with force limited collaborative robots, force and contact pressure measurements must also be taken to confirm the effectiveness of the protective measures. This requires the measurement of forces and contact pressures. A force measurement device (FMD) is used to collect these measurements.

The force measurement device is an electronic device that can measure the force exerted by a robot during a contact event. This data is captured and plotted against time as a force curve. This is needed in order to determine the duration of the force. The duration of the force dictates which biomechanical limit must be applied. Recall that a force level that has a duration of less than 0.5 s must be below the transient force limit. Where the force duration is equal to or greater than 0.5 s, the quasi-static biomechanical limit must be applied.

The second essential aspect of the FMD is that it mimics the human body region so that the forces measures approximate the forces felt by a human. In a3333 contact event, the forces exerted by the robot are affected by the characteristics of the object that is struck. The object's physical properties of mass, stiffness and damping affect the amount of energy absorbed by the object and the amount reflected back to the robot. This then affects the force curve and, therefore, the duration and magnitude of the force. Therefore, the FMD must mimic the human body in order to measure the approximate force felt by the human. In order to mimic the human body, the three physical properties must match the human body. The different parts of the human body will exhibit differing mass, stiffness and damping.

Different human body regions will present a different mass to the robot. For example, if a robot moving through space along its programmed path contacts the lower arm of a human, the mass of the lower arm is needed. Therefore, the FMD would need to have the same mass as the human arm. Also, the arm is free to be push back in space. So the FMD must also be free to recoil back as the contact occurs. In order to do this, the FMD must be on a slide that is positioned in the direction that the arm will initially move and with the same mass of the arm. It gets complicated to match the mass and direction of recoil.

Now let's look at a contact event where the robot crushes the lower arm against a machine. In this event, the mass of the lower arm is combined with the machine as far as the robot is concerned. In this case the mass of the arm doesn't matter. If the machine is bolted to the floor, then the robot will feel that mass of the arm, machine and the earth. So, in this case, the FMD can just be secured to the machine.

When a force is applied to a human body region, the bones and ligaments will bend and give. This is the stiffness characteristic of the body. Each region has a different stiffness. The FMD is set up with a spring to match the stiffness in order to mimic the force response of the body for a given collision. The spring constants range from 10 N/mm for the abdomen to 150 N/mm for the head.

The body also has soft tissue that affects the force response. The FMD contact face is fitted with a rubber pad that mimic the degree of damping effect of the soft tissue. A stiffer pad is used for regions with thin soft tissue such as the hand and a pad with more compliance such as the upper arm.

Guidance for make measurements is provided in a publish document from the RIA, Robotics Industries Association. The document is the ANSI RIA TR R15.806, Testing Methods for Power & Force Limited Collaborative Applications.

8 Additional Risk Reduction

Once measurements are made in the points identified in the risk assessment, the results are evaluated against the biomechanical limits of the standard. When the forces or contact pressures exceed the limits, the engineers need to modify the safety configuration, programming or construction on the system to bring those contact events down within the limits. The approach is the same as described in the design phase above.

1. Modify the safety parameters to lower the resulting forces and contact pressure.
    a. Speed limit
    b. Force limit
    c. Safety boundaries that limit speed or force when the robot enters the zone.
2. Modify the program to lower resulting forces and contact pressure.
    d. Defensive programming techniques
    e. Modification of the programmed path to provide more clearance
3. Make changes to the design and construction to lower the resulting forces and contact pressures.
    f. Design of layout to eliminate contact events
    g. Construction of tools and fixes to round edges
    h. Add padding to guard edges or absorb energy
    i. Add compliance to reduce force
    j. Add safety devices that detect persons and stop the robot
    k. Add presence sensing device to initiate a robot speed reduction

Once all the potential contact events have been addressed and brought into compliance, the validation is complete. The program and safety settings are protected by passwords. The configuration is recorded by electronic checksum or signature. Change control procedures are applied to periodically confirm that the settings have not been changed. Safety functions are periodically tested to confirm that these protective measures are still effective.

9 Conclusion

To take advantage of the use of collaborative robots can bring, the engineer must design in safety. The approach starts with a risk assessment to identify the design requirements. The design then applies protective measures to ensure that possible contact falls below the biomechanical limits of the human body. A validation is performed on the built system measuring forces and contact pressures. Finally, additional risk reduction measures, if necessary, are applied to reach an acceptable level of risk.

References

1. ISO/TS 15066:2016 (2016) Robots and robotic devices—collaborative robots. International Organization for Standardization
2. Melia M, Geissler B, König J, Ottersbach HJ, Umbreit M, Letzel S, Muttray A (2019) Pressure pain thresholds: subject factors and the meaning of peak pressures. Eur J Pain 23(1):167–182
3. ISO 14212–1:2007 (2007) Safety of machinery—risk assessment—part 1: principles. International Organization for Standardization
4. Dreyfuss H, Henry Dreyfuss Associates, Tilley, AR (1993) The measure of man and woman: human factors in design. Whitney Library of Design

Challenges in the Safety-Security Co-Assurance of Collaborative Industrial Robots

Mario Gleirscher, Nikita Johnson, Panayiotis Karachristou, Radu Calinescu, James Law, and John Clark

Abstract The coordinated assurance of interrelated critical properties, such as system safety and cyber-security, is one of the toughest challenges in critical systems engineering. In this chapter, we summarise approaches to the coordinated assurance of safety and security. Then, we highlight the state of the art and recent challenges in human–robot collaboration in manufacturing both from a safety and security perspective. We conclude with a list of procedural and technological issues to be tackled in the coordinated assurance of collaborative industrial robots.

Keywords Co-assurance · Cobot · Dependability trade-off · Cyber-security · Human–machine interaction · Risk management · Hazard identification · Threat analysis

---

M. Gleirscher (✉) · N. Johnson · R. Calinescu
Department of Computer Science, University of York, York, UK
e-mail: mario.gleirscher@york.ac.uk

N. Johnson
e-mail: nikita.johnson@york.ac.uk

R. Calinescu
e-mail: radu.calinescu@york.ac.uk

P. Karachristou · J. Law · J. Clark
Department of Computer Science, University of Sheffield, Sheffield, UK
e-mail: panayiotis.karachristou@sheffield.ac.uk

J. Law
e-mail: james.law@sheffield.ac.uk

J. Clark
e-mail: john.clark@sheffield.ac.uk

M. I. Aldinhas Ferreira and S. R. Fletcher (eds.), *The 21st Century Industrial Robot: When Tools Become Collaborators*, Intelligent Systems, Control and Automation: Science and Engineering 81, https://doi.org/10.1007/978-3-030-78513-0_11

1 Introduction

Collaborative robots (or *cobots*[1]) are expected to drive the robotics market in coming years [56], providing affordable, flexible, and simple-to-integrate robotic solutions to traditionally manual processes. This transformational technology will create new opportunities in existing markets such as food, agriculture, construction, textiles, and craft industries [20, 44], enabling more efficiency in production while reducing operator workloads and removing occupational hazards [20, 55].

Cobots will ultimately enable humans and robots to share physical spaces, and combine the benefits of automated and manual processes [57]. However, current applications are in the main limited to those requiring little physical collaboration, with humans and robots sharing spaces but working sequentially [21]; close, physical collaboration in the true sense (with robots responding in real-time to users) requires more complex sensing and control, resulting in highly Throughout this chapter we refer to *cobots* as robots, including their software and operational infrastructure, specifically designed for human–robot collaboration, although we accept that traditional robots may be used in collaborative ways when augmented with sufficient control and sensory systems complex safety cases. Whilst cobots may be designed to be inherently safe (when operating with limited capabilities), the process (and end effector or payload) often poses a greater threat than the robot itself [55].

ISO 10218 [22] set the standard for safety of industrial robots. In 2016, with a rapidly growing range of collaborative robots on the market, ISO 10218 was supplemented by ISO/TS 15066 [23], which specifies additional safety requirements for industrial *collaborative* robots. However, many manufacturing organisations have indicated that these do not go far enough in providing guidance, and that the lack of examples of good practice is hindering deployment of the technology. As a result companies are falling back on traditional segregational approaches to risk assurance, including physical or chronological isolation and barriers, which counteract many of the lauded benefits of collaborative robots.

The aim of this chapter is to explore existing approaches and best-practice for safety and security of collaborative robots, and to highlight the challenges. Section 1.1 provides an overview of safety and security approaches applicable to cobots. Sections 2 and 3 then elaborate on particular methodologies in the context of an industrial case study. Following the two perspectives—social and technical—of the Socio-Technical System (STS) design approach [5], Sect. 4 enumerates additional socio-technical and technical challenges arising from safety-security interactions.

[1] Throughout this chapter we refer to cobots as robots, including their software and operational infrastructure, specifically designed for human-robot collaboration, although we accept that traditional robots may be used in collaborative ways when augmented with sufficient control and sensory systems.

1.1 General Approaches to Safety and Security

One of the main sources of difficulty, which prompts the need for guidance, is the engineering complexity and diversity required to build a cobot, or indeed any complex system that involves interactions with humans. In order to reconcile multiple (often heterogeneous) goals and objectives, knowledge from multiple engineering disciplines is needed—mechanical, electrical, process, human-computer interaction, and safety and security. In this section, approaches to co-engineering and co-assuring safety and security that can be applied to cobots are explored. The discussion in this section will be an extension of previous work done in [24] with an emphasis on literature and application to industrial control, robotics and cobots. Throughout this section, *assurance* will refer to the process and outcome of identifying, reducing and arguing about the level of risk.

Pietre-Cambacedes et al. [45] give a comprehensive view of methods, models, tools and techniques that have initially been created in either safety or security engineering and that have then been transposed to the other quality attribute. However, one of the biggest challenges identified is the (mis-)use of risk language. Whilst both deal with the notion of risk, and aim to prevent situations which might result in negative consequences [11], their conditions and concepts of loss are sufficiently different to cause conflict to arise. A classical example of this is the conceptualisation of risk. Traditionally, the goal of safety is to prevent death or injury, therefore safety risk is concerned only with the hazards that might lead to an accident, and is calculated as a product of the likelihood of a hazard occurring and the severity of that hazard (i.e. safety risk = likelihood × severity). In contrast, the goal of security is to prevent loss of assets. These might include people, in which case the goal aligns with safety, however security encompasses a much larger scope so assets might include process, intellectual property, organisation reputation, and information. Thus many more factors must be included when analysing security risk (i.e. security risk = (threat × access) × (business impact, confidentiality, etc.)).

Figure 1, taken from previous work done on co-assurance, shows a model from the Safety-Security Assurance Framework (SSAF) [24]. In particular, it highlights the safety and security processes and interactions during the system lifetime. Each of the phases of the system will place different requirements on safety and security practitioners, for example early stages will have a focus on assisting the engineering process to lower the risk of the system, whereas during operation, the practitioner's focus will be on ensuring that operations are being performed as expected and no assurance claims made earlier are being violated.

The core idea underlying SSAF is that of *independent co-assurance* that allows for separate working but requires *synchronisation points* where information is exchanged and trade-off decisions occur. This allows practitioners to use specialised expertise and progress to occur within each domain because there is a shared understanding of *what* information will be required, and *where* and *when* it should be provided.

There are different modes of interaction ranging from silos (characterised by very few synchronisation points and little inter-domain communication) to unified

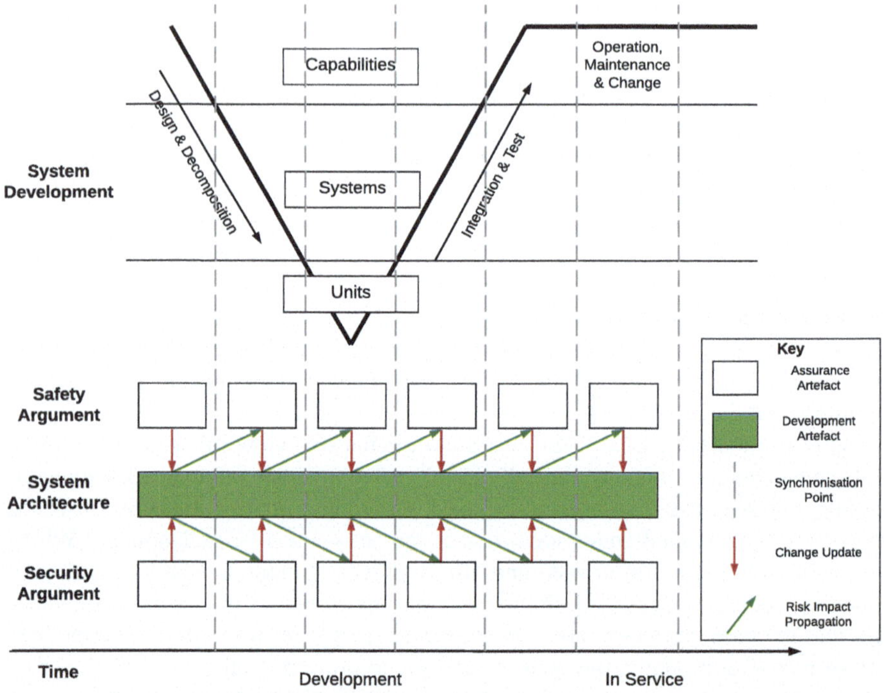

Fig. 1 SSAF interaction model with synchronisation points (taken from [24])

approaches (where the attributes are co-engineered and co-assured together, e.g. in [47]). A prerequisite for knowing information needs at synchronisation points is understanding the causal relationships within and between domains. This can be achieved using multiple approaches, some of which are shown in Table 1 which contains a subset of approaches that are commonly used. Approaches to safety and security risk management can be classified into several groups according to their level of formalism and their objectives. The classification framework of (1) structured risk analysis, (2) architectural methods, and (3) assurance and standards has been adapted from [45]:

(1) Structured Risk Analysis. These approaches are concerned primarily with understanding the cause-effect relationships for particular risks. In safety, hazards are analysed by experts using structured reasoning approaches such as Hazard and Operability Studies (HAZOP), Failure Modes Effects and Criticality Analysis (FMECA), and Bow Tie Analysis. These function by requiring the analyst to consider risk sources and outcomes systematically by using guide words over system functions. Example guidewords are *too much*, *too little*, *too late*, *etc.*; when applied to a cobot speed function would allow the analyst to reason about what would happen if the cobot was too fast/slow/late performing a task in a specific context. System-Theoretic Process Analysis (STPA) requires the system to be modelled as a controlled process rather than decomposing it into functions. Hazards relating to the control structure

Table 1 Approaches to safety and security assurance applicable to cobots

Approach	Safety	Security	Joint
(1) Structured risk analysis			
HAZOP	[10]	[53]	[59]
FME(C)A	[15] [6]		[48]
Bowtie analysis	[61]	[7]	[1, 8]
STPA	[35]		[60, 14]
STRIDE		[40]	
FTA	[17]		[25, 47]
Attack trees		[33, 49]	[47]
(2) Architectural approaches, testing & monitoring			
ATAM			[29]
Metrics			[31]
Failure injection	[43]	[4]	
(3) Assurance & standards			
Assurance cases	[30]	[19]	
Standards	IEC 61508 ISO 10218 ISO/TS 15066	IEC 62443 ISO 15408 (CC) ISO 27001	IEC TR 63069

are then reasoned about. The advantage of this approach is that hazards introduced by failure of intent are more readily identified than analysing functions, for example, where a cobot function does not fail and system requirements are satisfied, but it still leads to an unsafe state. There have been multiple adaptations of safety approaches to include security guidewords and prompts such as security-informed HAZOPs, FMVEA (FMEA and Vulnerabilities Analysis) and STPA-Sec.

Similar to its safety counterparts, STRIDE is a security threat analysis model that assists analysts to reason about types of threats relating to spoofing, tampering, repudiation, information disclosure, denial of service and elevation of privilege (more information is provided in Sect. 3). The main limitation of structured risk analysis is their over-reliance on expert opinion which makes the outcome only as good as the analyst performing the analysis [24].

Fault Tree Analysis (FTA) [17] is one of the earliest approaches used in safety analysis. Its process requires that the causes of functional failures and faults are decomposed; the outcome is then modelled in a directed acyclic graph as events connected by AND and OR gates. This tree-based approach has been extended to apply to both attacks and threats in security [33, 49]. Whilst the formality of these types of models allows for an improved review and audit of the risk analysis, they are limited by the assumptions made during the analysis process and whether or not they are validated subsequent to the analysis.

(2) Architectural Approaches, Testing & Monitoring. The third class of approaches seeks to understand the trade-offs between safety and security for the system architecture, and the effects of risk during operation. Architectural Trade-Off Analysis Method (ATAM) [29] is a methodology that allows structured negotiation of architectural strategies for quality attributes including safety and security. Dependability Deviation Analysis (DDA) [8] relies on applying Bow-Tie analysis to architectural models to understand the impact of dependability attributes on each other. In addition to these approaches, which are primarily applied early in the system development lifecycle, there are approaches that seek to test and verify the requirements output at even earlier stages. In addition to architectural model approaches such as ATAM and DDA, the concept of joint safety and security metrics [31] has been proposed to understanding safety and security functional assurance in industrial control systems. Examples of failure injection and testing include Hierarchically Performed Hazard Origin and Propagation Studies (HiP-HOPS) [43] which is a safety technique wherein Simulink models are used to generate fault trees to assess the impact of different faults. In security, penetration testing [4] is often used as an approach to understand the vulnerabilities in the engineered system.

(3) Assurance & Standards. The final class of approaches seeks to manage risk through safety and security processes and standards. The industrial control safety standard IEC 61808 is arguably the most influential safety standard created as many safety standards from other domains are its derivatives; examples include ISO 10218 [22] (robotics) and ISO/TS 15066 [23] (cobots). Similarly for security IEC 62443 (ICS product security), ISO 27001 (organisation security process), ISO 15408 (security requirements) have been applied to a wide range of industries. Standards from both domains have been largely isolated with very little alignment, however, as security has the potential to undermine the safety of a system, the two should be considered in conjunction. In an attempt to address this problem, joint standards are being created. One such standard is IEC TR 63069 (applied to ICS), which advocates the creation of a *security perimeter* within which safety analysis is performed. A fundamental part of the certification process for standards is presenting a structured argument with evidence to show that the assurance criteria have been satisfied. This argument takes the form of safety cases and security cases (examples of each in [30] and [19]).

Many of the approaches described above are general approaches or have been applied within an industry related to cobots, such as industrial control. There has been some preliminary application of many of these methods to cobots, for example, Lichte and Wolf [38] use an approach that combines graphical formalism and architectural methods to understand the safety and security of a cobot; this is a good start, however it still leaves many challenges unaddressed. In the following sections, we provide an illustrative cobot example and further details about the safety and security challenges.

1.2 Illustrative Example: The Cobot

In the previous section, the focus was on general approaches to safety and security that could be applied to a collaborative robotic system. However, there is a paucity of information about how these work in practice. The industrial case study discussed in the following Sects. 2 and 3, seek to provide greater clarity about how to undertake safety and security analysis for a cobot.

As many types of human-robot interactions, Kaiser et al. describe *scenarios of co-existence* [28]: (1) *Encapsulation*, with physically fenced robot work areas, (2) *co-existence*, without fencing but separation of human and robot work areas, (3) *cooperation*, with shared work areas but without simultaneous operation in the shared area, and (4) *collaboration*, with shared work areas and simultaneous, potentially very close interaction inside the shared area.

In our example, we consider scenarios that match (3) or (4) and a system that comprises primarily of a plant (i.e. a manufacturing cell including one or more robots), operators, and an automatic controller.

2 Cobot Safety

In this section, we (i) review safety risk analysis and handling in human-robot collaboration and (ii) highlight recent challenges in the development of controllers implementing the safety requirements.

2.1 Analysing Safety Risks in Cobot Settings

In safety analysis, one focuses on two phenomena: *accidents* and *hazards*. An *accident* is a more or less immediate and undesired *consequence* of a particular combination of the plant's state and the environment's state, which together form the *cause* of the accident. The plant's portion of such a cause is called a *hazard* and the environment's portion an *environmental condition* [36, p. 184], both first-class entities of safety risk analysis.

Safety Risks. Accidents, their causes, and hazards in human-robot co-existence in manufacturing have been discussed in the literature since the mid 1970s (e.g. [2, 27, 28, 54]). Notably, Nicolaisen [42] envisioned collaborative, potentially mobile, robots closely working together with humans in shared work areas in 1985. In 1986, Jones [27] distinguished two major classes of hazards relevant to human-robot collaboration:

- *Impact hazards*: Unexpected *movements* (e.g. effectors or joints reaching beyond the planned work area), for example due to failing equipment (e.g. valves, cables,

electronics, programs); dangerous work pieces handled (e.g. released) in an unexpected or erroneous manner; manipulations (e.g. welding, cutting, transport) with hazardous side effects (e.g. welding sparks, flying tools or work piece residuals, robots bumping into humans).
- *Trapping hazards*: Workers are trapped between a robot and a static object (e.g. a machine or cage wall) while the robot is active or in a programming, maintenance, or teaching mode.

Both classes of hazards can cause the robot to collide with, or catch, a human co-worker. This includes robot joints, end-effectors, or dangerous work pieces hitting and injuring co-workers.

2.2 Handling Safety Risks in Cobot Settings

Accidents or hazards can be *prevented* by employing measures to avoid their occurrence. More generally, accidents or hazards can be *mitigated* by employing measures to significantly reduce accident or hazard *probability* or reduce accident *severity*. An accident cause is considered *latent*[2] if there are sufficient resources (e.g. time, bespoke safety measures) to mitigate the accident (e.g. by removing the hazard or the environmental condition).

For example, a robot arm on its way to a shared area W (hazard) and the operator on their way to W (environmental condition) form a *cause* of an impact *accident* to happen. This cause seems *immediate* unless a measure (e.g. stopping the arm on a collision signal, or the operator jumping away) renders it to be *latent*, i.e. possible if this measure were to stay inactive.

Measures can be either *intrinsic* (not requiring control equipment) or *functional*[3] (requiring control equipment). A functional measure is said to be *passive* if it mitigates certain accidents (e.g. a car airbag system) and *active* if it prevents certain accidents (e.g. an emergency braking system). Functional measures focusing on the correctness, reliability, or fault-tolerance of the controller (nowadays a complex programmable electronic system) are called *dependability* measures [2].

Safety Measures. Experience from accidents and risk analyses has lead to many technological developments [2, 18, 28], partly inspired by precautions when interacting with machinery. Table 2 gives an overview of measures in use.

Concepts for Functional Measures. Jones [27, p. 100] describes safety monitors as an additional control device for safety actions like emergency stop and limiters (e.g. speed, force). He further suggests safety measures be designed specifically for each *mode of operation* (e.g. programming or teaching, normal working, and

---

[2] As opposed to immediate causes falling outside the scope of risk handling.

[3] Note that functional safety in IEC 61508 or ISO 26262 deals with the dependability,particularly, correctness and reliability, of critical programmable electronic systems.Safety functions or, here, "functional measures" form the archetype of such systems.

Table 2 Examples of safety measures in human–robot collaboration, classified by stage of escalation and by the nature of the underlying technical mechanism

Stage	Type	Intrinsic	Functional
Hazard prevention	(1) Safety barrier, physical safeguard	Fence	Interlock
	(2) Avoidance of development mistakes	Controller verification	Controller verification
Hazard mitigation & accident prevention	(3) Improved reliability and fault-tolerance		Fault-tolerant scene interpretation
	(4) Intrusion detection (static, dynamic, distance-based)		Speed & separation monitoring; safety-rated monitored stop
	(5) Hand-guided operation		
Accident mitigation	(6) Power and force limiting	Lightweight components flexible surfaces	Variable impedance control, touch-sensitive and force-feedback control
	(7) System halt		Emergency stop

maintenance), each to be examined for its designed and aberrant behaviour [27, p. 87]. Internal malfunction diagnostics (e.g. programming error detection, electronic fault detection, material wear-out monitoring) can also inform such a safety monitor and trigger mode-specific actions.

Alami et al. [2] highlight advantages of *interaction-controlled* robots over *position-controlled* robots. The former are easier for planning and fewer assumptions on the structure of the work area and the robot's actions need to be made, while the latter need extensive pre-planning and require stronger assumptions.

Based on ISO 15066, Kaiser et al. [28] summarise design considerations (e.g. work area layouts) and collaborative *operation modes* beyond traditionally required emergency stop buttons (also called dead man's controls):

- *Safety-rated monitored stop*: A stop is assured while powered, thus prohibiting simultaneous operation of robot and operator in the shared area.
- *Hand-guiding operation*: The robot only exercises zero-gravity control and is guided by the operator, hence, no actuation without operator input.
- *Speed and separation monitoring*: Robot speed is continuously adapted based on several regions of distance between robot and operator.
- *Power and force limiting*: To reduce impact force on the human body, the robot's power and applied forces are limited.

Guidelines. Standardisation of safety requirements for industrial robots started in Japan with the work of Sugimoto [54]. In the meantime, ANSI/RIA R15.06, ISO

10218, 13482, and 15066 have emerged, defining safety requirements and providing guidance for safety measures. The aforementioned safety modes are part of the recommendations in ISO 15066.

2.3 *Risk Analysis and Handling in a Cobot Example*

In our example, an operator and a cobot undertake a spot-welding process. The open-sided work cell consists of a robot positioned between a spot-welding machine and a shared hand-over area. During normal operation an operator sits outside the cell and uses this area for exchanging work pieces. The layout allows staff to enter the active cell when needed.

In Table 3, we exemplify results of a preliminary safety analysis (i.e. hazard identification, assessment, requirements derivation, cf. Sect. 2.1) of the cell along with the identified safety requirements compliant with state-of-the-art measures as summarised in Sect. 2.2. The right column specifies the *overall safety requirements* for each *accident* and the technical safety requirements (e.g. the mode-switch requirements) for each *latent cause* (left column) indicating how the *hazard* can be removed from the critical state in due time. Each safety requirement specifies the conditions of a corresponding measure to mitigate any escalation towards one of the listed accidents.

2.4 *Recent Challenges in Cobot Safety*

Because robots and human co-workers have complementary skills, it has long been an *unfulfilled desire for both actors to simultaneously and closely work in shared areas* [27, p. 70]. Consequently, complex guarding arrangements, established for safety reasons, interfere with efficient production workflows. For mobile robots, fencing in the traditional sense is rarely an option. Back in the 1980s, another problem was that robots often had to be programmed through a teaching-by-demonstration approach with workers in the cage while robots fully powered. However, this rather unsafe method has been superseded by more sophisticated programming techniques (e.g. simulation, digital twins).

From the viewpoint of co-workers, close interaction has proven to be difficult in some settings as invisible joint motion planning procedures not rarely result in *unpredictable joint/effector movement* patterns. Assuming collaborative scenarios with complex tasks shared between humans and robots, Hayes and Scassellati [18] raise further conceptual issues, such as how cobots recognise operator intent, how they autonomously take roles in a task, how they include operators' intent in their trade-offs, and how they self-evaluate quantities such as risk during operation.

Unfortunately, more sophisticated robots incorporate a larger variety of more *complex failure modes*. Sensors need to sufficiently inform the robot controller for the

Table 3 Results of hazard identification, assessment, and mitigation analysis referring to measures recommended in ISO 15,066

Critical event	Safety requirement†
Accident (undesired, to be mitigated)	
The robot arm collides with the operator	The robot shall avoid active collisions with the operator
Welding sparks cause operator injuries (i.e. burns)	The welding machine avoids sparks injuring the operator
Latent Cause (to be reacted upon in a timely manner)	
The operator and the robot use the shared hand-over area at the same time	(m) The robot shall perform a safety-rated monitored stop and (r) resume normal operation after the operator has left the shared hand-over area
The operator approaches the shared hand-over area while the robot is away from the hand-over area (undertaking a different part of the process)	(m) If the robot is transferring a work piece to the hand-over area then it shall switch to power and force limiting mode and (r) resume normal operation after the opera-tor has left the shared hand-over area
The operator has entered the safeguarded area of the cell while the robot is moving or the welding process is active	(m) The welding machine, if running, shall be switched off and the robot shall switch to speed and separation monitoring mode. (r) Both robot and welding machine shall resume normal operation after the operator has left the area and acknowledged the safety notification provided via the operator interface
The operator is close to the welding spot while the robot is working and the welding process is active	(m) The welding machine shall be switched off and the robot shall perform a safety-rated monitored stop. (r)Both robot and welding machine shall resume normal operation or idle mode with a reset procedure after the operator has left the area and acknowledged the safety notification provided via the operator interface

†(m) mitigation, (r) resumption

estimation and *timely reduction of the impact on the human body*. For example, "speed and separation monitoring" requires *accurate and reliable sensors* (e.g. stereo-vision systems, laser scanners), particularly, when multiple safety zones are considered, to be able to control several modes of operation. Unknown faults, complex failure modes, and additional sensor inputs lead to *security* as a more than ever important prerequisite to assuring cobot safety, as pointed out by Kaiser et al. [28]. Finally, the overarching challenge is to *provide practical safety and performance guarantees of the behaviour of a manufacturing cell* based on realistic assumptions and leading to certifiable assurance cases.

3 Security

As indicated in Sect. 2, an important aim of research has been to facilitate the release of cobots from highly constrained caged environments to enable greater productivity [41]. Ensuring their security is a critical aspect. The security of robots has received less attention than safety concerns and cobot-specific security issues have received hardly any attention at all. We can learn from extant attempts to secure robots in general (though this area itself is still a challenge) and seek to identify new challenges for cobots.

Some security issues have direct safety implications, for example any attack that can gain control over the physical actions of a cobot may cause the robot to behave in a physically dangerous fashion, even to the point of launching a physical attack on the co-worker. Some, for example the leakage of personal data, may not have any obvious safety effect, but must be addressed in a manner that satisfies security related regulations (e.g. GDPR). Below we examine elements of security whose challenges have particular relevance to cobot security.

3.1 *Threat analysis*

Threat Modelling is the use of abstractions which aid the process of finding security risks. The result of this operation is often referred to as a threat model. In the case of robotics, threat modelling defines risks that relate to the robot and its software and hardware components while offering means to resolve or mitigate them [58].[4] In general terms, threat modelling is figuring out what might go wrong in terms of security with the system being built, and helps address classes of attacks, resulting in the delivery of a much more secure product [51].

An example of threat modelling used in various systems is STRIDE. Kohn-fekder and Garg [32] introduced the STRIDE threat modelling approach in 1999 and it has since become one of the most commonly used threat modelling methods. The STRIDE acronym stands for Spoofing, Tampering, Repudiation, Denial of Service and Elevation of Privilege. This methodology has recently been used in a document concerning the threat modelling of robots [52] (at the time of writing this is still a draft document).

Moreover, a novel way of identifying attacks/threat modelling is used in a paper by Trend Micro [39]. They identify five robot specific attack classes and their concrete effects, as depicted in Table 4.

[4] We would add that analysing the wider system in which a cobot exists is essential,since for example linking OT and IT—e.g. supporting accounting and other businessfunctions—is a common practice and greatly enlarges the threat landscape.

Table 4 Types of attacks to robots by Trend Micro [39]

Robot attack class	Requirements violated			Concrete effects
	Safety	Integrity	Accuracy	
1. Altering the control loop parameters	*V*	*V*	*V*	Products are modified orbecome defective
2. Tampering with calibration parameters	*V*	*V*	*V*	Damage to robot
3. Tampering with the production logic	*V*	*V*	*V*	Products are modified orbecome defective
4. Altering the user perceived robot state	*V*	X	X	Injuries to robot operator
5. Altering the robot state	*V*	X	X	Injuries to robot operator

3.2 *Review of Existing Security Approaches*

Security Policies. For a system to be *secure* there needs to be an unequivocal statement of what it means to be secure, what properties the system must uphold, what must happen and what must not happen. Such a statement is typically provided by a *security policy*. The policy may cover physical (e.g. access to a physical robot environment may be required to be via specific doors accessible only to authorised users), logical (e.g. access control policy), and procedural aspects (e.g. vetting of staff). For example, Zong, Guo and Chen [62] propose a policy-based access control system that enables permission management for robot applications developed with ROS (Robot Operating System). They introduce Android-like permissions so that each application in ROS can be controlled as to whether a certain resource can be accessed or performed.

Authentication. underpins security in various forms. It can be thought of as the verification of one or more claims made about system agents or elements to be valid, genuine or true. Most commonly, authentication applies to a user or a process that wishes to gain access to a system and its resources. Authentication may be required from the user to the system (*user identification*) or from the system to the user.[5] Dieber et al. [9] introduces a dedicated Authentication Server (AS) that tracks which ROS node subscribes or publishes in a topic. The aim of ROS is to provide a standard for the creation of robotic software that can be used on any robot. A node is a sub-part of this robotic software. The software contains a number of nodes that are placed in packages, two example nodes are the camera driver and the image processing software. These nodes need to communicate with each other [46]. The proposed AS also handles node authentication and produces topic-specific encryption keys.

[5] Though in many applications guarantees of authenticity of the system will be provided by physical and procedural means—for example, it may be practically veryhard to replace an authentic robot with a malicious or fake one.

User authentication is typically provided in three ways: (i) *something you know*, (ii) *something you have*, and (iii) *something/someone you are*. Passwords are the most common means of user authentication. A *claim* is first made by supplying a specific user identifier, or *user id* (the person supplying that user id is claiming to be the person associated with it by the system). Accompanying such a claim with a password recognised by the system is an example of (i). (ii) is also a major means of authentication. Possession of a 'token' (or physical identifier) is deemed to establish the link with an authorised user. Thus, Radio Frequency Identification Tags (RFIDs) may be programmed with the identities of specific users to whom they are given. Biometric approaches are examples of (iii). These make use of physical (biological) or behavioural features of a user. Fingerprints, voiceprints, retinal or iris images are all examples of biometrics that can be used to identify a individual.

We would observe that most user authentication is *one off*: the user authenticates at the beginning of a session and the privileges which go with such authentication prevail throughout that session. Problems ensue when a user walks away from a terminal and another user takes over, or when an authorised user deliberately hands over to an unauthorised one. A variety of attempts have been made to counter this using *continuous authentication*. Such approaches may prove of significant use where cobots are concerned. Hass, Ulz and Steger [16] approach this problem with expiring passwords and smart cards equipped with fingerprint readers, so only authorised users are authenticated to use the robot.

Intrusion Detection Systems (IDS). Intrusion Detection Systems are effective counter-measures for detecting attacks or improper use of systems by detecting any anomalies or inappropriate use of the host machines or networks. The concept of IDSs was first introduced by Anderson in 1980 [3]. There are three main intrusion detection approaches: signature-based, anomaly-based, and specification-based [34, 37, 50].

Signature-based systems recognise basic pre-packaged patterns of misbehaviour. For example, three consecutive failed attempts to log into a cobot management system might be regarded as a 'signature' of a potential attack. Another example includes the use of specific payloads in service requests recognised by the presence of specific bit strings in code. The bit string would form a signature of malfeasance.

An *anomaly-based approach* typically seeks to profile 'normal' behaviour in some way and measure how closely current behaviour is to that previously ascertained profile. Often, the underlying measure is a statistical one, and an anomaly is raised when current behaviour veers outside the historically established profile of normal behaviour. Some attributes lend themselves to monitoring in this way, for example many events incurred, such as the number of page faults or cache-misses. Often an anomaly-based system is required to classify current behaviour as 'normal' or 'anomalous'. Thus, it is no surprise that a wide variety of pattern classification approaches can, and have, been brought to bear on intrusion detection problems. A major claim for anomaly-based approaches is that they have good potential for detecting previously unseen attacks. The degree to which this is true is controversial; strictly speaking, they will flag as anomalous monitored behaviours that are not reasonably close to historic trends. However, some malicious attacks may have behaviours that are plausibly consistent with normal behaviours. If the symptoms of an attack are similar to the symptoms of another anomalous attack then the new attack will also be flagged as anomalous. Furthermore, most behaviours are unique if you look into the detail. Thus, such approaches require that, at some level of detail, attacks look different to 'normal' behaviour. This may or may not be true.

Specification-based intrusion detection [50], being a niche application, assumes that the behaviour follows a specified protocol. An intrusion is any deviation from this protocol. A variety of attacks on communication or application protocols can be detected this way. For example, a malicious application running on a cobot may attempt to make use of currently unused fields in a communication protocol to leak information. Such leaking will be detectable.

An example of the application of intrusion detection techniques to a robotic system is given in [26], where the authors implement an IDS using deep learning models. There are two components in the system proposed: one signature-based and one anomaly-based. The signature-based component is intended to detect misuse by identifying known malicious activity signatures, whilst the anomaly-based component detects behavioural anomalies by analysing deviations from the expected behaviour. The approach has two significant challenges: processing time and detection accuracy. In short, the IDS must be very fast with a low false positive rate. This is acceptable when using the signature-based model, but the anomaly detection component requires more time and is more prone to falsely indicating an attack.

In another instance, Fagiolini, Dini and Bicchi [12] propose an intrusion detection system that detects misbehaviour in systems where robots interact on the premise of event-based rules. They refer to misbehaving robots as intruders that may exhibit uncooperative behaviour due to spontaneous failure or malicious reprogramming. The proposed IDS is applied in an industrial scenario where a number of automated forklifts move within an environment that can be represented as a matrix of cells and macro-cells.

3.3 Application to Cobots

To the best of our knowledge, there would appear to be no published research specifically targeting security for cobots. However, the authors are currently undertaking work into the application of security concepts described above. In the next section we identify some of these and explain why the application to cobotic security may require some sophistication.

Security Policy. Since the rise of cobots is quite recent we should not be surprised to find no security policy work specifically targeting cobots. This area, however, is an intriguing and subtle one. A security policy will generally have to maintain some notion of privacy, as personal data may be stored in the system (e.g. a staff capabilities database). But the area is much more subtle than is usual for security policy. Issues of who should get access to what and under what circumstances, may have significant effect on co-workers. What is a reasonable policy on access when there is contention between co-workers? What is suitable ‘etiquette’ in such cases? Some data will be less privacy-sensitive and safety-relevant and the motivations for security policy elements will be different. A security policy and its implementation may be able to enforce elements of safety practice. For example, overly long shifts that break working time directives can be readily policed via security policy. Requirements for suitable skills or training can be enforced too. Ultimately, a great deal of reasonable co-working constraints boil down to access control of one form or another and such control can be specified via policy.

Authentication. Our current work is carried out in collaboration with industrial users of cobots. Workers generally operate a single robot cell and work on a continual basis, punctuated by scheduled breaks. It seems highly unlikely that the robots in these cells can be faked or replaced in some way without notice (though we accept that software could be maliciously updated). Consequently, it is only one-way user-to-robot authentication that needs to be considered. However, this authentication needs to be persistent, to account for changes in staffing. Some notion of initial authentication by traditional means, for example a standard password or token based approach, can be supplemented by an appropriate continuous authentication method. We are currently investigating the use of physical interaction properties (i.e. forces applied to the robot by the co-worker) and other biometrics (accelerometry applied to co-worker hands, eye movements and ECG monitoring) for user authentication.

The resources and capabilities made available by a cobot may vary across co-workers. Part of the appeal of cobots is, however, that they can be readily reprogrammed to do a variety of tasks. As co-workers may vary in their skills, experience and qualifications, we might quite legitimately allow certain tasks to be done only by co-workers with specific skills and training. A suitably configured system with access to the capabilities of specific users can easily enforce such policies. Thus, issues of cobot policy compliance can readily be handled via suitable access control, provided there is credible user authentication.

3.4 Challenges for Cobot Security

Many aspects of cybersecurity analysis and system development remain unaltered. But our work so far reveals that cobots also present either specific challenges in their own right, or else require considerable subtlety of thought to engage with cobot-specifics. Below we highlight areas we believe are in need of security focus:

Developing security policies and related elements that fully take into account the human, including both single co-worker cases and multi-co-worker cases. Policies must be sympathetic of user needs, well-being, and diversity issues. Crafting a policy for a mobile cobot working in different company domains is as much an exercise in occupational psychology as it is an issue in cyber-security. Particularly, in an industrial work setting, the co-working modus operandi and co-worker well-being must be considered. There are significant possibilities of couching a variety of working practice constraints in terms of security policy.

Development of practical templates for authentication requirements. Authentication needs will vary a great deal between deployment domains. Thus, a single worker collaborating with an industrial robot on a welding task might lend itself to initial password logon followed by continuous authentication by a variety of means (e.g. biometrics or token). Manufacturing with several on-site suppliers and changing external staffing is markedly different, perhaps requiring on-going speech recognition for continuous authentication. Guidance will need to be developed about the pros and cons of authentication approaches in specific cobot environments. These will include guidance regarding specific approaches in harsh electromagnetic environments. As an example of the latter, one of our industrial collaborators uses cobots to perform elements of arc-welding; the arc-ing could affect any authentication scheme based around wireless communications (including, e.g. distance bounding protocols).

Development of a plausible strategy for cobot forensics, which as far as we are aware, has received no attention in the literature. Digital forensics is largely concerned with the generation of credible evidence as to who did what, where and when (and also why). This is important not just for cyber-security reasons, for example to investigate security breaches, but also for health and safety reasons, where reconstructing an accident from event logs may be required.

Developing a credible IDS for use in cobotic environments. Again, we know of no work in this area. It is clear that domain specifics will need to be handled. Mobile cobots will present challenges over and above those for fixed cobots.

4 Cobot Co-Assurance

In previous sections, we have discussed how to reduce safety and security risks in cobots, and the challenges that arise. However, reducing risk within the in-dividual domains alone is insufficient to claim the acceptability of overall risk. There are multiple factors that can affect co-assurance and the confidence in the reduction of overall risk. These factors can be divided into two categories:

Socio-technical Factors—these are concerned with the processes, technology, structure and people required for co-engineering and the *co-assurance process*. Such factors play an important role, e.g. in complex information systems engineering [5], and so co-assuring socio-technical factors encompasses a large scope that includes organisational, regulatory, and ethical structures, management practices, competence, information management tools, *etc.*

Technical Factors—these are primarily concerned with the causal relationships between risk conditions and artefacts of the *engineered system*, *e.g.* accidents, hazards, attacks, threats, safety and security requirements.

Below, we consider these factors for cobots, and discuss the challenges that arise with respect to the interactions between safety and security risks.

4.1 Socio-technical Challenges

The many socio-technical challenges include:

Ethics. Interaction with co-workers is at the core of cobot use and ethics issues will inevitably arise. For example, ethical issues related to operator monitoring within the cobot will have to be resolved, for example by getting informed consent from co-workers.

Risk prioritisation. The principled handling of trade-offs across competing risks in cobots (safety, security, economic etc.) requires care in both 'normal' and exceptional circumstances. For example, when a security problem has been published with potential safety implications but a vendor patch is not yet available (possibly a common occurrence given the heavy use of commercially available components in cobots), what should be done immediately? Different actions may have different economic consequences and organisations are rarely equipped to judge such situations rapidly.

Risk representation and communication. To facilitate better decision making, the *profile* of various risks in particular circumstances needs to be made apparent, e.g. using structured risk analysis (Sect. 1.1). This may encompass risk attributes outside the usual safety and security (or further dependability).

Coordinating and integrating risk analyses. The system design may be modified, as the design and analyses progress, to incorporate necessary security and safety mechanisms. Since neither security nor safety are preserved under refinement (i.e. moving to a more concrete representation) the *specific way* a safety or security measure is implemented matters and so safety and security synchronisation will be required at what are believed to be at feasibly stable points in the development. Table 1 on page 4 summarises several candidate approaches to be adopted to prioritise, align, or integrate safety/security analyses.

Ensuring quality and availability of third party evidence. Cobots will usually integrate components from several vendors. Those components may draw on further vendors' products. Ensuring the quality or availability of relevant documentation throughout the supply chain to support safety and security assurance arguments, both individually and together, due to interdependencies, will often prove challenging. A vendor patch to a vulnerability may, for example, be available only as an executable.

Handling dynamic threat and impact landscapes. A cobot's threat landscape may vary greatly over time, e.g. when new vulnerabilities in critical components emerge, and this will likely affect safety. Rapidly assessing the safety impact of security flaws (and vice versa) is not a mainstream activity for manufacturing organisations and the skills required may well be lacking.

Resourcing safety and security interactions. There are many interactions between safety and security in cobots. Resourcing the investigation of such aspects will present significant challenges in the workplace.

Increasing automation of assurance. Cobot assurance is complex and dynamic, requiring repeated analyses and risk judgements to be frequently revisited. Increasing automation of assurance will be essential to ensure feasibility of any sound process. Table 1 (rightmost column) on page 4 lists approaches that, used individually or in combination, can be a basis for continuous alignment of safety and security assurance of cobots in manufacturing.

4.2 Technical Challenges

Some technical challenges in cobot co-assurance are given below:

Deadly conflicts. Cobots may be used in dangerous environments with human presence. Protecting against dangerous, and possibly deadly, conflicts is a challenge. For example, ensuring access control policy does not cause problems, i.e. by prohibiting essential data access for maintaining safety.

Consolidated audit policy. Logging events are an important, if somewhat prosaic, component of cobots. Many events *could* be logged but the potential volume is huge. We need to determine for what purposes we would want logged information and seek to craft a fitting policy. For security purposes, we might wish to access logs for determining situational awareness, for identifying specific behaviours, or for contributing towards evidence in a criminal court case. For safety purposes, we might wish to reconstruct an accident and its causes. We must be aware of the possibility of logging causing problems (e.g. affecting real-time performance of critical services or, even worse, causing deadlocks). Of course, the logs themselves must be protected from attack, as a successful security attack might destroy forensic log evidence. The topic of *attribution* is difficult in almost any system, however the complexity of cobots and their wider connectedness will make this aspect even hard for cobots.

Unifying anomaly detection. There are challenges regarding the *efficient* use of data to inform anomaly detection for safety and security as the data that can inform safety and security decisions may overlap. For example, the presence of sophisticated malware might reveal itself in degradation trends of operational performance. Thus, traditional health monitoring for safety and predictive maintenance reasons might highlight the presence of malware (one of Stuxnet's remarkable attack goals was to cause centrifuges to wear out more quickly [13]).

Reacting to compromise. What should be done when some elements of a cobot are perceived to be compromised in some way (e.g. via physical component failure or the presence of malware)? Having detected something is potentially 'wrong', management may wish to take appropriate action. This requires significant situational awareness (which may be difficult in cobots) and we need to be very careful to ensure the cure (response) is not worse than the problem, i.e. does not itself cause unpalatable security or safety issues. Some degree of degradation may be inevitable.

Ensuring resilience to direct access. Cobots will come into close proximity with humans providing an opportunity for interference or damage, both physical and digital. We must find useful and preferably common mechanisms to either resist attack or else make it apparent when it has occurred.

Coping under failure. The failure modes of a cobot and their consequences must be thoroughly understood. A degraded system may be compromisable in ways a fully operational system is not. Co-assurance approaches from process automation such as, e.g. FACT [47], could cross-fertilise cobot-specific co-assurance procedures.

Adversarial attacks. Systems whose activities are underpinned by machine learning or other AI[6] may be susceptible to *adversarial attack*. This is where the robot is fooled into malfunctioning in some way because it is presented with specially crafted input. This is a common worry from both a security and safety perspective.

Testing of cobots. Current robot testing approaches will need to be adapted to the new requirements of cobot settings. Moreover, the development of more rigorous testing approaches for cobots is essential.

Software and system update. Cobot and wider system update is a common concern and we need to ensure updates are made with appropriate authority, integrity, and that the updates do not cause unpalatable harm. From a security perspective, patches are a commonplace, however the safety implications of such changes are rarely considered. Other domains understand the importance of maintaining integrity of updates (e.g. updating of remote satellites or automotive software in the field). In some cases updates may need to be 'hot', i.e. at runtime. Again, a compromised update process will play havoc with safety and security. Furthermore, the effects of a host of run-time attacks are likely to have safety implications (e.g. attacks on a database via SQL injection, or similar).

4.3 Conclusion

With this chapter, we summarise general approaches to safety and security assurance for cobots. Further detail is provided in the form of risk-based assurance approaches for both safety and security of cobots (Sect. 1.1), illustrated by an example (Sects. 2.3 and 3.3). Finally, we describe the challenges that arise when attempting to assure and align the two quality attributes for a cobot. These challenges are not limited to technical factors about how and when to relate risk conditions between safety

[6] Cobots, as with other manufacturing technologies, will include increasingly complexartificial intelligence software.

and security, but include significant socio-technical factors that influence the co-assurance process itself. These challenges form the crux of the contribution of this work. Although no claims are made with regards to completeness, we believe that the challenges identified in Sect. 4 form the basis of a preliminary research roadmap, and that addressing them is essential for the safe and secure deployment of cobots in an industrial setting.

Acknowledgements Contribution to this chapter was made through the Assuring Autonomy International Programme (AAIP) Project—CSI:Cobot. The AAIP is funded by Lloyds Register Foundation.

References

1. Abdo H, Kaouk M, Flaus JM, Masse F (2018) A safety/security risk analysis approach of industrial control systems: a cyber bowtie–combining new version of attack tree with bowtie analysis. Comput Secur 72:175–195
2. Alami R, Albu-Schaeffer A, Bicchi A, Bischoff R, Chatila R, Luca AD, Santis AD, Giralt G, Guiochet J, Hirzinger G, Ingrand F, Lippiello V, Mattone R, Powell D, Sen S, Siciliano B, Tonietti G, Villani L (2006) Safe and dependable physical human-robot interaction in anthropic domains: State of the art and challenges. In: 2006 IEEE/RSJ international conference on intelligent robots and systems. IEEE. https://doi.org/10.1109/iros.2006.6936985
3. Anderson JP (1980) Computer security threat monitoring and surveillance. Technical report, James P. Anderson Company
4. Arkin B, Stender S, McGraw G (2005) Software penetration testing. IEEE Secur Priv 3(1):84–87
5. Bostrom RP, Heinen JS (1977) Mis problems and failures: a socio-technical perspective. Part I: the causes. MIS Quarterly, pp 17–32
6. Bouti A, Kadi DA (1994) A state-of-the-art review of fmea/fmeca. Int J Reliab Qual Saf Eng 1(04):515–543
7. Cook A, Smith R, Maglaras L, Janicke H (2016) Measuring the risk of cyber attack in industrial control systems. In: 4th international symposium for ICS & SCADA cyber security research 2016 (ICS-CSR). BCS eWiC. https://doi.org/10.14236/ewic/ics2016.12
8. Despotou G, Alexander R, Kelly T (2009) Addressing challenges of hazard analysis in systems of systems. In: 2009 3rd annual IEEE systems conference. IEEE, pp 167–172
9. Dieber B, Breiling B, Taurer S, Kacianka S, Rass S, Schartner P (2017) Security for the robot operating system. Robot Auton Syst 98:192–203
10. Dunjó J, Fthenakis V, Vı´lchez JA, Arnaldos J (2010) Hazard and operability (HAZOP) analysis. a literature review. J Hazard Mater 173(1–3):19–32
11. Eames DP, Moffett J (1999) The integration of safety and security requirements. In: International conference on computer safety, reliability, and security. Springer, Berlin, pp 468– 480
12. Fagiolini A, Dini G, Bicchi A (2014) Distributed intrusion detection for the security of industrial cooperative robotic systems. IFAC Proc Vol 47(3):7610–7615
13. Falliere N, Murchu LO, Chien E (2011) W32. stuxnet dossier. White Pap Symantec Corp Secur Response 5(6):29
14. Friedberg I, McLaughlin K, Smith P, Laverty D, Sezer S (2017) STPA-SafeSec: safety and security analysis for cyber-physical systems. J Inf Secur Appl 34:183–196
15. Gilchrist W (1993) Modelling failure modes and effects analysis. Int J Qual Reliab Manag

16. Haas S, Ulz T, Steger C (2017) Secured offline authentication on industrial mobile robots using biometric data. In: Robot world cup. Springer, Berlin, pp 143–155
17. Handbook FT, Roberts N, Vesely W, Haasl D, Goldberg F (1981) Nureg-0492. US Nuclear Regulatory Commission
18. Hayes B, Scassellati B (2013) Challenges in shared-environment human-robot collaboration. In: Proceedings of the collaborative manipulation workshop at HRI
19. He Y, Johnson C (2012) Generic security cases for information system security in healthcare systems. In: 7th IET international conference on system safety, incorporating the cyber security conference 2012, IET (2012). https://doi.org/10.1049/cp.2012.1507
20. Hinojosa C, Potau X (2018) Advanced industrial robotics: taking human-robot collaboration to the next level. Eurofound & European Commission, Brussels, Belgium
21. IFR (2018) Demystifying collaborative industrial robots. International Federation of Robotics, Frankfurt, Germany
22. ISO 10218 (2011) Robots and robotic devices—safety requirements for industrial robots. Standard, robotic industries association (RIA). https://www.iso.org/standard/51330.html
23. ISO/TS 15066 (2016) Robots and robotic devices—collaborative robots. Standard, robotic industries association (RIA). https://www.iso.org/standard/62996.html
24. Johnson N, Kelly T (2018) An assurance framework for independent co-assurance of safety and security. In: Muniak C (ed) Journal of system safety. International system safety society (January 2019), presented at: the 36th international system safety conference (ISSC). Arizona, USA
25. Johnson N, Kelly T (2019) Devil's in the detail: through-life safety and security co-assurance using ssaf. In: International conference on computer safety, reliability, and security. Springer, Berlin
26. Jones A, Straub J (2017) Using deep learning to detect network intrusions and malware in autonomous robots. In: Cyber sensing 2017, vol 10185. International Society for Optics and Photonics, p 1018505
27. Jones RH (1986) A study of safety and production problems and safety strategies associated with industrial robot systems. Ph.D. thesis, Imperial College
28. Kaiser L, Schlotzhauer A, Brandstötter M (2018) Safety-related risks and opportunities of key design-aspects for industrial human-robot collaboration. In: Lecture notes in computer science. Springer International Publishing, pp 95–104. https://doi.org/10.1007/978-3-319-99582-3_11
29. Kazman R, Klein M, Barbacci M, Longstaff T, Lipson H, Carriere J (1998) The architecture tradeoff analysis method. In: Proceedings. fourth IEEE international conference on engineering of complex computer systems (Cat. No. 98EX193). IEEE, pp 68–78
30. Kelly TP (1999) Arguing safety: a systematic approach to managing safety cases. Ph.D. thesis, University of York York, UK
31. Knowles W, Prince D, Hutchison D, Disso JFP, Jones K (2015) A survey of cyber security management in industrial control systems. Int J Crit Infrastruct Prot 9:52–80
32. Kohnfelder L, Garg P (1999) The threats to our products. Microsoft Interface, Microsoft Corp 33
33. Kordy B, Mauw S, Radomirovic´ S, Schweitzer P (2010) Foundations of attack–defense trees. In: International workshop on formal aspects in security and trust. Springer, Berlin, pp 80–95
34. Lee W, Stolfo SJ, Mok KW (1999) A data mining framework for building intrusion detection models. In: Proceedings of the 1999 IEEE symposium on security and privacy (Cat. No. 99CB36344). IEEE, pp 120–132
35. Leveson NG (2003) A new approach to hazard analysis for complex systems. In: International conference of the system safety society
36. Leveson NG (2012) Engineering a safer world: systems thinking applied to safety. Engineering Systems, MIT Press. https://doi.org/10.7551/mitpress/8179.001.0001
37. Liao HJ, Lin CHR, Lin YC, Tung KY (2013) Intrusion detection system: a comprehensive review. J Netw Comput Appl 36(1):16–24
38. Lichte D, Wolf K (2018) Use case-based consideration of safety and security in cyber physical production systems applied to a collaborative robot system. In: Safety and reliability–safe societies in a changing world. CRC Press, pp 1395–1401

39. Maggi F, Quarta D, Pogliani M, Polino M, Zanchettin AM, Zanero S (2017) Rogue robots: testing the limits of an industrial robot's security. Technical report, Trend Micro, Politecnico di Milano
40. Martı´n F, Soriano E, Cañas JM (2018) Quantitative analysis of security in distributed robotic frameworks. Robot Auton Syst 100:95–107
41. Matheson E, Minto R, Zampieri EG, Faccio M, Rosati G (2019) Human–robot collaboration in manufacturing applications: a review. Robotics 8(4):100
42. Nicolaisen P (1985) Occupational safety and industrial robots. In: Bonney Y (ed) Robot safety. IFS (Publications) Ltd., pp 33–48. https://doi.org/10.1007/978-3-662-02440-9
43. Papadopoulos Y, McDermid JA (1999) Hierarchically performed hazard origin and propagation studies. In: International conference on computer safety, reliability, and security. Springer, Berlin, pp 139–152
44. Pawar VM, Law J, Maple C (2016) Manufacturing robotics: the next robotic industrial revolution. UK-RAS White Papers, UK-RAS Network. https://doi.org/10.31256/wp2016.1
45. Piètre-Cambacédès L, Bouissou M (2013) Cross-fertilization between safety and security engineering. Reliab Eng Syst Saf 110:110–126
46. Quigley M, Conley K, Gerkey B, Faust J, Foote T, Leibs J, Wheeler R, Ng AY (2009) ROS: an open-source robot operating system. In: ICRA workshop on open source software, vol 3.2. Kobe, Japan, p 5
47. Sabaliauskaite G, Mathur AP (2015) Aligning cyber-physical system safety and security. In: Cardin MA, Krob D, Lui PC, Tan YH, Wood K (eds) Complex systems design & management asia. Springer, Berlin, pp 41–53. https://doi.org/10.1007/978-3-319-12544-2_4
48. Schmittner C, Ma Z, Smith P (2014) FMVEA for safety and security analysis of intelligent and cooperative vehicles. In: International conference on computer safety, reliability, and security. Springer, Berlin, pp 282–288
49. Schneier B (1999) Attack trees. Dr. Dobb's J 24(12):21–29
50. Sekar R, Gupta A, Frullo J, Shanbhag T, Tiwari A, Yang H, Zhou S (2002) Specification-based anomaly detection: a new approach for detecting network intrusions. In: Proceedings of the 9th ACM conference on computer and communications security. IEEE, pp 265–274
51. Shostack A (2014) Threat modeling: designing for security. John Wiley & Sons
52. Special interest group on next-generation ROS: ROS 2 robotic systems threat model. https://design.ros2.org/articles/ros2_threat_model.html
53. Srivatanakul T, Clark JA, Polack F (2004) Effective security requirements analysis: Hazop and use cases. In: International conference on information security. Springer, Berlin, pp 416–427
54. Sugimoto N (1977) Safety engineering on industrial robots and their draft standards for safety requirements. In: Proceedings of the 7th international symposium on industrial robots. IEEE, pp 461–470
55. Tang G, Webb P (2019) Human-robot shared workspace in aerospace factories. In: Human-robot interaction: safety, standardization, and benchmarking. CRC Press, pp 71–80
56. Tantawi KH, Sokolov A, Tantawi O (219) Advances in industrial robotics: from industry 3.0 automation to industry 4.0 collaboration. In: 2019 4th technology innovation management and engineering science international conference (TIMES-iCON). IEEE, pp 1–4 (2019)
57. Vanderborght B (2019) Unlocking the potential of industrial human-robot collaboration. Publications Office of the European Union, Brussels, Belgium
58. Vilches VM (2019) Threat modeling a ros 2 robot. https://news.aliasrobotics.com/threat-modeling-a-ros-2-robot/
59. Winther R, Johnsen OA, Gran BA: Security assessments of safety critical systems using HAZOPs. In: International conference on computer safety, reliability, and security. Springer, Berlin, pp 14–24 (2001)

60. Young W, Leveson NG (2014) An integrated approach to safety and security based on systems theory. Commun ACM 57(2):31–35
61. Yu M, Venkidasalapathy J, She Y, Quddus N, Mannan SM, et al (217) Bow-tie analysis of underwater robots in offshore oil and gas operations. In: Offshore technology conference
62. Zong Y, Guo Y, Chen X (2019) Policy-based access control for robotic applications. In: 2019 IEEE international conference on service-oriented system engineering (SOSE). IEEE, pp 368–3685

Task Allocation: Contemporary Methods for Assigning Human–Robot Roles

Niki Kousi, Dimosthenis Dimosthenopoulos, Sotiris Aivaliotis, George Michalos, and Sotiris Makris

Abstract Human Robot Collaboration (HRC) is considered as a major enabler for achieving flexibility and reconfigurability in modern production systems. The motivation for HRC applications arises from the potential of combining human operators' cognition and dexterity with the robot's precision, repeatability and strength that can increase system's adaptability and performance at the same time. To exploit this synergy effect on its full extent, production engineers must be equipped with the means for optimally allocating the tasks to the available resources as well as setting up appropriate workplaces to facilitate HRC. This chapter discusses existing approaches and methods for task planning in HRC environments analysing the requirements for implementing such decision-making strategies. The chapter also highlights future trends for progressing beyond the state of the art on this scientific field, exploiting the latest advances in Artificial Intelligence and Digital Twin techniques.

1 Introduction

The latest trend in today's globalized market expresses an evident shift from mass production to mass customization and personalization [1]. In fact, product personalization is the global trend driving the development of modern production systems [2]. However, EU factories struggle to follow the market demand for new products due to the performance limitations of current production systems [3]. Manufacturing sectors such as the automotive and white goods, maintain the serial production paradigm where the processes are predetermined and resources (robots and machines) are in fixed positions with pre-programmed operations. For this kind of setups a significant amounf of cost, time and effort is needed to introduce new product variants [4]. On the other hand, various EU industries such as the aeronautics and aerospace mainly rely on manual assembly due to high complexity on integrating robot based solutions in the existing workplaces [5]. Nevertheless, fully manual production imposes strain

N. Kousi · D. Dimosthenopoulos · S. Aivaliotis · G. Michalos · S. Makris (✉)
Laboratory for Manufacturing Systems & Automation, Patras University, Patras, Greece
e-mail: makris@lms.mech.upatras.gr

M. I. Aldinhas Ferreira and S. R. Fletcher (eds.), *The 21st Century Industrial Robot: When Tools Become Collaborators*, Intelligent Systems, Control and Automation: Science and Engineering 81, https://doi.org/10.1007/978-3-030-78513-0_12

to human workers since they need to lift, carry and handle heavy loads in several occasions. The generated strain, in the long term, may create musculoskeletal disorders (MSD) leading to high rate of absenteeism in work and thus frequent production down times [6].

Driven by the abovementioned challenges, European robotics industry is moving towards enabling hybrid, more flexible production systems aiming at: (a) making the factories of the future more cost effective and (b) restoring the competitiveness of the EU manufacturing industry [7]. In this context, the latest trends in EU research foster the deployment of HRC applications [8]. New types of factories exploiting the capabilities of multiple resources such as human operators and mobile and/or stationary robot assistants are emerging [9]. The deployment of this new production paradigm raises a set of technical challenges oriented around: (a) the safety related issues that occur from the HR co-existence in fenceless environment [8, 10–13], (b) the industrial requirement for easy robot programming techniques so for non-robot experts to be able to efficiently re-program the robots [14–17], (c) the need for new, intuitive human side interfaces for seamless interaction and communication with the robot system [18–21], and (d) the call for scalable mechanisms for monitoring and control in HRC execution including human operators in the execution loop [8, 22–24].

The above are technical aspects that mainly concern the low—level HRC process execution focusing on robot control methods and user interfaces development. However, to enable a full scale assembly scenario, it is critical to implement methods that will answer the question *of how to plan the activities between the available humans and robots in an optimal way* [25] and *how to re-organize the work when changes are needed or unexpected events occur* [7].

The research topic of Task Allocation has been extensively investigated over the last decades for answering those questions [26]. The core part of these methods employs intelligent decision-making tools that analyze the received input for generating optimal task assignments to the human robot collaborative teams. On top of that, the latest trends in EU research foster the integration of these tools with digital instances of the factory that are dynamically updated based on the status of the physical factory. This enables the continuous monitoring of the execution and the dynamic work re-organization to adapt in variations of the process. This task planning lifecycle (Fig. 1) raises a group of specific technical challenges that are analyzed in this chapter.

The chapter structure is organized as follows: Sect. 2 provides a thorough investigation on the current approaches implemented for the task allocation in HRC environments, including scientific methodologies as well as industrial case studies. Section 3, based on the identified limitations of existing methods, highlights the future trends towards deploying more robust and closer to industrial standards solutions. Finally, Sect. 4 summarizes the discussed points drawing the conclusions as well as an outlook on the next steps.

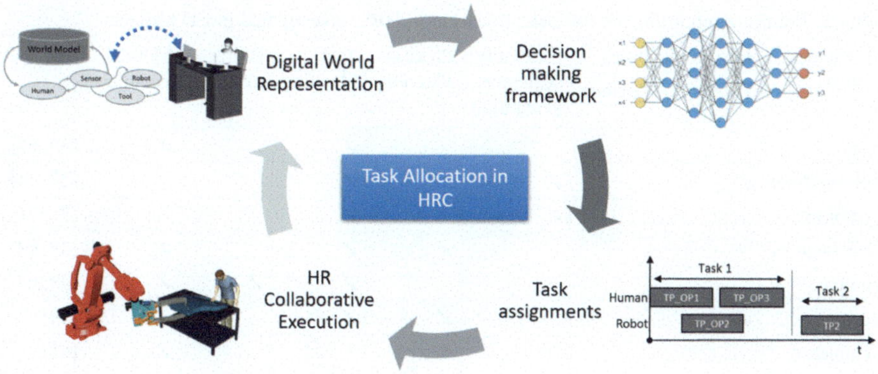

Fig. 1 Task Allocation in HRC environments

2 Current State

The previous section provided an introduction on the topic of task allocation in human and robot resources as well its importance for exploiting the collaborative teams' capabilities to their full extend. This chapter is dedicated to providing a thorough insight into the current state of the art regarding the technical aspects involved in deploying a task allocation decision-making system. Table 1 provides an overview of the core elements that are discussed in the following subsections.

2.1 Planning Requirements for HRC Environments

Task allocation in manufacturing can be defined as the process of distributing responsibilities and workloads in different production units in an optimal way [26]. In

Table 1 Overview of task allocation core elements

Task Allocation aspects	Approaches	References
Planning requirements	• Layout planning • Assembly sequence and Process planning • Task planning • Action/Motion planning	[27–29] [30–34] [7, 25–27, 35–43] [7, 37, 43–45]
Modeling	• Hierarchical model • Petri Net modeling • STRIPS-like representation • Knowledge manager-ontologies	[7, 25, 27, 28, 38, 43, 46] [47, 48] [49, 50] [11, 51]
Decision-making	• Operations research methods • Artificial intelligence methods • Simulation-based methods	[26, 50, 52] [14, 28, 38, 53–55] [27, 28, 44, 47, 56–58]

Table 2 Planning requirements for task allocation in HRC investigated in EU projects

EU Funded projects	Layout planning		Assembly sequence generation	Process Planning	Task planning	Path planning	Motion Planning	Ref
	Fixed	Dynamic						
X-act					✓		✓	[35, 61]
Robo-partner	✓		✓		✓			[27, 54]
Symbio-tic			✓		✓			[7, 62]
Four by three				✓	✓		✓	[45, 63]
Thomas		✓			✓	✓	✓	[43, 64]
Recam	✓			✓	✓			[29]
Symplexity							✓	[65]
Sharework				✓	✓		✓	[66, 67]
Sherlock		✓	✓		✓	✓	✓	[68]
Factory in a day					✓		✓	[69]
Pick place				✓	✓	✓	✓	[70]

traditional, fully automated, mass production systems, where the robots are operating behind fences, the working environment is rendered highly predictable and the workload is pre-planned in a deterministic way [59]. However, when it comes to the co-existence of human operators and multi-purpose robot workers, the complexity in the planning process increases, involving multiple decisions to be made [28]. The questions raised range from the organization of the layout so to effectively support HRC, to the adaptive motion planning of robots so to ensure collision free and human aware trajectories generation in real time.

In an attempt of the research community to unlock the potential of HRC on industrial environments, the topic of task allocation has been investigated through multiple EU funded projects [60] trying to answer the above questions. Table 2 provides an overview of the identified planning requirements for HRC based on the industrial requirements posed in a group of the most relevant European funded projects.

When employing multiple types of resources in a single production system, including stationary and / or mobile robots and human operators, it is critical to ensure that the workplace will be able to facilitate the collaboration between them [54]. To this extent, the workplace layout design has been gaining a lot of attention [71] regarding the optimization of HR collaborative environments against criteria related to human ergonomy and robot reachability aspects [27]. The fixed layout planning provides the optimal locations for stationary passive and active resources [28] while the dynamic layout planning refers to the real time efficient re-location of mobile robots which dynamically changes the working environment [43].

Nevertheless, to be able to validate a generated layout it is critical to identify which tasks will be performed by the robots and which ones by the humans. Considering this, the combination of layout planning and workload allocation has emerged in the last years [25]. To automate the planning process in a wider extent, the investigation starts from methods to automate the assembly sequence generation [27, 62] in order to generate process plans [31] that can be executed by HR teams. The target is to analyse part, product and resource characteristics to automatically generate the list of required assembly tasks as well as the human operators' and robot resources' suitability for each of these tasks [63]. The next step is to use this information for distributing the tasks to the resources in an optimal way against criteria such as human ergonomy, investment cost and production efficiency [54]. In parallel, in an environment where robots are working with humans in free space, it is important to ensure collision free trajectories for the robot resources. The integration of the motion planning for stationary resources and path and motion planning for mobile manipulators in the task planning process has been proved beneficial in terms of selecting the optimal HRC task allocation [43, 45].

2.2 Modeling Aspects of the Planning

An important aspect of developing an automated task allocation system is the design of a unified modeling approach incorporating the particularities of the involved multi-purpose resources as well as the process specifications and characteristics [72]. In order to model the industrial environment, a knowledge manager composed of ontologies to represent the environment of the robot can be used [51]. Ontologies are very reusable and flexible at adapting to dynamic changes, eliminating the need to re-compile whenever a change is needed. Through ontologies, industrial scenarios can also be modelled, in which robots collaborate with humans, in terms of robot behaviors, tasks that they can accomplish, and the objects they can manipulate/handle from an interaction point of view. A similar approach [11] proposes a method to combine semantic knowledge representation with classical AI techniques to build a framework that can assist robots in task planning at the high symbolic level. A semantic knowledge ontology is developed for representing the environmental description and the robot primitive actions, while a recursive back-trace search algorithm handles the task planning process.

STRIPS-like operators are also used because of their simplicity and their capability to provide a compact representation of a domain [49]. A STRIPS-like planning operator (PO) is composed of two parts. The first part is the precondition part, which is a logic clause with the required attribute-values to yield changes that need to be made. The other part of a PO, called effect part, specifies attribute-values based on the precondition part because of the execution of an action. With the generated Pos, the robot is able to generate plans that permit to cope with all the restrictions of its environment [50].

A model for assembly plan generation and selection for human and robot coordinated (HRC) cell, called Dual Generalized Stochastic Petri Net (GSPN) model, has been suggested by [47]. Based on this model, the Monte Carlo method was used to study the time relevant cost and payment cost for different alternatives, and Multiple-Objective Optimization (MOOP) related Cost-effectiveness analysis is

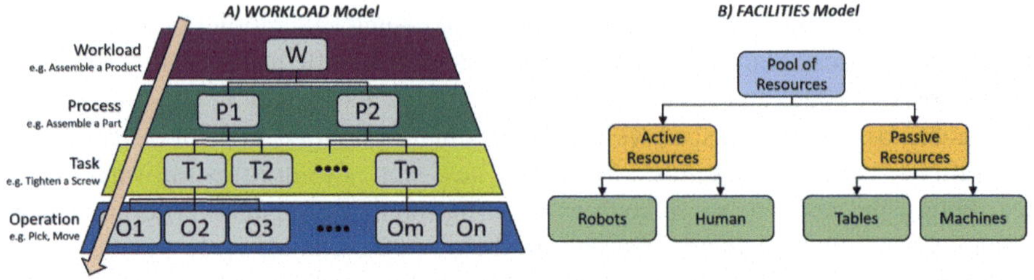

Fig. 2 Hierarchical **a** Workload model, **b** Facilities model [28]

adopted to select the optimal ones. However, similar to the previous approaches, only the existence of stationary robots is considered. Also, in [48] a plan generating tool for robotic applications is described. The tool, called ROPES (robot plan generating system), is based on modeling by a special type of Petri nets called predicate/transition (Pr/T) nets and follows an analysis technique of the Pr/T nets, called the T-invariant method, to generate robot plans.

Another proposed modeling framework follows the hierarchical modeling for the available resources. Under these principles, the resources have been classified in active and passive resources. Human and robotic operators represent the active resources, while the passive resources are actors such as assembly tables and compression machines. This hierarchical modeling is also used to model the HRC workload (Fig. 2). Based on these principles, the workload is decomposed into several processes. A process involves a set of tasks (high-level activities), which constitute a group of operations (low-level robot activities) [28]. This approach provides the flexibility of performing the task allocation decision in the level of activities, either in Task or Operation level that fits best to the requirements of the investigated case study [7].

2.3 Decision Making Approaches

A fundamental aspect of designing an optimal manufacturing system is the decision-making process. In this section, a group of relevant decision-making methods and tools is presented, being divided in three main categories: operations research, artificial intelligence, and simulation [1].

2.3.1 Operations Research Methods

Mathematical programming can be defined as the use of techniques for optimizing (minimizing or maximizing) a given objective function of a number of decision variables [1]. A decision-making algorithm for distributing the work in industrial assembly systems is implemented in [73]. In this approach, human operators handle the needed changes in the process plan. Thus, the additional investment to the system could be reduced while benefiting from its cost reduction effects. In this case, the robotic operators take over processes that do not need a large amount of additional investment to adapt to the product changes. However, this method cannot address the requirements of unstructured/ dynamic environments.

A decision-making framework, integrated into a planner to enable task allocation between human and robotic resources, based on STRIPS-like operators [49] has been described by [50] and [52]. Under this approach, a symbolic definition of the initial status is generated. The initial situation, combined with specifications about the target, is used to search for plans to match this target.

A different approach uses the evaluation of attributes that are assigned to each assembly task [26]. These attributes are divided into factors from 0 to 1, where 1 defines tasks that can be fully automated and 0 defines tasks that cannot be automated. The automation value that emerged for each task defines whether the task can be handled by human operators or robots.

2.3.2 Artificial Intelligence Methods

The field of Artificial Intelligence can be defined as the use of techniques that maximize the usefulness of computers by exploiting their intelligence. A dynamic AI-based task planning system for HRC is proposed in [38]. This approach uses Artificial intelligence methods to cope with the uncertainty that is inserted into the system by the active presence of humans and to adapt the generated task plans in a dynamic way to the real-time behavior of the operators.

Neuro-cognitive mechanisms [53], have been implemented and used in industrial applications. Under this approach, an experiment-based model is used for action preparation and decision-making in HRC tasks. From this perspective, the coordination of actions and goals among the partners is considered as a dynamic process that integrates contextual cues, shared task knowledge, and predicted the outcome of others' motor behavior. Reinforcement learning has been used in [74] to predict the distribution of the priority of move selections and whether a working sequence is the one resulting in the maximum of the HRC efficiency.

A search-based framework has been proposed to generate multiple alternative production schedules by allocating each task to a suitable active resource and evaluates them using multiple criteria to select the best plan. An HR task allocation plan is achieved through a set of decisions and the evaluation of multiple criteria, such as the human operator's needs [36], human ergonomics [25] and manufacturing system performance [7, 43]. A set of decision steps has been considered that allows for the evaluation of the generated task assignments to the resources. These approaches represent the task allocation problem in a form of search tree, as indicatively illustrated in Fig. 3, and use heuristic functions for the ranking of alternatives at each brach of the tree [7].

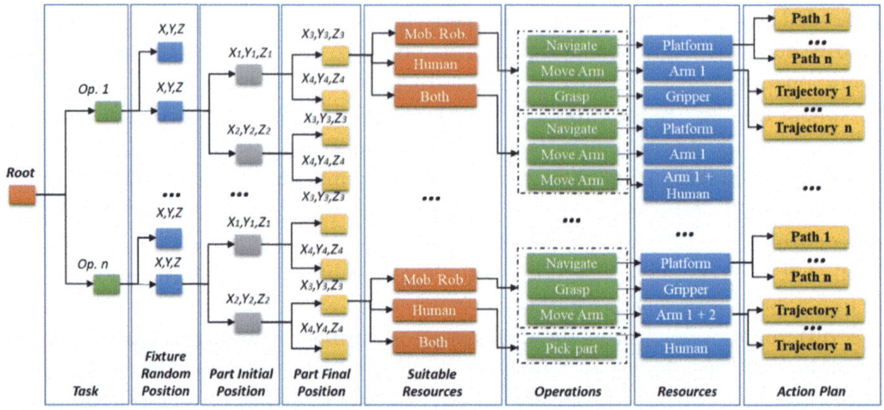

Fig. 3 Indicative search tree formulation of the HRC Task Allocation problem [64, 75]

2.3.3 Simulation-Based Methods

Computer simulation is a wide term that describes a variety of computer software, which simulates the operation of a manufacturing system. The input of such computer simulation software tools are decision variables, which specify the main characteristics of a manufacturing system. A simulation-based approach for task allocation is presented in [56]. Under this approach, the planning system covers all necessary aspects from identifying the manual until the final evaluation of the plan. The data needed for the task allocation are obtained via simulations. Also, after the task allocation process, realistic simulations are used to validate the feasibility of the generated task plans.

A different approach presented in [27] uses CAD models to generate alternative schedules. Under this method, data about the part's characteristics are extracted from the existing CAD models of the product. After defining the product's main characteristics, the assembly sequence is defined, while an intelligent algorithm is used to generate alternative task allocation plans.

Simulation tools are used for motion planning of mobile robots in [44]. Under this approach, 2D or 3D simulations are used to provide the information needed for safe and accurate motion planning, while a Mapping sub-module is responsible for providing data related to the structure of the shop-floor that are needed for the navigation of the robot.

Simulation-based methods can also be used for validating and evaluating the emerging Task plans. Discrete event simulation methods are commonly used for task planning validation in collaborative scenarios, as they can prove to be a helpful tool in quantifying the processing times for both the operator and the robot [58]. These techniques combined with multi-objective optimization methods for assigning tasks to resources have also been implemented [47].

A different Task plan validation method has been implemented in [28, 54]. Under this approach, there is an intelligent search-based algorithm [55] that executes the assignment procedure. This decision-making framework is responsible for the alternatives' generation and the evaluation and validation of these alternatives are based on multi-criteria values retrieved from 3D simulations. A similar approach in [57] uses 3D simulations for optimizing and validating layout planning and task allocation among the human operators and the robots in decentralized manufacturing systems (DMS).

The aforementioned methods are limited only in offline planning and, therefore, they lack the ability to perform online changes in the generated task plans during the process. A framework for planning shared human–robot activities in a dynamic way within an industrial environment has been developed in [7], where a multi-criteria decision-making module (following the hierarchical modeling approach) is being implemented for offline scheduling and online re-allocation of the remaining tasks in case of abnormal events.

2.4 Industrial Examples

As mentioned in Sect. 2.1, the task allocation problem attracted the interest of multiple EU funded research projects over the last years. Through these research projects, decision-making applications for task planning and work re-organization have been implemented in multiple human–robot collaboration use case demonstrators. Indicative examples concerning different industrial sectors are provided below.

Automotive: The automotive industry has exhibited a high interest in adopting HRC applications in the production facilities. This has motivated extensive research in the field of task planning around use cases driven by this manufacturing sector's requirements.

In [27], the HRC assembly of the rear axle of a passenger vehicle is being investigated employing a high payload industrial robot and one human operator. A 3D simulation-based decision-making tool has been deployed generating an optimal layout embedded with an effective task allocation plan against human ergonomy and manufacturing performance criteria.

The benefit of employing dual arm robot capabilities working along with humans has been analyzed in [35] using a search-based task planning method to optimally allocate the tasks for a vehicle's dashboard assembly against resource utilization and cycle time related criteria.

The task planning in collaborative assembly of a turbocharger has been tackled by [7]. The method was implemented for re-scheduling of the plan when a resource breakdown occurred focusing on minimizing the products' flow time.

In [76] a skill-based approach has been implemented in the assembly process of a set diesel engine proving that through the human robot interaction the human active time may be reduced up to 30% of the total manual assembly operation and human robot collaboration enable heavy tasks' assignments to industrial robots.

Aeronautic: A task planning approach for the assembly of aircraft fuselages is introduced in [77]. This approach is based on the comparison of robot and human skills for performing several tasks involving joining processes.

White goods: A task planning method was implemented for optimizing the sealing process of a refrigerator performed in an HRC cell [25]. The discussed HRC cell employed two human operators and two robot medium payload industrial manipulators (COMAU). The implemented task planner, integrated with Process Simulate 3D simulation software by SIEMENS [78], generated different alternative layout plans combined with optimal task plans. The different alternatives were evaluated based on user defined criteria such as human ergonomic analysis, and total cycle time's estimation from the simulated process. The selected layout and task plan assignment is characterized by 18% less floor space, 45% less Human muscle strain required and shorter cycle time by 2 s than the second alternative in the ranking.

Linear actuator production: A skill-based task planning approach for the assembly of linear electric actuators is presented in [26]. The task planning result shown that above 70% of the assembly process investigated may be automated but the minimization of the available resources idle time should be considered as well.

Polyethylene blown film production: A 3D simulation-based decision-making tool for the collaboration of mobile units and human operators inside the production line of polyethylene blown films presented in [44]. The proposed algorithm was based on a digital twin of the factory designed in Visual Components [79] to provide the infrastructure for generating the path plans of the mobile robot.

Injection machine wax pattern production: A dynamic task planning algorithm for the assembly and disassembly operations of a metal die through the collaboration of human and robot resources is presented in [63]. The proposed planner is responsible for meeting the production requirements and for handling the temporal uncertainty of the execution system to supervise the human operations.

3 Future Trends and Transformations

The investigation presented in the previous section shows that the research community has posed a considerable effort over the past years to analyse the industrial requirements for developing decision-making tools to support production designers and engineers to incorporate HRC systems in manufacturing. However, the existing approaches exhibit limitations to cover the challenge of continuous monitoring of the execution and real time workload re-organization in order to adapt in: (a) the continuously changing production requirements, (b) in the competences and skills of the operations, (c) unforeseen events such as machine breakdown. To be able to cover these requirements, future trends are expected to integrate the latest advances and capabilities provided by the Digital Twins, Artificial Intelligence and High-Performance Computing. This topic is discussed in the following subsections.

3.1 Digital Twins and Task Planning

In the context of Industry 4.0 the use of virtual models of the production system, the so-called Digital Twins, has been emerging. These are commonly used for the surveillance, evaluation, planning and monitoring of a production shop floor. As presented in [80] and [81], digital twins may be defined as virtual representations of the physical factory dynamically updated with real time data coming from sensors installed in the actual shop floor. A digital twin for testing human–robot collaboration scenarios including mobile robots for maintenance and assembly operations in the energy and the automotive industry is presented [82] and [64] respectively. The use of a digital twin for task planning and re-planning operations in human robot collaborative environments is pointed out in several scientific approaches. As presented in [43], a digital twin can have the main role in the evolution of decision-making mechanisms. A search-based task planning algorithm is capable to produce feasible task schedules tested in a virtual environment which will be evaluated by a multi-criteria decision-making module using data from their simulated execution. A digital twin-based decision-making tool is presented in [44] focusing on the collaboration of mobile units and human operators. An online decision-making and optimization approach without interrupting the physical production process is introduced in [83].

Even though several digital twins have been designed for different industrial use cases, many limitations are detected and require further developments from having accurate simulation models and task planning results:

- Lack of controllers able to control the motion of both physical and simulated resources. It is very difficult to have a simulated human motion close enough to a real human motion due to the different personalization and characteristics of each operator. Also, several platforms support controllers able to manipulate both the physical and simulated robotic resources using the same interface. However, these are quite new and there are only a few robot platforms supported.
- Limited simulation of objects' physics characteristics (type of surface, weight, etc.). In this way, the simulation of these objects' manipulation by the operator and the ergonomic analysis of his motion are inaccurate.
- Lack of safety sensors' drivers for simulation tools. Safety sensors' developers may provide a simulation driver for their products but usually these drivers are not compatible with a simulation tool supporting human and robotic resources.
- Lack of sensors drivers able to predict any mechanical failure of robotic resources. Nowadays, it is very common to install different types of sensors on the robotic resources to predict mechanical failures and accordingly maintain them.

Despite the fact that there are many simulation tools for robotic resources [84–86], their limitation to support either human operators or sensing devices makes these tools unable to be used in an HRC digital twin. Simulation models should be further tested in different industrial use cases to be more stable and ready to be included in the task planning operations. It is expected that through the IoT development, multiple sensors with wireless connection will be added in the libraries of different simulation tools allowing the designing of big-sized production lines' digital twins [87]. In parallel with future High Performance Computing (HPC) ability to receive and analyze big data [88, 89], it is expected that task planning systems will be able to be deployed in bigger scale industrial applications. In addition, HPC will be able to address the existing limitations in real time decision-making, facilitating online execution monitoring and task replanning close to industrial requirements.

3.2 AI for Resource Suitability Calculation

Having both human and robot resources working in shared workstations creates the need for methods to analyze the suitability and the capability of the available resources to execute each task. The possible candidates able to execute a task inside a human–robot collaborative environment might be a human operator, a robotic resource or a combination of them depending on the type and the complexity of the task [7]. Task planning approaches based on the capabilities of robot resources are presented in [90] and [91]. According to [92], the workspace and the reachability of a robot arm may be taken under consideration during the task planning process. Another approach taking into consideration the capabilities of both human and robot resources (robot reach, robot payload, human ergonomic analysis, etc.) and product's characteristics (weight, size, flexibility, etc.) is presented in [27]. In addition, the previous and next task to be executed based on product's specifications may be taken under account through the scheduling process [35]. All these characteristics may be used in case task of the initial planning or emergency re-planning operation too. Depending on the efficiency of the robot and human resource to execute a specific task, the task sequence might be changed by exchanging the responsible resource for executing this task [42].

Until now, these algorithms have been tested in laboratory level and further testing and modifications are required for their insertion in physical manufacturing lines. The limitations include:

- No generic approach to identify the suitable resources for the execution of each action. Resources' capabilities based on the configuration of the resources (mobile or stationary, single or multi-arm) and several environmental parameters (temperature, light brightness, etc.).
- Lack of simulation controllers for human resources. The suitability of human resources for different tasks of a production line based on the ergonomic analysis of human operations from a simulation model. Due to the different personalization and physical characteristics of each operator (weight, strength, etc.), it is difficult to create an accurate generic human simulation model and the ergonomics analysis might be inaccurate.

It is expected that in the next years, new enhanced methods will be implemented for modeling human and robot resources providing more information about these resources' abilities. In addition, the real time status and behavior may be monitored in order to detect online changes on the human operators' preferences as well as machines' capabilities. In this way, the tasks can be re-organized in order to adapt in the resources' current state. In parallel, the robotics community makes significant efforts to create unified models to be used for resources' description in order to support this activity [93, 94].

3.3 Ergonomy and Safety Considerations in Task Planning

The HRC systems of the future are envisioned as human-centered collaborative robotic cells that will exploit human capabilities such as dexterity and flexibility from the one hand, while from the other hand they will improve the quality of the working environment (how the characteristics of work impact on worker wellbeing and satisfaction), by reducing human's physical load and stress. This goal could be achieved through automating the non-value adding activities and allocating less physically demanding tasks to the human operators.

There are several simulated ergonomic analysis tools (OWAS, NIOSH, NORDIC body map, etc.) [95] used for task planning alternatives' evaluation in laboratory level [25], but more detailed ergonomic analysis tools dedicated in the human motion analysis [96] is expected to be integrated into the task planning process over the next years. From the safety point of view, a limitation of the current task planning approaches is the lack of tools to adapt the safety design of a production process to the alternative task plans. New methods that will incorporate the dynamic update with the risk assessment in the planning process are expected to provide more robust and effective solutions, being able to meet the safety regulations' (ISO 10,218–1, 10,218–2 and ISO/TS 15,066:2016) requirements.

4 Conclusion

The latest trends in manufacturing foster the deployment of reconfigurable assembly lines in order to accommodate personalized products, increase production rate, and decrease cycle time and costs. To meet these challenges, human operators and robot resources are aspired to work together under collaborative workspaces. To get the most out of their synergy effect, methods for effectively planning and allocating the tasks are of high importance. Driven by this need, this chapter presented currently used methods for task planning as well as some future trends regarding this topic.

The direct benefits deriving from the effective distribution of tasks as well as the optimal HRC workplace involve the support of production managers in planning of shared assembly tasks, the optimal exploitation of the available space in the factory as well as the reduction of the time needed to reconfigure the production system. It is evident in the literature that great advances have been achieved during the last decade in the deployment of HRC task planning systems and validation through industrial cases. Nevertheless, several further steps need to be made in order for the technology to be ready for actual industrial use:

- **Incorporation of powerful Digital Twins** that will provide real time information on the shopfloor status by seamlessly integrating multiple sensing and control devices,
- **Deployment of powerful AI based decision-making systems in HPC machines** that will allow the real time analysis of big amount of raw sensor as well as simulation data,
- **Integration of detailed ergonomic analysis tools** so for the planning process to optimize human operators job quality and healthiness during HRC operations,
- **Integration with dynamic risk assessment tools** in the decision-making process so to ensure human operators' safety during real time re-organization of the workload as well as the compliance of the generated task plans to safety regulations.

In the era of the 4th industrial revolution that is accompanied with significant progress in Robotics and AI, the exploitation of human robot collaboration potential in industry is closer to reality than ever. Future systems are expected to position human operators in the centre of the system aiming to optimize their working conditions and well-being under the factories of the future.

References

1. Chryssolouris G (2006) Manufacturing systems: theory and practice. Springer-Verlag, New York
2. Michalos G, Kousi N, Makris S, Chryssolouris G (2016) Performance assessment of production systems with mobile robots. Proced CIRP 41:195–200. https://doi.org/10.1016/j.procir.2015.12.097
3. Michalos G, Makris S, Chryssolouris G (2015) The new assembly system paradigm. Int J Comput Integr Manuf 28:1252–1261. https://doi.org/10.1080/0951192X.2014.964323
4. Michalos G, Makris S, Papakostas N, Mourtzis D, Chryssolouris G (2010) Automotive assembly technologies review: challenges and outlook for a flexible and adaptive approach. CIRP J Manuf Sci Technol 2:81–91. https://doi.org/10.1016/j.cirpj.2009.12.001
5. Kousi N, Michalos G, Aivaliotis S, Makris S (2018) An outlook on future assembly systems introducing robotic mobile dual arm workers. Proced CIRP 72:33–38. https://doi.org/10.1016/j.procir.2018.03.130
6. Hall A, Tiemann M, Herget H, Rohrbach-Schmidt D, Seyfried B, Troltsch K, Inez Weller S, Braun U, Leppelmeier I, Martin P, Dorau R (2012) BIBB/BAuA-Erwerbstätigenbefragung Arbeit und Beruf im Wandel, Erwerb und Verwertung beruflicher Qualifikationen
7. Nikolakis N, Kousi N, Michalos G, Makris S (2018) Dynamic scheduling of shared human-robot manufacturing operations. Proced CIRP 72:9–14. https://doi.org/10.1016/j.procir.2018.04.007
8. Michalos G, Makris S, Spiliotopoulos J, Misios I, Tsarouchi P, Chryssolouris G (2014) Robo-partner: seamless human-robot cooperation for intelligent, flexible and safe operations in the assembly factories of the future. Proced CIRP 23:71–76. https://doi.org/10.1016/j.procir.2014.10.079

9. Gkournelos C, Kousi N, Christos Bavelos A, Aivaliotis S, Giannoulis C, Michalos G, Makris S (2019) Model based reconfiguration of flexible production systems. Proced CIRP 86:80–85. https://doi.org/10.1016/j.procir.2020.01.042
10. Michalos G, Makris S, Tsarouchi P, Guasch T, Kontovrakis D, Chryssolouris G (2015) Design considerations for safe human-robot collaborative workplaces. Proced CIRP 37:248–253. https://doi.org/10.1016/j.procir.2015.08.014
11. Ji Z, Qiu R, Noyvirt A, Soroka A, Packianather M, Setchi R, Li D, Xu S (2012) Towards automated task planning for service robots using semantic knowledge representation. In: IEEE 10th international conference on industrial informatics. IEEE, Beijing, China, pp 1194–1201
12. Bicchi A, Peshkin MA, Colgate JE (2008) Safety for physical human-robot interaction. In: Siciliano B, Khatib O (eds) Springer handbook of robotics. Springer, Berlin Heidelberg, Berlin, Heidelberg, pp 1335–1348
13. Kulić D, Croft EA (2006) Real-time safety for human–robot interaction. Robot Auton Syst 54:1–12. https://doi.org/10.1016/j.robot.2005.10.005
14. Makris S, Tsarouchi P, Surdilovic D, Krüger J (2014) Intuitive dual arm robot programming for assembly operations. CIRP Ann 63:13–16. https://doi.org/10.1016/j.cirp.2014.03.017
15. Matthaiakis SA, Dimoulas K, Athanasatos A, Mparis K, Dimitrakopoulos G, Gkournelos C, Papavasileiou A, Fousekis N, Papanastasiou S, Michalos G, Angione M, Makris S (2017) Flexible programming tool enabling synergy between human and robot. Modena, Italy
16. Surdilovic D, Yakut Y, Nguyen T-M, Pham XB, Vick A, Martin-Martin R (2010) Compliance control with dual-arm humanoid robots: design, planning and programming. In: 2010 10th IEEE-RAS international conference on humanoid robots. IEEE, Nashville, TN, pp 275–281
17. Schmidt B, Wang L (2013) Contact-less and programming-less human-robot collaboration. Proced CIRP 7:545–550. https://doi.org/10.1016/j.procir.2013.06.030
18. Makris S, Karagiannis P, Koukas S, Matthaiakis A-S (2016) Augmented reality system for operator support in human–robot collaborative assembly. CIRP Ann 65:61–64. https://doi.org/10.1016/j.cirp.2016.04.038
19. Gkournelos C, Karagiannis P, Kousi N, Michalos G, Koukas S, Makris S (2018) Application of wearable devices for supporting operators in human-robot cooperative assembly tasks. Proced CIRP 76:177–182. https://doi.org/10.1016/j.procir.2018.01.019
20. Cherubini A, Passama R, Crosnier A, Lasnier A, Fraisse P (2016) Collaborative manufacturing with physical human–robot interaction. Robot Comput-Integr Manuf 40:1–13. https://doi.org/10.1016/j.rcim.2015.12.007
21. Goodrich MA, Schultz AC (2007) human-robot interaction: a survey. Found Trends® Hum-Comput Interact 1:203–275. https://doi.org/10.1561/1100000005
22. Argyrou A, Giannoulis C, Sardelis A, Karagiannis P, Michalos G, Makris S (2018) A data fusion system for controlling the execution status in human-robot collaborative cells. Proced CIRP 76:193–198. https://doi.org/10.1016/j.procir.2018.01.012
23. Karagiannis P, Giannoulis C, Michalos G, Makris S (2018) Configuration and control approach for flexible production stations. Proced CIRP 78:166–171. https://doi.org/10.1016/j.procir.2018.09.053
24. Michalos G, Kousi N, Karagiannis P, Gkournelos C, Dimoulas K, Koukas S, Mparis K, Papavasileiou A, Makris S (2018) Seamless human robot collaborative assembly—an automotive case study. Mechatronics 55:194–211. https://doi.org/10.1016/j.mechatronics.2018.08.006
25. Tsarouchi P, Michalos G, Makris S, Athanasatos T, Dimoulas K, Chryssolouris G (2017) On a human-robot workplace design and task allocation system. Int J Comput Integr Manuf 1–8. https://doi.org/10.1080/0951192X.2017.1307524
26. Malik AA, Bilberg A (2019) Complexity-based task allocation in human-robot collaborative assembly. Ind Robot Int J Robot Res Appl 46:471–480. https://doi.org/10.1108/IR-11-2018-0231
27. Michalos G, Spiliotopoulos J, Makris S, Chryssolouris G (2018) A method for planning human robot shared tasks. CIRP J Manuf Sci Technol 22:76–90. https://doi.org/10.1016/j.cirpj.2018.05.003

28. Tsarouchi P, Spiliotopoulos J, Michalos G, Koukas S, Athanasatos A, Makris S, Chryssolouris G (2016) A decision making framework for human robot collaborative workplace generation. Proced CIRP 44:228–232. https://doi.org/10.1016/j.procir.2016.02.103
29. Järvenpää E, Siltala N, Hylli O, Lanz M (2019) Implementation of capability matchmaking software facilitating faster production system design and reconfiguration planning. J Manuf Syst 53:261–270. https://doi.org/10.1016/j.jmsy.2019.10.003
30. Zhang H-C, Alting L (1994) Computerized manufacturing process planning systems, 1st edn. Chapman & Hall, London ; New York
31. Bikas C, Argyrou A, Pintzos G, Giannoulis C, Sipsas K, Papakostas N, Chryssolouris G (2016) An automated assembly process planning system. Proced CIRP 44:222–227. https://doi.org/10.1016/j.procir.2016.02.085
32. Shoval S, Efatmaneshnik M, Ryan MJ (2017) Assembly sequence planning for processes with heterogeneous reliabilities. Int J Prod Res 55:2806–2828. https://doi.org/10.1080/00207543.2016.1213449
33. Wang L (2007) Process planning and scheduling for distributed manufacturing. Springer, London
34. Pintzos G, Matsas M, Papakostas N, Mourtzis D (2016) Disassembly line planning through the generation of end-of-life handling information from design files. Proced CIRP 57:740–745. https://doi.org/10.1016/j.procir.2016.11.128
35. Tsarouchi P, Makris S, Chryssolouris G (2016) On a human and dual-arm robot task planning method. Proced CIRP 57:551–555. https://doi.org/10.1016/j.procir.2016.11.095
36. Alami R, Clodic A, Montreuil V, Sisbot EA, Chatila R (2005) Task planning for human-robot interaction. In: Proceedings of the 2005 joint conference on Smart objects and ambient intelligence innovative context-aware services: usages and technologies—sOc-EUSAI '05. ACM Press, Grenoble, France, p 81
37. Bidot J, Karlsson L, Lagriffoul F, Saffiotti A (2017) Geometric backtracking for combined task and motion planning in robotic systems. Artif Intell 247:229–265. https://doi.org/10.1016/j.artint.2015.03.005
38. Cesta A, Orlandini A, Umbrico A (2018) fostering robust human-robot collaboration through AI task planning. Procedia CIRP 72:1045–1050. https://doi.org/10.1016/j.procir.2018.03.022
39. Lozano-Perez T, Jones JL, Mazer E, O'Donnell PA (1989) Task-level planning of pick-and-place robot motions. Computer 22:21–29. https://doi.org/10.1109/2.16222
40. Mishra A, Bhuyan S, Agrawal S, Deb S, Sen D Task level planning and implementation of robotic assembly under machine vision guidance
41. Muñoz P, R-Moreno MD, Barrero DF, (2016) Unified framework for path-planning and task-planning for autonomous robots. Robot Auton Syst 82:1–14. https://doi.org/10.1016/j.robot.2016.04.010
42. Müller R, Vette M, Geenen A (2017) Skill-based dynamic task allocation in human-robot-cooperation with the example of welding application. Proced Manuf 11:13–21. https://doi.org/10.1016/j.promfg.2017.07.113
43. Kousi N, Dimosthenopoulos D, Matthaiakis A-S, Michalos G, Makris S (2019) AI based combined scheduling and motion planning in flexible robotic assembly lines. Proced CIRP 86:74–79. https://doi.org/10.1016/j.procir.2020.01.041
44. Seder M, Petrović L, Peršić J, Popović G, Petković T, Šelek A, Bićanić B, Cvišić I, Josić D, Marković I, Petrović I, Muhammad A (2019) open platform based mobile robot control for automation in manufacturing logistics. IFAC-Pap 52:95–100. https://doi.org/10.1016/j.ifacol.2019.11.055
45. Pellegrinelli S, Orlandini A, Pedrocchi N, Umbrico A, Tolio T (2017) Motion planning and scheduling for human and industrial-robot collaboration. CIRP Ann 66:1–4. https://doi.org/10.1016/j.cirp.2017.04.095
46. Michalos G, Kaltsoukalas K, Aivaliotis P, Sipsas P, Sardelis A, Chryssolouris G (2014) Design and simulation of assembly systems with mobile robots. CIRP Ann 63:181–184. https://doi.org/10.1016/j.cirp.2014.03.102

47. Chen F, Sekiyama K, Sasaki H, Huang J, Sun B, Fukuda T (2011) Assembly strategy modeling and selection for human and robot coordinated cell assembly. 2011 IEEE/RSJ international conference on intelligent robots and systems. IEEE, San Francisco, CA, pp 4670–4675
48. Zhang D (1991) Planning with Pr/T nets. In: Proceedings 1991 IEEE international conference on robotics and automation. IEEE Computer Society Press, Sacramento, CA, USA, pp 769–775
49. LaValle SM (2006) Planning algorithms. Cambridge University Press, Cambridge
50. Agostini A, Torras C, Wörgötter F (2011) Integrating task planning and interactive learning for robots to work in human environments. In: Proceedings of the 22nd international joint conference on artificial intelligence. Barcelona, Catalonia, Spain
51. Maurtua I, Ibarguren A, Kildal J, Susperregi L, Sierra B (2017) Human–robot collaboration in industrial applications: safety, interaction and trust. Int J Adv Robot Syst 14:172988141771601. https://doi.org/10.1177/1729881417716010
52. Agostini AG, Torras C, Wörgötter FA. General strategy for interactive decision-making in robotic platforms. Institut de Robotica i Informatica Industrial
53. Bicho E, Erlhagen W, Louro L, Costa e Silva E, (2011) Neuro-cognitive mechanisms of decision making in joint action: a human–robot interaction study. Hum Mov Sci 30:846–868. https://doi.org/10.1016/j.humov.2010.08.012
54. Tsarouchi P, Matthaiakis A-S, Makris S, Chryssolouris G (2017) On a human-robot collaboration in an assembly cell. Int J Comput Integr Manuf 30:580–589. https://doi.org/10.1080/0951192X.2016.1187297
55. Michalos G, Sipsas P, Makris S, Chryssolouris G (2016) Decision making logic for flexible assembly lines reconfiguration. Robot Comput-Integr Manuf 37:233–250. https://doi.org/10.1016/j.rcim.2015.04.006
56. Weßkamp V, Seckelmann T, Barthelmey A, Kaiser M, Lemmerz K, Glogowski P, Kuhlenkötter B, Deuse J (2019) Development of a sociotechnical planning system for human-robot interaction in assembly systems focusing on small and medium-sized enterprises. Procedia CIRP 81:1284–1289. https://doi.org/10.1016/j.procir.2019.04.014
57. Bochmann L, Bänziger T, Kunz A, Wegener K (2017) Human-robot collaboration in decentralized manufacturing systems: an approach for simulation-based evaluation of future intelligent production. Proced CIRP 62:624–629. https://doi.org/10.1016/j.procir.2016.06.021
58. Malik AA, Bilberg A (2017) Framework to implement collaborative robots in manual assembly: a lean automation approach. In: Katalinic B (ed) DAAAM Proceedings, 1st edn. DAAAM International Vienna, pp 1151–1160
59. Unhelkar VV, Dorr S, Bubeck A, Przemyslaw AL, Perez J, Siu HC, Boerkoel Jr JC, Tyroller Q, Bix J, Bartscher S, Shah JA (2018) introducing mobile robots to moving-floor assembly lines: design, evaluation and deployment
60. (2020) Unlocking the potential of industrial human–robot collaboration: a vision on industrial collaborative robots for economy and society. Publications Office, LU
61. Kaltsoukalas K, Makris S, Chryssolouris G (2015) On generating the motion of industrial robot manipulators. Robot Comput-Integr Manuf 32:65–71. https://doi.org/10.1016/j.rcim.2014.10.002
62. Kardos C, Kovács A, Váncza J (2017) Decomposition approach to optimal feature-based assembly planning. CIRP Ann 66:417–420. https://doi.org/10.1016/j.cirp.2017.04.002
63. Cesta A, Orlandini A, Bernardi G, Umbrico A (2016) Towards a planning-based framework for symbiotic human-robot collaboration. In: 2016 IEEE 21st international conference on emerging technologies and factory automation (ETFA). IEEE, Berlin, Germany, pp 1–8
64. Kousi N, Gkournelos C, Aivaliotis S, Giannoulis C, Michalos G, Makris S (2019) Digital twin for adaptation of robots' behavior in flexible robotic assembly lines. Proced Manuf 28:121–126. https://doi.org/10.1016/j.promfg.2018.12.020
65. Magrini E, De Luca A (2017) Human-robot coexistence and contact handling with redundant robots. In: 2017 IEEE/RSJ international conference on intelligent robots and systems (IROS). IEEE, Vancouver, BC, pp 4611–4617
66. URL SHAREWORK EU Project. https://sharework-project.eu/. Accessed 9 Aug 2020

67. Orlandini A, Cialdea Mayer M, Umbrico A, Cesta A (2020) Design of timeline-based planning systems for safe human-robot collaboration. In: Vallati M, Kitchin D (eds) Knowledge engineering tools and techniques for AI planning. Springer International Publishing, Cham, pp 231–248
68. URL SHERLOCK EU Project. https://sherlock-project.eu/. Accessed 9 Aug 2020
69. Cencen A, van Deurzen K, Verlinden JC, Geraedts JMP (2014) Exploring human robot coproduction. In: Proceedings of the 2014 IEEE emerging technology and factory automation (ETFA). IEEE, Barcelona, Spain, pp 1–4
70. Faroni M, Beschi M, Pedrocchi N (2019) An MPC framework for online motion planning in human-robot collaborative tasks. In: 2019 24th IEEE international conference on emerging technologies and factory automation (ETFA). IEEE, Zaragoza, Spain, pp 1555–1558
71. Aghazadeh S, Hafeznezami S, Najjar L, Huq Z (2011) The influence of work-cells and facility layout on the manufacturing efficiency. J Facil Manag 9:213–224. https://doi.org/10.1108/14725961111148117
72. Michalos G, Makris S, Rentzos L, Chryssolouris G (2010) Dynamic job rotation for workload balancing in human based assembly systems. CIRP J Manuf Sci Technol 2:153–160. https://doi.org/10.1016/j.cirpj.2010.03.009
73. Takata S, Hirano T (2011) Human and robot allocation method for hybrid assembly systems. CIRP Ann 60:9–12. https://doi.org/10.1016/j.cirp.2011.03.128
74. Yu T, Huang J, Chang Q (2020) Mastering the working sequence in human-robot collaborative assembly based on reinforcement learning. ArXiv200704140 Cs Eess
75. URL CORDIS THOMAS EU project. https://cordis.europa.eu/project/id/723616. Accessed 31 Aug 2020
76. Dianatfar M, Latokartano J, Lanz M (2019) Task balancing between human and robot in mid-heavy assembly tasks. Proced CIRP 81:157–161. https://doi.org/10.1016/j.procir.2019.03.028
77. Müller R, Vette M, Mailahn O (2016) Process-oriented task assignment for assembly processes with human-robot interaction. Proced CIRP 44:210–215. https://doi.org/10.1016/j.procir.2016.02.080
78. URL Siemenes process simulate. https://www.plm.automation.siemens.com/en_gb/Images/7457_tcm642-80351.pdf. Accessed 9 Jul 2020
79. URL visual components. https://www.visualcomponents.com/. Accessed 9 Jul 2020
80. Wang XV, Kemény Z, Váncza J, Wang L (2017) Human–robot collaborative assembly in cyber-physical production: classification framework and implementation. CIRP Ann 66:5–8. https://doi.org/10.1016/j.cirp.2017.04.101
81. Bilberg A, Malik AA (2019) Digital twin driven human–robot collaborative assembly. CIRP Ann 68:499–502. https://doi.org/10.1016/j.cirp.2019.04.011
82. Pairet E, Ardon P, Liu X, Lopes J, Hastie H, Lohan KS (2019) A digital twin for human-robot interaction. In: 2019 14th ACM/IEEE international conference on human-robot interaction (HRI). IEEE, Daegu, Korea (South), pp 372–372
83. Negri E, Ardakani HD, Cattaneo L, Singh J, Macchi M, Lee J (2019) A digital twin-based scheduling framework including equipment health index and genetic algorithms. IFAC-Pap 52:43–48. https://doi.org/10.1016/j.ifacol.2019.10.024
84. Metta G, Natale L, Nori F, Sandini G, Vernon D, Fadiga L, von Hofsten C, Rosander K, Lopes M, Santos-Victor J, Bernardino A, Montesano L (2010) The iCub humanoid robot: an open-systems platform for research in cognitive development. Neural Netw 23:1125–1134. https://doi.org/10.1016/j.neunet.2010.08.010
85. Diankov R, Kuffner J (2008) OpenRAVE: a planning architecture for autonomous robotics
86. (2020) URL OpenHRP3. http://fkanehiro.github.io/openhrp3-doc/en/about.html
87. Mois G, Folea S, Sanislav T (2017) Analysis of three IoT-based wireless sensors for environmental monitoring. IEEE Trans Instrum Meas 66:2056–2064. https://doi.org/10.1109/TIM.2017.2677619
88. Usman S, Mehmood R, Katib I (2018) big data and hpc convergence: the cutting edge and outlook. In: Mehmood R, Bhaduri B, Katib I, Chlamtac I (eds) Smart societies, infrastructure, technologies and applications. Springer International Publishing, Cham, pp 11–26

89. Qi Q, Tao F (2018) Digital twin and big data towards smart manufacturing and industry 4.0: 360 degree comparison. IEEE Access 6:3585–3593. https://doi.org/10.1109/ACCESS.2018.2793265
90. Kootbally Z (2016) Industrial robot capability models for agile manufacturing. Ind Robot Int J 43:481–494. https://doi.org/10.1108/IR-02-2016-0071
91. Pedersen MR, Nalpantidis L, Andersen RS, Schou C, Bøgh S, Krüger V, Madsen O (2016) Robot skills for manufacturing: from concept to industrial deployment. Robot Comput-Integr Manuf 37:282–291. https://doi.org/10.1016/j.rcim.2015.04.002
92. Zacharias F, Borst C, Hirzinger G (2007) Capturing robot workspace structure: representing robot capabilities. In: 2007 IEEE/RSJ international conference on intelligent robots and systems. IEEE, San Diego, CA, USA, pp 3229–3236
93. Robot operating system (ROS). http://www.ros.org/
94. Muñoz Arancón M, Montano G, Wirkus M, Hoeflinger K, Silveria D, Tsiogkas N, Hugues J, Bruyninckx H, Bensalem S, Alanen J (June 20–22) ESROCOS: a robotic operating system for space and terrestrial applicationS. Netherlands
95. Setyanto NW, Efranto RY, Lukodono RP, Dirawidya A (2015) Ergonomics analysis in the scarfing process by OWAS, NIOSH and nordic body map's method at slab steel plant's division. IJIRSET 4:
96. URL Ergonomics eTool (2020). https://www.osha.gov/SLTC/etools/electricalcontractors/materials/heavy.html

Implementing Effective Speed and Separation Monitoring with Legacy Industrial Robots—State of the Art, Issues, and the Way Forward

Alberto Moel, Scott Denenberg, and Marek Wartenberg

Abstract Collaborative applications using traditional industrial robots and Speed and Separation Monitoring (SSM per ISO/TS 15066) rely on safe stopping if a Protective Separation Distance (PSD per ISO/TS 15066) is violated. However, larger industrial robots have longer stopping times and their control architectures are not designed for flexible external interaction. Robot manufacturers also provide stopping time and distance data for calculating the PSD, but these data are often fragmented, hard to interpret, and overly conservative. Hence, the "worst-case" PSD calculation for SSM is generally more conservative than warranted. We believe truly fluid human–robot collaboration will be possible but will require a closer interlocking between the robot controller and the safety system and a more precise characterization of robot stopping times and distances. In this paper we describe techniques to improve latencies and response times using SSM with existing robot control architectures. We also propose longer-term alternatives for consideration by the industry.

1 Introduction

Current manufacturing trends towards mass customization and faster product cycles means that manufacturers cannot amortize the costs of fully automated workcells; instead, they need the flexibility of human labor. As these end-market trends continue, the industry has gingerly moved towards allowing robots and humans to safely be in the same space while the robot is operating. In that way the judgment, dexterity, and flexibility of humans can begin to be paired with the strength, reliability, and precision of robots. The first solution for simple human–robot collaboration to gain acceptance has been robots meeting the ISO/TS 15066 Power and Force Limited (PFL) [1] specifications. The growing popularity of PFL robots has shown that collaborative applications can raise productivity, provide faster fault recovery, and increase unit production rates.

A. Moel (✉) · S. Denenberg · M. Wartenberg
Veo Robotics, Waltham, USA
e-mail: am@veobot.com

M. I. Aldinhas Ferreira and S. R. Fletcher (eds.), *The 21st Century Industrial Robot: When Tools Become Collaborators*, Intelligent Systems, Control and Automation: Science and Engineering 81, https://doi.org/10.1007/978-3-030-78513-0_13

However, standards-driven risk assessments (such as those required by ISO 10218–2 [2]) require that PFL robots be speed- and force-limited when operating in proximity to humans or otherwise safety-guarded if there are other sources of risk in the application, such as a dangerous workpiece or end effector. This greatly reduces the range of their application—manufacturing steps can either be completed by a human or via automation, but usually not both.

An alternative to PFL robots are collaborative applications using traditional industrial robots (or even PFL robots) and Speed and Separation Monitoring (SSM per ISO/TS 15066). In SSM, the robot system and operator may move concurrently in the collaborative workspace, and risk reduction is achieved by maintaining at least the Protective Separation Distance (PSD) between operator and robot. When the human–robot separation distance decreases to a value below the PSD, the robot slows down to decrease the PSD, or completely stops. When the human moves away from the robot, the robot system can resume motion automatically according to the SSM requirements.

Collaborative applications using SSM have different and less onerous limitations over PFL on robot speed and payload as well as end-effector design, leaving the robot and workcell in a safe state when the PSD is violated. A vision-based implementation of SSM provides a way to overcome the PFL limitations, making large industrial robots aware of humans, and opening new opportunities for human–robot collaboration.

However, larger and heavier traditional industrial robots have longer stopping times and their control architectures are not designed for flexible external interaction (with, for example, large command latencies), resulting in longer PSDs. We believe true, high performance, and safe human–robot collaboration is certainly possible, but will require a much closer interlocking between the robot controller and the safety system than what is currently allowed by existing control and safety architectures.

In this paper we review the state of the art in SSM implementations and describe several techniques to improve latencies and response times using SSM with existing robot control architectures. We also propose several longer-term alternatives for consideration by the industry.

2 Current State

Relative to the recent and rapid progress of PFL robot installations, practical SSM implementations are still in their infancy. In addition to related prior work on collision avoidance (see for example, [1] and references therein), some methods and prototypes focused particularly on dynamic SSM in the context of the relevant robot safety standards [2–6] have been disclosed. However, most of the published research prototypes are *functional* implementations only. In other words, they do not fulfill machinery safety requirements as per ISO 13849–1 [6]. A good review of the different forms of human–robot collaboration, including SSM, can be found in [7], and a survey of human–robot collaboration in industrial settings is available in [8].

As a sample of the extant research, [9] introduces an SSM system using a Microsoft Kinect V2 to continuously detect a human worker within a shared workspace. Obviously, the use of a non-safety-rated sensing element such as a Kinect V2 means such an approach cannot be used in production. Reference [10] introduces some practical SSM case examples that can be integrated into a transferable robotic platform. Some of the work is simulation-based and then implemented in the real world, such as [11], which presents an SSM manufacturing cell for automotive brake disc assembly. Virtual environment simulation is used to determine the SSM algorithm parameters for estimating the PSD.

Other work focuses on practical SSM algorithms, such as the trajectory dependent dynamic SSM proposed in [12], SSM metrics [13], PSD calculations [14], certification possibilities [15] or other design considerations [16] and [17]. Hardware and computation platform considerations have been also studied. For example, [18] discusses the use of unsafe devices and protocols for safety functions considering functional safety at the system level, while [19] reviews techniques for sensor fusion for human safety in industrial workcells.

Practical implementations that highlight the issues with legacy robot controllers, specifically high or unpredictable latencies are unusual. A notable exception is [20], which finds there is an inherent delay of 300–500 ms from the time the robot is ordered to reduce speed until it begins to execute the order [21].

However, most of the demonstrated applications rely on existing sensing elements and components that are not safety-rated and hence cannot be used in a production environment. The few that are meant for production environments using traditional safety sensing components such as light curtains and 2D scanners make for cumbersome and rigid implementations with little flexibility and poor performance.

"True" fluid human–robot collaborative workcells using SSM are a work in progress for several reasons. The novelty of the SSM standard specifications; computational difficulties in implementing the SSM standard; legacy robot safety architectures with incomplete and sparse safety data (in particular poorly-characterized stopping times and distances); and most importantly, limited availability of safety-rated sensing elements are all barriers to wide adoption of SSM collaborative applications in production environments. We deal with each issue below before discussing the path forward.

2.1 Novelty of SSM Standards Specifications

The key parameter in SSM implementations is the PSD, and is defined based on the concepts used to create the minimum distance formula in ISO 13855 [3], modified to take into account the dynamic nature of hazards associated with robot motion. The PSD is calculated between the robot and any area or volume that a person could potentially occupy and a safety-rated speed reduction or protective stop is initiated if the PSD is violated. The robot can automatically restart and continue its trajectory once the operator moves away and re-establishes the PSD.

2.1.1 The Protective Separation Distance Equation

The PSD, S_p, can be described by Eq. (1):

$$S_p(t_0) = S_h + S_r + S_s + C + Z_d + Z_r \tag{1}$$

where

$S_p(t_0)$ is the protective separation distance at current time t_0;

S_h is the contribution to the protective separation distance attributable to the operator's change in location;

S_r is the contribution to the protective separation distance attributable to the robot system's reaction time;

S_s is the contribution to the protective separation distance due to the robot system's stopping distance;

C is the intrusion distance, as defined in ISO 13855; this is the distance that a part of the body can intrude into the sensing field before it is detected;

Z_d is the position uncertainty of the operator in the collaborative workspace, as measured by the presence sensing device resulting from the sensing system measurement tolerance;

Z_r is the position uncertainty of the robot system, resulting from the accuracy of the robot position measurement system.

$S_p(t_0)$ allows the PSD to be calculated dynamically, allowing the robot speed to vary during the application. $S_p(t_0)$ can also be used to calculate a fixed value for the protective separation distance, based on worst-case values.

2.1.2 Estimating the PSD in Practice is not Easy

Following [22], we make the assumption that the robot can begin moving at any time directly towards the human at a speed v_r, which, in turn, is moving directly towards the robot at a speed v_h. The robot takes an amount of time T_r to begin stopping, and once it begins stopping it will do so at a constant deceleration $a_s > 0$. Under those assumptions, we can derive Eq. (6) in [22]:

$$S \geq S_p(v_r) = v_h\left(T_r + \frac{v_r}{a_s}\right) + v_r T_r + \frac{v_r^2}{2a_s} + (C + Z_d + Z_r) \tag{2}$$

A stop of robot motion must be issued at the instant the actual separation distance S falls below S_p since otherwise the operator would be able to reach the robot within the system's reaction time T_r.

A proper implementation of the SSM supervision criterion in (2) must include a safety-rated computation of the anticipated stopping behavior of the robot, based on its momentary pose, velocity and payload. Suitable algorithms should execute in a real-time manner on the time scale of the refresh rate of the workspace supervision. However, validating the speed estimates of the operator and robot is a difficult task. For example, it has been observed in some robot systems that the mechanisms by which the robot tracks velocities for control differ from the mechanisms by which velocities are reported externally. Independent (i.e. external) verification of velocities is then required to safely quantify discrepancies in reported versus actual values. No clear generalized method for reliably and safely determining robot velocities is readily available.

Another variable that needs to be estimated is the speed of the operator, v_h in Eq. (2). v_h is often assumed to be a maximum of 1.6 m/s based on the specifications in ISO 13855. Even though the 1.6 m/s is an accurate assessment of normal walking speeds, it may be even more prudent instead to use instead the worst-case assumption of 2.0 m/s from ISO 13855 to accommodate instantaneous rapid motions such as arm motions when sufficiently near the hazard, as emphasized in ISO 10218–2 [2]. But in reality, and depending on the application, v_h could actually be lower than required by the standard and still provide a safe working environment.

Because of these measurement and computational complexities, manufacturing engineers and system integrators will often calculate (using the maximum robot speed, load and extension) a static minimum distance as defined in ISO 13855 and install a laser scanner or light curtain at a distance larger than that maximum calculated value. This static calculation might not consider the application at all and often leads to wasted factory floor space and limited access to the robot for collaborative tasks.

2.2 *Difficulties in Implementing SSM in Legacy Robot Control and Safety Architectures*

Existing robot control and safety architectures were not designed for collaborative applications. In general, these are legacy architectures, developed decades ago for situations where the robots and humans were completely isolated from each other and all robot operation occurs autonomously. These limitations are of three types: poor integration of safety and control architectures and robot kinematics, unpredictable and large controller latencies, and poor and incomplete availability of stopping distance and time data for PSD calculations. We tackle each of these in turn.

2.2.1 Poor Integration of Safety and Control Architectures and Robot Kinematics

Due to technological limitations, a given robot system may be disqualified for use in a collaborative environment. Example disqualifying conditions include the robot being incapable of reporting positions or velocities with low latency, a lack of support for reporting positions or velocities of all joints, unreliable delays in stopping times, or the inability to receive and process external signals to trigger a controlled stop.

However, aware of the market potential of collaborative robotics, robot manufacturers and automation providers, have, over the last years, introduced simple but reliable safety sensing and interfaces that allow some level of human–robot collaboration. These systems can safely stop or slow down the robot on a signal, initiated, for example, by a human triggering an area scanner. These systems also allow for dynamically constraining the speed or reach of the robot in certain software-defined physical zones.

One such widely available is the use of functions such as a safety-rated Speed Limit. This functionality guarantees a robot's tool-center point (TCP) will not move above a set speed in mm/s, allowing for closer proximity between the human and robot. However, the safety-rated function does not govern the speed itself, that is left to the non-safety rated trajectory controller, instead it simply monitors TCP speed and initiates a Protective Stop if a speed over the set limit is achieved. Further, as a requirement of ISO 10218, protective stopping data is provided per joint, and not for TCP. Knowing how long each joint takes to stop is directly used in the calculation of the PSD and setting a TCP speed limit requires complex inverse kinematic computation to determine what that TCP speed corresponds to for each joint.

Additionally, safety-rated protective stops leave the robot in a state with servo power turned off. Restarting from this state requires any alarms to be cleared, servos to be reenergized, and a start command sent. This takes complex signaling and can take several seconds to restart robot motion. Although this is a great improvement over lockout-tagout procedures currently used on the factory floor, more desirable functionality is possible with a stop that leaves servos energized.

Hence, dynamic definition and computation of safe areas usually cannot be accomplished without a manual reset of safety-related part of control systems and cannot use a continuous value from sensors as thresholds [6]. Consequently, over-conservative safety distances required around the robot workspace could increase the size of the layout [7, 8].

Lastly, this functionality, along with communication of real-time ethernet directly to the controller is not safety-rated, so even positions reported by the robot cannot be directly used in safety applications. Slowing the robot via commanded velocities over ethernet can allow for much more fluid interaction but is likewise not safety rated and cannot be used on the factory floor.

2.2.2 Large and Unpredictable Robot Controller Latencies

There is a general need for low-latency safety-rated robot interfaces that can control robot speed and trajectory in real time through external signals. This means that the amount of time necessary to perform a complete cycle of the safety system, from sensing to robot speed adjustment, must be very low. An SSM decision to slow or stop the robot is made by a sensing system and the robot controller is responsible for applying the requested action.

There is some evidence of long and unpredictable legacy robot controller latencies in the literature. For example [9] describes a human–robot collaboration experimental setup using an ABB IRB120 robot programmed in RAPID [10]. To improve the sensing and computational latencies, they reduced the quality of the robot CAD model and optimized the network used to connect the different sensing and computation elements, including disabling network congestion protocols [11].

Although the resulting system is capable of robot speed adjustment in real-time, there are no hard guarantees for latency (i.e., it is not" hard real-time" as defined in [12]). Safety system latencies were recorded experimentally. The average latency of the sensing and control system was 6.13 ms, with a maximum latency of 389.6 ms, which would make it unfit for safety-rated applications requiring deterministic latencies. However, all of this is rendered moot. With the robot used in this implementation, there is an inherent delay of 300–500 ms from the time the robot is ordered to reduce speed until it begins to execute the order [10], making the approximately 6 ms latency of the safety system an insignificant contribution to overall system latency.

In other work, [13] identifies a related but equally important issue. They find a delay of about 220 ms between when the robot was a specific location and when it reported it was at that same position. Further, it was observed that, depending on the method used to stop the robot safely, the time necessary to bring the robot velocity to 1 mm/s varied considerably.

This controller latency is captured in T_r in Eq. (2), and we can model the impact of this latency on the PSD by looking at the comparative statics on how S_p changes as a function of T_r and v_r:

$$\frac{\partial S_p}{\partial T_r} = v_h + v_r \tag{3}$$

$$\frac{\partial S_p}{\partial v_r} = \frac{1}{a_s}(v_h + v_r) + T_r \tag{4}$$

An increasing T_r both increases the effect of robot speed v_r on the PSD S_p, and increases the minimum PSD from the higher latency. We can see this effect graphically in Figs. 1 and 2 where we model Eq. (2) using the parameters in Table 1.

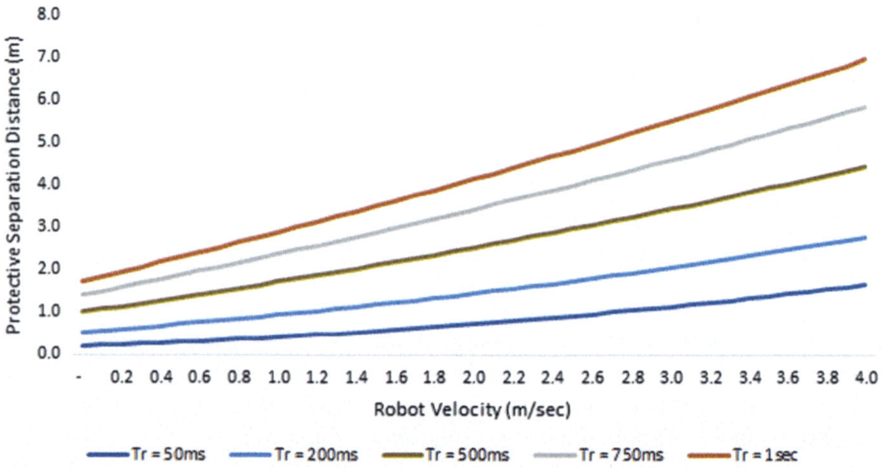

Fig. 1 PSD as a function of robot velocity v_r parametrized by robot latency T_r

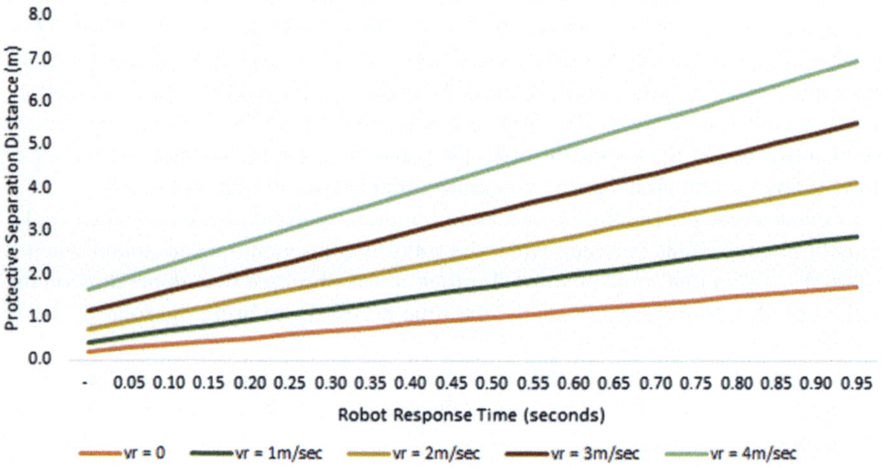

Fig. 2 PSD as a function of robot response time T_r parametrized by robot velocity v_r

Table 1 PSD calculation parameters used in Figs. 1 and 2

v_h	Human velocity	1.6 m/sec
a_s	Robot deceleration	10 m/sec^2
C	Intrusion distance	0.088 m
$Z_d + Z_r$	Robot and sensor uncertainty	0.13 m

2.2.3 Issues with Stopping Time and Distance Data for PSD Calculations

Robot manufacturers also provide graphs and data points for stopping times and distances as a function of speed, payload, and arm extension, which are necessary for the correct implementation of SSM and calculating the PSD.

ISO 10218–1 Annex B requires that the stopping time and distance for different dynamic states of the robot (at a minimum, 33, 66, and 100% of speed, load, and extension) be provided by robot manufacturers. However, these data are often fragmented, hard to interpret, overly conservative, and sometimes generated by simulation. The requirement itself is interpreted differently by different robot manufacturers—some provide this data in tables of different sparseness, while others provide it in graphs that vary in resolution quality.

There is also no guidance from the standards on how to extrapolate this data down to zero. In some cases the published stopping data indicates that it will take the same amount of time for the robot to stop whether it is moving at 33% speed with 33% load, 100% speed with 100% load, or any combination in between. This is likely because the worst-case number for 100% speed and 100% load was propagated throughout the table.

With the limited data available, the worst-case PSD calculation for SSM is functional but much less precise and more conservative than warranted by the robot dynamics and the control system latencies. Without access to planning information used by robot controllers (which uses such data to determine precisely how robot motion will cease), the uncertainty represented in the PSD prevents truly close and efficient collaborative operation.

As an example, Fig. 3 shows the extrapolated PSD as a percentage of TCP speed for actual robot data, with the calculated PSD from Eq. (2) for the parameters in

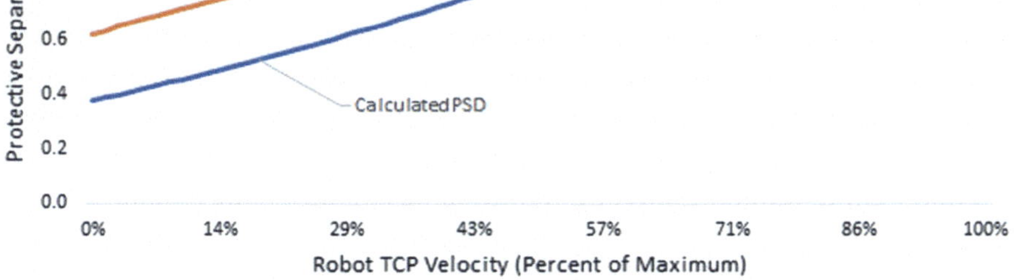

Fig. 3 Calculated and extrapolated PSD as a function of TCP maximum velocity

Table 1 and an estimated robot command latency T_r of 100 ms. The extrapolation to $v_r = 0$ implies that T_r is closer to 250 ms than the observed number closer to 100 ms. We believe that the reported numbers are very conservative as the assumption is that the robot will be in a fixed guarded space and hence variations on robot payload, velocity, or extension are not relevant to the safety calculation.

2.3 *Issues with Safe Sensing for SSM Implementations*

Although many technologies like 2D sensors and LIDAR have been available for several years, they are not currently sufficient to make traditional industrial robots safe in collaborative manufacturing settings. Recently, 3D active vision-based safety systems (off-the-shelf RGB, depth cameras and projectors, or 3D time-of-flight) have gained momentum due to their affordable price, flexible installation, and easy tailoring. However, these systems have not been widely adopted and standardized and their safety architectures and ratings have not been fully developed (see [14] for a comprehensive review).

2.3.1 Performance Issues with Existing and New Sensing Technologies

A vision-based implementation of SSM provides a way to overcome the PFL limitations, making large industrial robots aware of humans. However, current safety-rated approaches relying on 2D sensors (i.e. laser scanners) do not provide the richness of data required to implement dynamic SSM.

For example, the laser scanners are limited to identifying intrusions in 2D, which means any 3D elements of the intrusion are not captured. If a human steps into a space monitored by a floor-level laser scanner, it cannot determine if the human has stretched a hand into the workcell in a dangerous way. Further 2D scanners (and existing 3D scanning technology) have limited "intelligence" when dealing with occluded or shadowed volumes, which are likely to change as the robot and human move around the workspace.

3D time-of-flight (ToF) sensing allows for direct calculation of a PSD between the robot and operator, with well-understood failure modes and the best range versus resolution tradeoff. However, turning millions of 3D data points into a sufficiently rich (and Category 3 PLd safe) representation of the workcell requires dual-channel, multi-core parallel processing in real time (25 Hz or higher) with low latency from sensing to control and a safety watchdog monitoring the latency and communications with the robot.

2.3.2 Integration Issues in SSM Implementations Using Advanced Sensing

There are several challenges to the broad implementation of 3D sensing technologies for safe human–robot collaboration. First, to work with 3D sensors, a robot controller must have Category 0/1/2 protective stopping functions with low latency and dynamic characterizations, the ability to externally override robot speed with low latency, safety-rated speed-limiting functions, and functional safety zones. As we have discussed, these kinds of interfaces are not always available.

Second, when a sensing system reports the position of a human operator, it is vital to have an estimate of how much time has actually elapsed since the human was at the reported location. If there is a sufficiently long latency between the times when an event occurs and when the event is reported, the delay presents a significant safety risk regardless of the accuracy and precision of the reported data. For instance, if a human is walking at 1.6 m/sec is reported half a second late, the separation distance between the robot and the human may be 80 cm shorter than indicated.

This delay directly impacts the measurement uncertainty of the human detection/tracking system and affects the robot reaction time T_r. To compensate for this larger uncertainty, the separation distance may be increased. Currently, there is no standardized methodology for measuring the reaction time of a human-detecting sensor. Rather, it is treated as a constant value that the integrator must provide. Given the nonlinearity of the human tracking problem, accurately predicting the reaction time of a human-detecting system is difficult.

Lastly, and most importantly, unless the workcell is carefully configured, sensor-based safety devices may require larger safety distances than physical fences, because they do not prevent human access to the robot working area. Especially faster and larger robots with a longer stopping time require a longer PSD.

3 Future Trends and Transformations

As technology evolves, the development of reliable and production ready SSM implementations for collaborative robotics will continue. Progress will hinge on new safety-sensing architectures and novel algorithms for 3D SSM implementations. Over the long term, it will also require robot manufacturers to substantially modify their control and safety architectures for true and fluid human–robot collaboration. But as a temporary measure, we propose a couple of workable SSM implementations that rely on currently available safety architectures.

3.1 New Safety Sensing Architectures

3D ToF sensors that cost $75,000 in 2010 now cost less than $100 thanks to the mass-market appeal of ToF chipsets. Additionally, the embedded processors, powerful SoCs, and FPGAs that are needed to run 3D computing algorithms are now readily available. And robot manufacturers have introduced richer APIs for safety to their robot controllers, which means today's controllers can be equipped to handle and respond to commands from 3D computing platforms.

3D sensing enables a much closer interlock between the actions of machines and the actions of humans, which means industrial engineers will be able to design processes where each subset of a task is appropriately assigned to a human or a machine for an optimal solution that maximizes workcell efficiency and lowers costs, while keeping human workers safe.

As an example, we have developed a fail-safe dedicated hardware system for 3D sensing to implement dynamic SSM [23]. 4–8 ISO 13849-compliant sensors feed data into a redundant ISO 13849 compliant computing platform that calculates the PSD and generates safety-rated control signals to modify robot movement speed or issue a protective stop (Fig. 4).

Using dual-channel safety-rated interfaces it is possible to safely reduce robot speed to decrease the PSD allowing for closer collaboration or initiate a protective stop when the PSD is violated. Because our system perceives the state of the environment around the robot at 30 frames per second in 3D, it is able to analyze all of that data to: determine all possible future states of a robot's environment, anticipate possible undetected items and monitor for potentially unsafe conditions, and then communicate with the robot's control system to slow or stop the robot before it comes in contact with anything it's not supposed to contact.

3.2 Safe Vision and Perception Algorithms

Not only is SSM a perception problem, it is a safe perception problem. Systems that provide safeguarding functionality in industrial robot workcells are required to comply with functional safety standards as described in ISO 13849. These standards require that no single hardware failure can lead to an unsafe situation and that both hardware and software development follow a structured process with traceability from requirements to testing, including for third-party software.

3.2.1 Reliable Data and Reliable Algorithms

To create a safe perception system, we need reliable data and reliable algorithms. The system in [23] uses 3D ToF sensors that are positioned on the periphery of the workcell to capture rich image data of the entire space. The architecture of the sensors

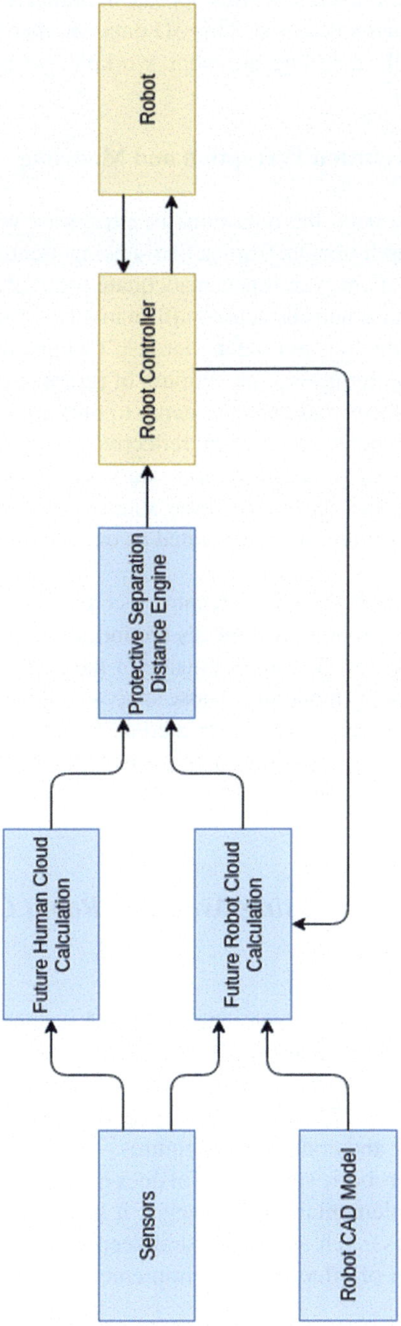

Fig. 4 Safety-rated sensing architecture for reliable SSM implementation

ensures reliable data with novel dual imagers that observe the same scene so the data can be validated at a per-pixel level. With this approach, higher level algorithms will not need to perform additional validation. This 3D data can then be used to identify key elements in the workcell, including the robot, workpieces, and humans.

3.2.2 Occupancy and Occlusion Perception and Modeling

In addition to using reliable data, the data must be processed with safety in mind. Most algorithms that use depth images from active sensing identify regions of space as either empty or occupied. However, this is inadequate for a safety system because a safety system requires that humans be sensed affirmatively: a part of a human body not showing up in sensor data does not mean there isn't a human there.

Because all active sensing requires some amount of return to detect objects, variability in reflectivity of surfaces can cause systems to output false negatives. Dark fabrics, for example, sometimes have very low reflectivity, so active IR sensors may not be able to "see" the legs of someone wearing dark jeans. This is unsafe, so the system in [23] classifies spaces as one of three states: empty (something can be seen behind it), occupied, or unknown and treated as occupied until the system can determine it to be otherwise.

This approach also addresses static and dynamic occlusions. In a workcell with a standard-size six axis robot moving workpieces around, there will always be some volumes of space that are occluded from or outside of the field of view of all of the sensors, either temporarily or permanently. Those spaces could at some point in time contain a human body part, so they are also treated as occupied for SSM purposes: a human could be reaching their arm into a space near the robot that none of the sensors can observe at that moment.

3.3 SSM Implementations Using Available Robot Control Functions

Even though work continues in the development of SSM systems with the requisite safety architecture, the level of achievable SSM performance of the robot-safety system as a whole is critically dependent on the robotic systems—not just the physics of the manipulator and latency of robot controllers and safety systems, but also on the capabilities of the safety and control architectures.

Current state of the art in robotic safety control does not allow for true low-latency high-performance SSM implementations. However, it is possible to utilize existing functional safety capabilities, such as keep-in and keep-out zones, and non-safety-rated speed overrides as part of effective SSM implementations.

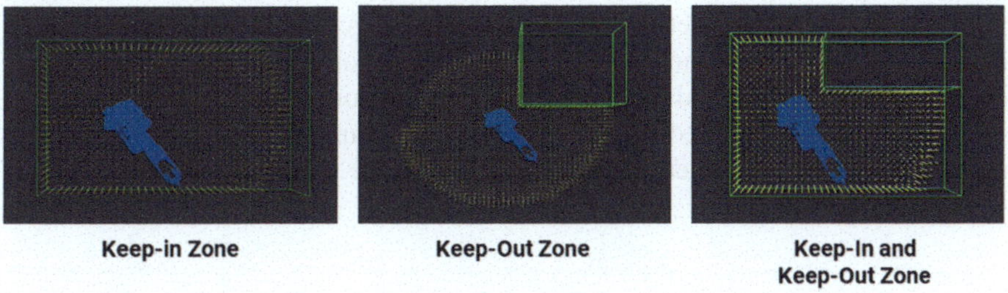

Fig. 5 Illustration of a keep-in and keep-out zone, and a joint keep-in/keep-out zone

3.3.1 Keep-In and Keep-Out Zones

Keep-in and keep-out zones are software-defined volumes that can be programmed using safety-rated functions in existing robot controllers. Once a keep-in or keep-out zone is defined, the robot is safely constrained within a keep-in zone, or prevented from entering a keep-out zone, reducing the future possible range of motion (the Robot Future Cloud, or RFC) of the robot (Fig. 5).

Path optimization may include dynamic switching of zones throughout the task, creating multiple RFCs of different sizes, in a similar way as described for the operator. And switching of these dynamic zones may be triggered not only by a priori knowledge of the robot program, but also by the instantaneous detected location of the robot or the instantaneous detected location of the operator.

For example, if a robot is tasked to pick up a part, bring it to a fixture, then perform a machining operation on it, the RFC can be dynamically updated based on safety-rated axis limiting at different times within the program.

3.3.2 Using Speed Overrides as Safety Functions

It is possible to use the PSD calculation to govern the speed of the robot as well, e.g., in an application where the robot path cannot deviate from its original programmed trajectory. In this case the PSD between the human and RFC are dynamically calculated and continuously compared to the instantaneous measured distance between them.

However, instead of a system which alters the robot path, or simply initiates a protective stop when the PSD is violated, speed can be governed to a set point at a distance larger than the PSD. At the instant when the robot reaches a lower set point, not only will the robot's RFC be smaller, but the distance the operator is from the new RFC will be larger. In one approach, governing to a lower set point can be achieved with a precomputed safety function already present on the robot controller.

As the measured distance between the future positions decreases, the commanded speed of the robot decreases accordingly, but the RFC size does not change. Once the robot has slowed down to the particular set point (at a distance larger than the PSD), the velocity at which the safety-rated joint monitor triggers an emergency stop can be decreased in a stepwise manner, shrinking the RFC accordingly. The decreased RFC corresponding to a decreased PSD may allow the operator to work in closer proximity to the robot in a safety-rated way. This approach is functional but cumbersome in practice. However, other methodologies relying on the speed override can be used for more fluid interaction.

3.4 Future Work and Collaboration with Robot Manufacturers

We believe true, high performance, and safe human–robot collaboration is certainly possible, but will require a much closer interlocking between the robot controller and the safety system than what is currently allowed by existing control and safety architectures. We propose several alternatives for consideration by the industry, in rough order of difficulty of implementation: better and more granular stopping time and distance data for safety calculations, improved existing robot safety functions, a safety-rated Category 2 stop, faster control response times for lower latency, safety-rated robot position data and speed overrides, and a safety-rated robot trajectory planner.

3.4.1 Improve Published Category 1 Stop Data

As discussed in Sect. 2.2.3, published robot stopping time and distance data, although hewing to the standards' disclosure requirements, leaves much to be desired. Higher collaborative productivity (from a better-estimated and smaller PSD) is possible if stopping data more accurately reflects the true dynamics of the robot. Using the current worst-case conservative published numbers for all cases does guarantee a safe state no matter the robot's speed or load but doing so also greatly limits the functionality of dynamic SSM.

Ideally, we would like to see much more granularity on the disclosed data for a broad range of data points along payload, speed, and robot pose and extensions, empirically-tested equations for estimating stopping time and distance data for a broad range of parameter values, or explicit manufacturer-approved guidance for the end user to experimentally determine the data.

3.4.2 Improved Existing Robot Safety Functions

In general, as described in Sect. 2.2.1, the robot control and safety functions are not well-integrated, with safety stopping functions and robot trajectory control not really "talking" to each other. Current implementations provide safety-rated Speed Limit functionality at the TCP, but the ISO 10218–1 standard only requires that stopping data be characterized and provided at the joint level. This situation is very likely driven by the fact that safety functions beyond a simple e-stop were grafted onto robot controllers many years after the control architecture was defined.

If a Speed Limit could be enforced at the joint level, the provided stopping data could be directly applied. Instead, it is currently necessary to compute a complex inverse kinematic function to map TCP speed limits to specific joint speed limits. It is likely that such a change in the modality on how robot controllers enforce speed limits (particularly the introduction of safety-rated speed governors at the robot joint level) will require additional specialized and safety-rated hardware.

3.4.3 Safety-Rated Category 2 Stop

Building an end-to-end safe SSM implementation requires that every element of the sensing, computation, and actuation loop be safety-rated. However, the standards currently require that only the Category 0 (uncontrolled safety stop) and Category 1 (controlled stop with motor shutdown) be safety-rated. Category 2 stops (controlled stop with motors powered on) are not required to be safety-rated. And since they are not safety-rated, there is rarely any stopping distance or time data provided.

This means all practical SSM implementations must rely on, at best, Category 1 stops. The issue is that requiring the motors to be powered off means that automatic restart is cumbersome and time consuming. Ideally, it would be preferred to use a Category 2 stop, as the motors would remain engaged and restart after a Category 2 stop would be in the order of milliseconds. This would provide much more fluid human–robot interaction.

3.4.4 Improved Robot Controller Latencies

As presented in Sect. 2.2.2, robot controller command latencies are both longer than desired and not always deterministic. In particular, the command latencies in the execution of a Protective Stop (for example, a Category 1 stop) are "bundled" in the published stopping time and distance data needed for the practical calculation of S_r in Eq. (1). And, because of the "worst-case" approach to published stopping time and distance data, these data are likely to incorporate the worst possible command latency.

What is needed is better data disclosure; outlining in more detail command latencies (so that the command latency can be separated from the actual robot stopping kinematics); deterministic, fully repeatable and fully-characterized command latencies; and most important (but most difficult), much lower absolute command latencies, on the order of tens of milliseconds at most. We believe upgrading robot controllers to minimize latencies to external commands will be difficult, as control architectures were not designed for extensive external interaction, which would be needed for fluid human robot collaboration.

3.4.5 Safety-Rated Robot Position Data and Speed Overrides

As described in Sect. 2.2.1, robot position data is generally not safety-rated, is not deterministic (as it is transmitted over UDP), and can have very high latency, with reported position and actual position lagging, in some instances, by hundreds of milliseconds. And because the position data is not safety-rated, other (likely more costly, computationally intensive, and slow) means of confirming the robot position are needed.

A related issue is the functionality of robot speed overrides. Most robot controllers include the ability to override TCP speed as a percentage of a programmed value. This functionality is key for human robot collaboration, especially for slowing down (or speeding up) the robot in fluid response to human approach (or separation). By slowing down the robot, we can both dynamically reduce the PSD and signal to the human that the robot "sees" and is "responsive" to the human approach. However, this speed override is generally not safety-rated and can have long and non-deterministic command latencies. Building a safety-rated, low-latency, and deterministic Ethernet channel (for example, using a black channel approach [24]), for both reading robot position and providing robot speed overrides would be a major step in updating robot controllers for safe and fluid SSM.

3.4.6 Safety-Rated Robot Trajectory Planner

At this point, we know of no robot manufacturer where the robot trajectory planner is safety rated. Even a simple repetitive trajectory cannot be "guaranteed" safe and cannot be used to predict the future robot state in compliance with the ISO 13849 and ISO 10218–1 standards. If the trajectory was safety-rated, it would obviate the need for external safety monitors, minimize all the uncertain terms in Eq. (1) and allow for the calculation of the smallest PSD possible and the tightest integration for human–robot collaboration. Developing a safety-rated trajectory planner would possibly involve a bottom-up redesign of control architectures, but it would "solve" the problem of sub-optimal SSM implementations.

4 Conclusion

In spite of its current limitations, major manufacturers and robotics end users are trying out collaborative robotics using SSM as it has the potential to greatly improve productivity, lower setup time and costs, as well as provide better ergonomics for an aging production labor workforce. In this paper we provided an overview of SSM, the technical and commercial issues preventing widespread adoption, and some practical implementations given the current state of robot safety controllers.

We have also identified key ideas where robot manufacturers and industry participants can work together to improve the performance of all sensor-based industrial robot control. In lieu of directly improving stopping functionality, dynamic SSM can become much more efficient if, as an industry, we simply report more accurate and detailed stopping performance data. Next steps on the side of the robot manufacturers might focus on the reliable estimation of stopping distances and providing simulation tools to support system integrators in the design of efficient collaborative applications.

Longer term, however, there is a general need for low-latency safety-rated robot interfaces that can control, in real time, robot speed and trajectory through outside signals. Robot manufacturers and industry participants need to develop and execute on a technology roadmap that shifts the mindset to a more open model of external robot control.

References

1. Schmidt B, Wang L (2014) Depth camera-based collision avoidance via active robot control. J Manuf Syst 33(4):711–718
2. ISO (2016) ISO/TS 15066:2016 robots and robotic devices—collaborative robots. ISO, Geneva
3. ISO (2010) ISO 13855:2010 safety of machinery—positioning of safeguards with respect to the approach speeds of parts of the human body. ISO, Geneva
4. ISO (2011) ISO 10218–1:2011, robots for industrial requirements—safety requirements—part 1: robot. ISO Copyright Office, Geneva
5. ISO (2011) ISO 10218–2:2011 robots and robotic devices—safety requirements for industrial robots—part 2: robot systems and integration. ISO, Geneva
6. ISO (2015) ISO 13849–1:2015 safety of machinery—safety-related parts of control systems—part 1: general principles for design. ISO
7. Lasota P, Fong T, Shah J (2014) A survey of methods for safe human-robot interaction. Found Trend Robot 5(4):261–349
8. Villani V, Pini F, Leali F, Secchi C (2018) Survey on human-robot collaboration in industrial settings: safety, intuitive interfaces and applications. Mechatronics 55:248–266
9. Rosenstrauch M, Pannen T, Krüger J (2018) Human robot collaboration—using kinect v2 for ISO/TS 15066 speed and separation monitoring. Proced CIRP 76:183–186
10. Salmi T, Väätäinen O, Malm T, Montonen J, Marstio I (2014) Meeting new challenges and possibilities with modern robot safety technologies. In: Zaeh M (ed) Enabling manufacturing competitiveness and economic sustainability. Springer International Publishing, Munich, Germany, pp 183–188
11. Belingardi G, Heydaryan S, Chiabert P (2017) Application of speed and separation monitoring method in human-robot collaboration: industrial case study. In: 17th International scientific conference on industrial systems. Novi Sad, Serbia
12. Vicentini F, Giussani M, Molinari Tosatti L (2014) Trajectory-dependent safe distances in human-robot interaction. In: Proceedings of the 2014 IEEE emerging technology and factory automation (ETFA). Barcelona, Spain
13. Zanchettin A, Ceriani N, Rocco P, Ding H, Matthias B (2015) Safety in human-robot collaborative manufacturing environments: metrics and control. IEEE Trans Autom Sci Eng 13(2):882–893
14. Safeea M, Mendes N, Neto P (2017) Minimum distance calculation for safe human robot interaction. Proced Manuf 11:99–106
15. Matthias B, Kock S, Jerregard H, Kallman M, Lundberg I, Mellander R (2011) Safety of collaborative industrial robots: certification possibilities for a collaborative assembly robot

concept. In: 2011 IEEE international symposium on assembly and manufacturing (ISAM). Tampere, Finland
16. Bdiwi M, Pfeifer M, Sterzing A (2017) A new strategy for ensuring human safety during various levels of interaction with industrial robots. CIRP Ann Manuf Technol 66:453–456
17. Michalos G, Makris S, Tsarouchi P, Guasch T, Kontovrakis D, Chryssolouris G (2015) Design considerations for safe human-robot collaborative workplaces. Proced CIRP 37:248–253
18. Vicentini F, Pedrocchi N, Giussani M, Molinari Tosatti L (2014) Dynamic safety in collaborative robot workspaces through a network of devices fulfilling functional safety requirements. In: ISR/Robotik 2014; 41st international symposium on robotics. Munich, Germany
19. Rybski P, Anderson-Sprecher P, Huber D, Niessl C, Simmons R (2012) Sensor fusion for human safety in industrial workcells. In: 2012 IEEE/RSJ international conference on intelligent robots and systems. Vilamoura, Portugal
20. Lasota P, Rossano G, Shah J (2014) Toward safe close-proximity human-robot interaction with standard industrial robots. In: 2014 IEEE international conference on automation science and engineering (CASE). Taipei, Taiwan
21. ABB Group (2014) Rapid reference manual. ABB, Sweden
22. Byner C, Matthias B, Ding H (2019) Dynamic speed and separation monitoring for collaborative robot applications: concepts and performance. Rob Comput Integr Manuf 58:239–252
23. Veo Robotics. www.veobot.com
24. Creech G (2007) Black channel communication: what is it and how does it work? Measurement + Control, pp 304–309

Ethical Aspects of Human–Robot Collaboration in Industrial Work Settings

Aimee van Wynsberghe, Madelaine Ley, and Sabine Roeser

Abstract In this chapter, we review and expand on the current ethical research on Human–Robot Collaboration in industrial settings. To date, the ethical issues discussed include: job loss, reorganization of labour, informed consent and data collection, user-involvement in design, hierarchy in decision-making, and coerced acceptance of robots. These wide-ranging issues are a useful starting point for discussion, yet as the number of robots designed and deployed as collaborators in industrial settings grows, ethical research must evolve to allow for more nuance in the previously listed issues as well as a recognition of novel concerns as they arise. In this paper, we suggest new ethical aspects related to collaborative robots in industrial settings, including: emotional impact on workers; effects of limited movement; the potential effects of working with one's replacement; the 'chilling effects' of performance monitoring; the possibility for disclosure of new and unintended information through data collection; and the inability to challenge computerized decisions. Taken together these thoughts are meant to open the door towards new forms of moral learning necessary for assessing the ethical acceptability of human–robot collaborations on the factory floor.

1 Introduction

The future of human–robot collaboration is changing as robots are no longer considered mere tools; rather, robots are now seen as collaborative partners in completing a task. Under the umbrella of Industrie 4.0 (I4.0), a term coined by the German government and widely adopted to describe the future of high-tech manufacturing (e.g. 'smart' factories and warehouses) [1], robots are being developed to be more adaptable, lightweight and flexible [1, 2]. These new qualities allow robots to "work with

M. Ley · S. Roeser
Delft University of Technology, Delft, Netherlands

A. van Wynsberghe (✉)
University of Bonn, Bonn, Germany
e-mail: aimee@uni-bonn.de

M. I. Aldinhas Ferreira and S. R. Fletcher (eds.), *The 21st Century Industrial Robot: When Tools Become Collaborators*, Intelligent Systems, Control and Automation: Science and Engineering 81, https://doi.org/10.1007/978-3-030-78513-0_14

people hand in hand on a common task (with a specific goal). They use their mechanical and sensory skills to make decisions with regard to products and processes" [3], p. 277. The close dynamic between human workers and robots within the setting of I4.0 is a radical shift from workers being required to avoid hazardous robots [2]. With the increase in collaborative robots comes the need for 'moral learning' [4] about the ethical impact resulting from these new kinds of human–robot interactions. To date, public debate concerning robots in the workplace have focused on fundamental ethical issues of job loss when robots replace humans, and/or physical safety of robots and humans in close collaboration. The purpose of this chapter is to expand the current ethical discussion on collaborative robots in industrial settings to include new concerns that have only recently come into view.

Despite the considerable change that will occur in workers' relationships with robots, there is only preliminary research into the ethics of collaborative robots in industry. To date, researchers have pointed to potential job loss [5], reorganization of labour [6, 7, 8], psychological harm [2], informed consent concerning data collection [2, 3, 7], the need for user involvement in technology design [7], hierarchy of decision-making [3], and coerced acceptance of robots and/or lack of ownership over new work [3, 7]. Each of these issues remain relevant in today's discussion of the ethics of collaborative robots, however, as the technology becomes more pervasive in factories across the globe it is important to take stock of other underexplored ethical issues.

The following chapter begins the process of 'moral learning' about the new ethical issues raised by collaborative robots in industrial contexts by building on the ethical issues listed above. In particular, adding a broader set of ethical considerations going beyond physical harm and informed consent to include a focus on emotions and embodied experiences. This provides more nuance to the previously listed 'psychological harm' to include other considerations such as: how robots impose ways of working on humans and how this affects people's experience of the meaning of work; potential effects of working with one's replacement; 'chilling effects' of performance monitoring; disclosure of new and unintended information through data collection; the inability to challenge computerized decisions; and the potential for bias in algorithmic decision-making. We conclude by arguing that to understand the nuanced and complex ethical issues in future factories and warehouses, attention must expand beyond the one-to-one of human and robot interactions and consider the human–robot system interaction (HRSI) when collaborative robots are placed within the context of the digital-physical systems of the industrial setting [9].

2 Ethics of Collaborative Robots in Industrial Settings: Current State of the Art

While the intention of this chapter is to begin the process of unpacking and reflecting on the variety of unintended, and potentially unethical, risks associated with collaborative robots in industrials settings, it should be noted that there are at the same time many positive effects of the technology. The new wave of collaborative industrial robots promises to help alleviate the burdens of dull, dangerous, and dirty work. For example, robots can take on the task of heavy and repetitive lifting, thereby preventing long-term health consequences of such work. Collaborative robots, otherwise called co-bots, are praised for their ability to assist workers in the factory often replacing workers from dangerous tasks. A recent BBC piece showed the health benefits of robots that could replace humans from tasks in a glass factory that resulted in the inhalation of dangerous substances [10]. Even others argue that the increase of robotics in industry will make it possible for humans to stop mind numbing repetitive tasks, and focus on using their creative and interpretive capacities [6].

It should also be noted that while the intention to remove humans from the tasks labelled as dull, that in itself represents an evaluation of certain forms of work currently being done and that there are individuals currently engaged in said 'dull' tasks that are entirely comfortable with jobs of that sort. As such, the first label of dull tasks as negative and/or in need of being alleviated, should itself be critically examined.

While preventing immediate harm and long-term physical or mental burnout certainly has a positive effect on quality of work, radically shifting workers' relationships with robots will inevitably impact quality of work, and therefore is deserving of sustained ethical attention. The ethical issue that receives the most attention, public and academic, is job loss. In a 2017 study, for example, Frey and Osborne estimated: "[...] 47% of total US employment is in the high risk category, meaning that associated occupations are potentially automatable over some unspecified number of years, perhaps a decade or two" ([5], p. 265). In response to the threat of massive unemployment and its potential societal effects, researchers argue for careful reorganization of labour, where robots take on the above mentioned dull, dirty and dangerous tasks, as well as redeployment, where people trained to take on new tasks that "require relatively more judgement, skills, and training". Multinational consultancy firm, Mckinsey & Company, for example, urges employers to ready their employees with the skills needed to work in the future, to invest in "jobs related to developing and deploying new technologies... workers globally will have to master fresh skills as their current jobs evolve alongside the rise of automation, robotics, AI, and the capable machines thereby enabled" [11], p. 27. Despite best efforts to shift focus from unemployment to redeployment within industry, the threat and fear of job loss is never fully eradicated and often acts as the negative motivator for taking seriously some of the ethical issues listed below.

Physical safety has long been a main ethical concern for humans and robots collaborating on the factory floor and accordingly a high priority for roboticists. The earliest robots in the factory were kept behind cages to prevent physical interactions with human workers, thereby minimizing harm as much as possible. Today, factory robots are working towards the capability to adapt velocity, movement and routes to avoid collision with other robots, artefacts or humans in their path, with the time of the highest level of safety possible.

Despite the efforts to maintain physical safety when humans and robots collaborate, studies of industrial applications have shown an insufficient effort to integrate worker's needs and wellbeing into the design and implementation of collaborative robots [2, 12], p. 163. Fletcher and Webb argue that the potential 'psychological harm' of workers is unduly neglected. They claim that asking an individual to work beside robots may be psychological wearing because he or she may have been trained to perceive the machines as untrustworthy or hazardous, "Simply asking individuals to change their assumptions of danger or accept prescribed levels of risk acceptability may infuse various psychological concerns and anxieties, and we should not assume they are able to adopt new concepts that directly conflict with long-held beliefs" [2], p. 165. The collaborative robots built for the I4.0 context are explicitly designed for increased physical interaction and/or closeness with humans [1], p. 270 and the effects of this proximity may well go beyond mere physical safety.

As stated above, the collaborative human–robot dynamic is made possible by advances in "algorithms, metrology, sensory capacity and actuators" [2], p. 162. These technological advances open up new ways for data on an individual to be collected, analyzed and stored [3], p. 282. Fletcher and Webb point out that this will not merely be used to ensure a worker's immediate safety, say to stop the robot from knocking him or her over, but also to record and analyze his or her performance [2], p. 166. When concerning the data collection resulting from working with a collaborative robot, the existing ethics literature tends to focus on informed consent. Questions arise about how data strategies will be communicated to workers and how 'informed' is to be measured [2], p. 166. For example, to what extent does a worker need to understand the technical aspects of data collection? This falls within a wider ethical discussion on best practices to explain to users how their information is being used and most often draws on the work done in medical settings and the concept of informed consent to structure said involvement [13–15]. Lengthy technical explanations, for example, are criticized for oversaturating people with information and discouraging them from reading carefully, thereby compromising their understanding. Gaining consent in a work setting is further complicated by the fact that consent may not be freely given if/when a person's job and livelihood are at stake, for example, if workers are asked to give consent by their employer and are worried about job loss if they decline consent.

Users' input in design and implementation is another ethical issue that has been raised in so far as worker's participation is often omitted in the design and implementation process [2], p. 162. In an effort to mitigate potential negative consequences of collaborative robots, some researchers have argued that end-users (the workers) should be involved in the design of the technology they work with [2], p. 163, [7], p. 23. If a robot is meant to complement human workers, then workers themselves have valuable insights into what would help them with their work, or what might simply be burdensome or dangerous. Workers' involvement in how robots are implemented also contributes to an individual's sense of meaningful control over his or her job, which can increase his or her quality of work [7], p. 25. The lack of worker involvement in design and/or implementation is itself seen an ethical issue worthy of attention but it also points towards another issue, an asymmetry in power between robot workers and human workers.

It has been suggested that 'as robots' abilities for autonomous decision-making increases the workers' autonomy decreases' [11], p. 27. Smart factories, for example, will need to remain competitive and will rely more and more on computers to make fast decisions that can be implemented quickly. As artificial intelligence (AI) is increasingly used in robots, robots may not just take over the physical labour, but also the cognitive functions as well, meaning "humans could indeed become obsolete as partners and teachers" [3], p. 287. The consequences of robots being primary decision-makers is not only potential job loss, but a shift in authority: "the robot is no longer a mere tool; instead it uses mankind as a tool" [3], p. 284.

Finally, the existing literature on the ethics of collaborative robots in industry both explicitly and implicitly suggests that workers may be coerced into accepting new robotic technologies. Some points listed above, namely fear of job loss, potential psychological harm, exclusion from the design process, and lack of technical knowledge, point to a lack of freedom when workers are choosing to work with robots. When discussing the use of exoskeletons, which can be used to support worker's bodies in heavy lifting, Bendel claims that a worker may *have* to put on the machine, due to economic motivations rather than intrinsic desire for human enhancement [3], p. 284. Again, we acknowledge that requiring workers to adapt their labour and bodies to an incoming technology is not a new phenomenon; however, we suggest that the technologies of I4.0 have a more dynamic relationship with a worker's body, taking a more active role as a collaborator rather than a mere tool. Concern about a human's sense of autonomy seems also to underlie discussions on the meaning of, and ownership of, work.

As the technology of robotics and AI are advancing at a nearly exponential rate [6], while at the same time still used in society in a relatively experimental manner [16], we suggest that those companies who are deploying collaborative robots should aim at moral learning, i.e. learning about the ethical issues that arise when workers have increased exposure to the technology. Moral learning, requires continually taking stock of the known and unknown ethical issues and is necessary in order to mitigate or prevent the previous ethical concerns related to human–robot collaboration. Our goal for the next section is to flag plausible risks for which more moral learning is required to fully understand their scope and impact.

3 Identifying Gaps to Address in the Future of Human–Robot Collaboration

In this section we expand the landscape of ethical issues that have already been raised by adding nuance to previously discussed issues as well as introducing new areas altogether for ethical attention. In both instances we arrive at this expanded list of ethical concerns by reflecting on the growing applications of collaborative robots in industrial contexts and the capabilities these robots have, as found in the academic literature. Given that there is little literature available on the topic, our goal is also to bring together reflections from other disciplines that address technologies with similar or overlapping technical capabilities. For example, the use of machine learning (ML) algorithms in banking, finance and/or healthcare, have raised concerns about the kind and disclosure of new information that can be derived about individuals when multiple streams of data are combined.

Furthermore, the existing ethical approaches discuss commonly acknowledged issues such as physical harm, privacy, autonomy and security. However, there are also other dimensions of ethical relevance that deserve attention, and which relate to the embodied nature of human–robot interactions and emotional impacts on people. These kinds of impacts are easily neglected because they are harder to pinpoint and formalize, often considered unquantifiable. They may even be considered to be irrelevant by some, but as we will argue below, they are actually of vital importance for a discussion of the ethics of human–robot collaboration and deserve immediate attention to safeguard the wellbeing of workers who will collaborate with robots.

3.1 Imposing on Work Routines

To begin, we take a closer look at the ethical issue of 'psychological harm', already raised by Fletcher and Webb; however, we suggest going into greater detail and depth about the forms that psychological harm may manifest itself. First, we wish to suggest the possibility that robots may impose a way of working of individuals, one that may feel restrictive and/or machine-like. Examples of this include having to adhere to specific routes, work at exact speeds, or place items with routine accuracy so that the robot's work is not interrupted. Being forced to move in such a precise way in order for the collaborative robot to function correctly may lead to frustration, or a sense that one is easily replaceable (with another person or robot). The latter can result in further feelings of being disengaged or insecure in one's jobp.

3.2 The Need to Focus on Emotions

Emotions are frequently neglected in decision making about technology, even in ethical, sociological and participatory approaches to technology assessment and responsible innovation. However, emotions and embodied experiences can point out important ethical issues that need to be addressed [17], and in this case can point towards underexplored issues in need of greater attention while collaborative robots are being deployed. We will argue that these aspects are ethically important and need full consideration.

The smart devices used in Amazon warehouses, as an example, result in workers attempting to be efficient in the extreme, wearing adult diapers to maximize efficiency otherwise prevented by bodily functions and the need to go to the washroom. The threat of job loss prohibits many people from challenging productivity expectations, even though meeting these standards may lead to physical or emotional discomfort. One worker describes: “You had to process the items in seconds and then move on. If you didn’t meet targets, you were fired” [18]. The effects of sustaining this quality of work can be dehumanizing: “After a year working on the floor, I felt like I had become a version of the robots I was working with” [18]. As people work hand-in-hand with collaborative robots, human workers may be further forced to move in machine-like ways. We put forward here that the scope of potential effects resulting from humans working alongside robots range from harmless adaptation of the body to a greater sense of dehumanization.

Another aspect of psychological harm, beyond a concern of dehumanization to human workers, may occur when the robot is perceived as a threat to one’s job. Collaborating with a robot may negatively affect a person’s quality of work because she feels as though she is working beside her replacement. The threat of job loss has been a popular subject of debate for years, but the integration of robots into dynamic work spaces marks a shift from a theoretical threat to one that is more concrete. Already within factories, warehouses, and stores, workers are starting to feel the need to compete with the productivity levels of the robots. For example, when Walmart introduced a robot, the workers expressed a sense of stress when stocking alongside a robot because they felt the need to become hyper-efficient and even ‘robotic’ [19]. When the task is repetitive and boring, humans compete in vain with robots because the machines can work tirelessly and without breaks. Still, people try to compete in order to keep their job, which results in horrifying and inhumane working conditions. For example, UK trade union GMB reported: “workers using plastic bottles to urinate in instead of going to the toilet, and pregnant women have been forced to stand for hours on end, with some pregnant women being targeted for dismissal” [20].

3.3 Performance Monitoring

In addition to psychological harm with an emphasis on emotional responses by workers, we also wish to include an additional set of ethical concerns related to ‘performance monitoring’ [2], p. 159, [3], p. 282. Data collection and analysis of the physical environment is a main goal of I4.0 and collaborative robots will play a role in this. However, if performance monitoring becomes surveillance for the sake of surveillance then freedom of workers is at risk. Places of work have never been spaces that allow unbridled freedom, factories and warehouses even less so. Managers oversee workers and cameras have long-since been installed. However, digital documentation of workers “enables monitoring, prioritization and judgement to occur across widening geographical distances and with little time delay; second, it allows the active sorting, identification, prioritization and tracking of bodies, behaviours and characteristics of subject populations on a continuous, real-time basis” ([21], p. 228). In addition, the data may be kept indefinitely without an official process to have the information deleted. As such, performance monitoring should be considered on a short term basis as well as a long term basis in which ones performance is stored and used for future evaluations and/or predictions of productivity.

Implicit or explicit surveillance in the workplace may have an additional unintended consequence known as the 'chilling effect'. Traditionally, the 'chilling effect' is made reference to "when governmental regulation and policy not directed at certain activities deters individuals from carrying out protected activities" [22], p. 585. In other words, the fear of surveillance may have a negative impact on one's ability to function normally in one's day-to-day life. A worker may experience the chilling effect when working alongside a collaborative robot because he or she does not understand or feel in control of the data being collected. Robbins and Henschke call this an 'information deficit', i.e. a lack of knowledge on the part of the worker pertaining to the quantity and/or quality of data that the robot is collecting on the worker. The information deficit, in this situation, exists when the employer has knowledge of, and access to, all the data being collected and the way it is used while the worker does not.

A survey done by the Pew Research Center showed that people experienced this chilling effect even when policies were put in place to protect citizen's rights. Thus, as Robbins and Henschke suggest, it is not only about the existence of policies but also about whether or not citizens, and in this case workers, have the 'assurance and ensurance' that their rights are protected [22]. The potential information deficit of workers links to the above mentioned concerns of Fletcher and Webb, Bendel, and Went et al., which grapple with the ethics of informed consent. In this case, we argue that ethics attention on informed consent should not only focus on what a worker can or should understand, but also on how it affects the choices of how people act and move in work. It can have significant impact on people's experience of belonging and commitment and being valued as agents in a work setting.

Another concern related to performance monitoring and surveillance is that of disclosure of new kinds of information when machine learning (ML) algorithms are used to 'make sense' of the data provided by workers. In other words, with outwardly mundane or meaningless data, a ML algorithm may be able to find (seemingly) meaningful correlations in the form of disclosure of new kinds of information. For example, when determining an individual's credit score a bank may analyse her online social presence. In this case her social media presence is being taken out of its original context and used to generate other information. Collaborative robots will contribute novel data streams that may be used to disclose information about individuals. The kinds of information will be dependent on the level of collaboration: passing items back and forth on the line; pressing buttons to get robots to do what you need; wearing an exoskelton. The results of such collaborations could be giving employers unexpected and private medical information of their employees. For example, sensors may indicate that a worker had developed a hand tremor. The vast amount of data collected in a factory or warehouse using smart technology may be used in various combinations with online search history or social media presence to reveal information about workers that was not originally intended or consented to.

3.4 Asymmetry in Power

Additionally, we add to Bendel's brief discussion on the prioritizing of technology over humans, in order to discuss how workers could feel a sense of competition or voicelessness alongside robots. Information collected by collaborative robots adds to the decentralized decision-making of modern industrial systems. When a manager doles out a task, they may be challenged (though not always without consequences). When given a task by a robot collaborator, however, it is difficult and maybe even dangerous to ask for more information in real-time. Further, explaining why one went against the robot's orders may prove challenging as computerized data is often considered objective and therefore correct. Yet, the very notion of objectivity is philosophically contentious. In the case of I4.0, computer analyses are not neutral or objective, but designed with particular perimeters in place. For example, the computer is designed to gather information on worker's speed and length of breaks, but not emotional wellbeing or relationship with a coworker. Therefore, when making decisions or delegating tasks, the computer will only analyse a part of the human experience of work.

The decentralized decision-making also obscures the process of determining responsibility and accountability. Assigning blame may soon be left to computer analysis, rather than to the interpretation and evaluation by humans. The first steps towards this have already begun, as an Amazon worker reports: "You're being tracked by a computer the entire time you're there. You don't get reported or written up by managers. You get written up by an algorithm" [20]. The ethical concern here is that algorithms can be rife with human biases. Countless studies today reveal how biases present in historical data are exacerbated in the resulting algorithm with extreme detrimental consequences for society when used in predictive policing and other public services like child maltreatment assessments by governmensts [23–26]. It would be narrow sighted to assume that industrial settings are impervious to similar algorithmic biases. Humans are biased, of course, but prioritizing computer systems under the guise of neutrality risks overlooking harmful prejudices.

3.5 Collaborative Robots and Sociotechnical Systems

The introduction of collaborative robots in industrial settings has focused largely on the ergonomics, safety, psychological safety, and efficiency of human–robot interaction. It is our contention, based on the analysis presented above, that the impact of the robot must be assessed or evaluated based on its influence on the system and not only on individuals, and based on other impacts on individuals than traditionally seen, including impact on their emotions and embodied experiences. It has been argued in other work that robots integrated into a healthcare system should be assessed not only for their impact between robot and patient or robot and surgeon but should be understood as having an impact on the healthcare system as a whole—robots will change the way resources are allocated, how surgeons are trained, or what is considered expertise [9]. As such, robots should be understood as a socio-technical system—their introduction impacts more than the human with whom it interacts, rather, the robot impacts the entire system to which it enters.

In the same way that robots introduced into healthcare systems should be assessed and evaluated according to their impact on the healthcare *system* rather than their impact on the individuals interacting with the robot, we suggest that collaborative robots in industry settings should be assessed according to their impact on the system as a whole. From this vantage point it becomes possible to understand concerns such as the chilling effect of the information deficit between workers and employers and how this can influence how a person interacts with a robot. This phenomenon, along with the other concerns raised in this paper, can only be fully understood by also taking the system into account. I4.0 demands this kind of multi-layered ethical thinking because the technology itself works within a system.

4 Conclusion

Current developments in industrial robots point towards a growing trend towards collaborative robots—robots and humans that collaborate to fulfil a task (or a series of tasks). While there are many positives to such collaborations, understood as an increase in efficiency and a decrease in physical strain on human workers, there are also a number of ethical concerns to take into consideration in the overall evaluation of such robots. This chapter was meant to review the current ethical issues identified to date and to point towards new and underexplored issues deserving of greater attention and deliberate study. In particular, we addressed: how robots impose ways of working on humans and how this affects people's experience of the meaning of work; the potential effects of working with one's replacement; the 'chilling effects' of performance monitoring; the possibility for disclosure of new and unintended information through data collection; the inability to challenge computerized decisions; and the potential for bias in algorithmic decision-making. Most notably we introduced the idea that the emotional impact robots will have on workers is an area in need of greater attention. As a final thought, we aimed to direct the reader's attention towards the need for evaluating robots as sociotechnical systems which demands recognition of the robots impact on all elements of a factory context (or other institutional setting) as a necessary requirement for grasping both the complexity and the breadth of unintended ethical consequences of a human and robot collaborating. Taken together these thoughts are meant to open the door towards new forms of moral learning necessary for assessing the ethical acceptability of human–robot collaborations on the factory floor.

Acknowledgements This research is supported by the Netherlands Organization for Scientific Research (NWO), project number 275-20-054 and as part of the Airlab TU Delft + Ahold Delhaize Collaboration.

References

1. Gilchrist A (2016) Industry 4.0: the industrial internet of things. Apress
2. Fletcher SR, Webb P (2017) Industrial robot ethics: the challenges of closer human collaboration in future manufacturing systems. In: Aldinhas Ferreira MI, Silva Sequeira J, Tokhi MO, Kadar EE, Virk GS (eds) A world with robots: international conference on robot ethics: ICRE 2015. Springer International Publishing, pp 159–169. https://doi.org/10.1007/978-3-319-46667-5_12
3. Bendel O (2018) Co-robots from an ethical perspective. In: Dornberger R (ed) Business information systems and technology 4.0: new trends in the age of digital change. Springer International Publishing, pp 275–288. https://doi.org/10.1007/978-3-319-74322-6_18
4. Van De Poel I (2018) Moral experimentation with new technology. In: New perspectives on technology in society: experimentation beyond the laboratory. (ibo van de Poel, Donna C. Mehos, and Lotte Asveld. Routledge, pp 59–79
5. Frey CB, Osborne MA (2017) The future of employment: How susceptible are jobs to computerisation?. Technol Forecast Soc Chang 114:254–280
6. McAfee A, Brynjolfsson E (2015) The second machine age: work, progress, and prosperity in a time of brilliant technologies. W. W. Norton & Company
7. Went R, Kremer M, Knotterus A (2015) Mastering the robot. The future of work in the second machine age. The Netherlands Scientific Council for Government Policy. https://english.wrr.nl/publications/investigation/2015/12/08/mastering-the-robot.-the-future-of-work-in-the-second-machine-age
8. Autor DH (2014). Skills, education, and the rise of earnings inequality among the "other 99 percent". Sci 344(6186):843–851
9. van Wynsberghe A, Li S (2019) A paradigm shift for robot ethics: from HRI to human–robot–system interaction (HRSI). Med Bioeth 9:11–20. https://doi.org/10.2147/MB.S160348
10. Politics of automation: Factory workers and robots (2019) In BBC news. https://www.bbc.com/news/av/uk-politics-48991461/politics-of-automation-factory-workers-and-robots
11. Teulieres M, Tilley J, Bloz L, Lugwig-Dehm PM, Wägner S (2019) Industrial robotics: insights into the sector's future growth dynamics. McKinsey & Company
12. Bartneck C, Kulic D, Croft E (2008) Measuring the anthropomorphism, animacy, likeability, perceived intelligence, and perceived safety of robots. Tech Report 8
13. Barocas S, Nissenbaum H (2014) Big data's end run around anonymity and consent. Privacy, big data, and the public good: frameworks for engagement, pp 44–75
14. Beauchamp TL, Childress JF (2001) Principles of biomedical ethics. Oxford University Press
15. Solove DJ (2012) Introduction: privacy self-management and the consent dilemma, vol 126, pp 1880–1893
16. van de Poel I (2013) Why new technologies should be conceived as social experiments. Eth Policy Environ 16(3):352–355. https://doi.org/10.1080/21550085.2013.844575
17. Roeser S (2017). Risk, technology, and moral emotions. Routledge
18. Yeginsu C (2018) If workers slack off, the wristband will know. (And amazon has a patent for it.). The New York Times. https://www.nytimes.com/2018/02/01/technology/amazon-wristband-tracking-privacy.html
19. Harwell D (2019) As Walmart turns to robots, it's the human workers who feel like machines—The Washington Post. https://www.washingtonpost.com/technology/2019/06/06/walmart-turns-robots-its-human-workers-who-feel-like-machines/

20. Drury C (2019) Amazon workers 'forced to urinate in plastic bottles because they cannot go to toilet on shift.' The independent. https://www.independent.co.uk/news/uk/home-news/amazon-protests-workers-urinate-plastic-bottles-no-toilet-breaks-milton-keynes-jeff-bezos-a9012351.html
21. Graham S, Wood D (2003) Digitizing surveillance: categorization, space, inequality. Crit soc policy 23(2):227–248
22. Robbins S, Henschke A (2017) The value of transparency: bulk data and authoritarianism. Surveill Soc 15(3/4):582–589. https://doi.org/10.24908/ss.v15i3/4.6606
23. Burrell J (2016) How the machine 'thinks': understanding opacity in machine learning algorithms. Big Data Soc 3(1):2053951715622512. https://doi.org/10.1177/205395171562 2512
24. Chouldechova A, Putnam-Hornstein E, Benavides-Prado D, Fialko O, Vaithianathan R (n.d.) A case study of algorithm-assisted decision making in child maltreatment hotline screening decisions, p 15
25. Corbett-Davies S, Pierson E, Feller A, Goel S, Huq A (2017) Algorithmic decision making and the cost of fairness. [Cs, Stat]. https://doi.org/10.1145/3097983.309809
26. Haussler D (1988) Quantifying inductive bias: AI learning algorithms and Valiant's learning framework. Artif Intell 36(2):177–221

Robots and the Workplace: The Contribution of Technology Assessment to Their Impact on Work and Employment

Tiago Mesquita Carvalho and Tiago Santos Pereira

Abstract In this chapter we address the challenges that automation/robotization poses for labor and employment issues. Our take is that technology assessment (TA) can provide a ground for both ethical reflection and social engagement towards participatory decision-making regarding the application of such technologies. The role of underlying narratives regarding robots and automation processes is also highlighted. The chapter debates labour substitution as a dominant narrative in economic analysis, while also stressing the need to contextualise technological change and innovation regarding robots and automation in the concrete work processes or tasks, bringing narratives closer to the ground. This discussion leads us to the second main theme of the chapter: the potential role of technology assessment in better exploiting the development and use of robots in the workplace, their unanticipated consequences and the ethical and social tensions arising therein. Such approaches do not aim at complete or sound predictions, but at building participatory and interdisciplinary processes. This chapter is then about how we ought to live and to relate to technology.

1 Introduction

The fields of ethics and philosophy of technology, STS and Technology Assessment (TA) have since long dealt with various topics concerning the use of robots at home, workplaces and in the battlefield [31]. Ongoing topics of research include more direct matters such as safety, privacy and responsibility and how far these concerns can be actively integrated into such artificial systems. The range and breadth of robotic innovations' impacts can give way not only to generic social conflicts, but also to policy, legal and ethical challenges as well [37, 36]. For evaluating these matters, it

T. M. Carvalho (✉) · T. S. Pereira
Collaborative Laboratory for Work, Employment and Social Protection (CoLABOR), Lisbon, Portugal
e-mail: tiago.carvalho@colabor.pt

T. S. Pereira
Centre for Social Studies (CES), University of Coimbra (UC), Coimbra, Portugal

M. I. Aldinhas Ferreira and S. R. Fletcher (eds.), *The 21st Century Industrial Robot: When Tools Become Collaborators*, Intelligent Systems, Control and Automation: Science and Engineering 81, https://doi.org/10.1007/978-3-030-78513-0_15

should also be accounted for that robotics has since decades been the center of several science fiction authors and other speculative works, which have undoubtedly helped raise questions about the nature of the interaction between humans and robots.

Our approach to these matters encompasses a broad criticism of technological determinism while acknowledging the role that related narratives play in constituting the plural social meaning of robotic applications [16]. The stories underlying the development of new and emergent science and technologies (NEST) like robotics, synthetic biology, nanotechnology, AI or geoengineering help shape collective imaginaries. The rejection of technological determinism and the importance of narrative conveys consequences for how TA is conducted. For instance, future visions of AI, automation and robots have been related with the dawn of a technological singularity [25], working almost as a spiritual beacon lightning the way to more promising expectations and applications. Regarding narratives, most popular is the fear of a complete or partial robot revolution or uprising. Even if it is highly unlikely that these fears materialize in an overnight upturn, it is important to underscore how collective imaginaries and visions of the future shape present relationships, either by showcasing anxiety or by unbridled hopes.

Concurrently, although progress in robotics currently proceeds at a relatively fast pace, in reality there are still a lot of technological thresholds and emerging practical problems that call into question the straightforwardness of such narratives towards images of the future. These difficulties relate, for instance, to solving apparently simple tasks that are obvious to humans but that currently thwart the best available scientific and technological knowledge [10]. Not all moral and intellectual capabilities that we usually relate to human beings can be programmed into robots or machines. Unlike humans, robots cannot as yet recognize ethically charged situations, as researchers struggle to give them a world and a sense of what matters and what's useless.

Our contention is that these thresholds and limits illustrate epistemic gaps in the development phase of robotics where TA and attention to ethical, legal and social concerns can be introduced and presented. Granted that the social impact of such products cannot be entirely predicted, there are still opportunities for interdisciplinary teams advancing new material ingenious solutions and circumvent troublesome friction points.

The role of TA should hence not be regarded as necessarily hindering or dampening technological innovation but as a thorough and engaged examination of the hurdles, issues and problems that may arise during and after developments. The point is to see that ethical concerns and TA methods reflect legitimate social, legal and moral concerns that can potentially be integrated by engineers and designers in the very core of robots or automated machines. Regulations for certain robotic performances are thus not necessarily opposed to innovation, but can even contribute to opening new markets where human laws and values are upheld. Some of the challenges that ethical and TA raise about new and upcoming technologies can also bolster industry itself to include such considerations in the design process or in later phases.

It might not be obvious in the context of liberal democracies that the importance of technological development for economic growth should be somewhat regulated and controlled. The underlying view is that the more institutions and professionals call forth a variety of questions and concerns, even apprehensions over the pace and impact of technological innovation, the more industry will be cautious about employing and developing marketable emerging technologies. Some institutional landscapes can appear to be cumbersome to companies that already face economic scarcities of various sorts in order to operate and finish projects with demanding deadlines and at the same time present profitable and socially useful products. In a worst case scenario, governmental regulations on innovative technological improvements can discourage business in a given geopolitical framework. Regulations may then be seen as part of a disadvantageous institutional landscape that affords competitors elsewhere an insurmountable advantage. There is thus a thin line of balance between avoiding or managing technological risks and uncertainties and the proper stimuli to a competitive and ingenious environment.

It must be also acknowledged that society is set to be rocked with many technological novelties emerging in the near future. Industrial robots have long been part of such developments and together with other groundbreaking technologies they are set to rise not only public hard questions related to economic growth, health, safety and employment but also other soft questions. Either way, the real effects of robotics will be bounded between social hopes and fears that are more or less realistic. The social meaning of such applications in public awareness sprouts opportunities for debate about their goals and on what grounds can and should their use be criticized. It is at such jointures that philosophically and ethically minded questions in TA about technology's role in general and industrial robots in particular can play a role regarding public narratives and the design of new forms of governance.

2 Technology, Robotics and Employment

Public concerns regarding robots in society have been largely framed by imaginaries of industrial processes, and the corresponding automation processes, going back to the very origin of the word 'robot' in Capek's play or to Chaplin's *Modern Times*. While the fears of robotic domination may be less founded, not giving rise to imaginaries of confrontation between humans and machines, concerns regarding the substitution of human labour by automated processes have proliferated, characterizing more aptly these tensions as forms of displacement, translation or articulation. While sociology, anthropology, or even management have been more concerned with the microlevel dynamics of such exchanges within organisations, the impacts of robots, and automation more generally, on employment has been an important concern in economic analysis.

Economists' interest in technical change has largely pointed to its contribution to the reduction of resource use, and in particular to its labour-saving effect. Such discussion emerges regularly as new waves of technological change [12] sweep across different sectors of the economy. The wider discussions on the impacts of innovation on employment have not reached clear conclusions. [34] reviewed different approaches to the 'compensation theory', which state that market forces should compensate the labour-saving effects resulting from technological change, including job losses, as well as its critiques, and concludes that job losses cannot be guaranteed to be counterbalanced ex-ante, highlighting the underlying risks of unemployment.

The current developments in automation, through the increased use of robotics or artificial intelligence (AI) technologies, have brought equally, if not more, intense debates on the potential displacement of jobs, and the risks to the employment of many workers. There are insights from previous studies which are relevant to better understand such processes. In particular, the type of innovation, whether it is a product or process innovation, is of central relevance to identify potential expected impacts, with the former having typically a positive impact on employment, while the latter having mostly a negative impact [28]. Another long stylized fact about innovation and employment states that there is a 'skill bias' in employment resulting from innovation processes, with unskilled jobs declining while more skilled jobs see an increase as technological change advances (idem).

These conclusions have come partly into question with the recent wave of automation processes. While the first generations of robots were largely deployed in heavy industrial processes, the more recent wave of automation has seen a wider use of robots and AI technologies that go well beyond the traditional industrial setting, reaching on to different organizational processes and, even, into the service sector, with a wider impact on process innovation, highlighting its potentially higher negative impact on employment. In particular with the rise of AI technologies and its implementation in the service sector, there is a wider potential impact across more sectors of the economy. This increased attention on the potential impact on employment of the current wave of technological innovation through automation has led to a number of forecasts of such impacts. In particular, work by Frey and [14, 13] has resonated widely, both based on its methodological approach as well as on its dramatic predictions regarding the impacts of automation, concluding that 47% of jobs in the US economy are at risk of automation in the near future.

Frey and Osborne's work follows the task-based approach to the study of work processes and labour markets which have characterized recent developments in the field [4]. By addressing task content, relevant differentiations can be made, for example, between routine tasks, often seen as unsatisfactory and with few rewards for workers, and more advanced tasks, with higher rewards but also requiring more advanced skills. Rather than simply considering jobs as the unit of analysis, research started looking at the concrete tasks that workers need to take in the course of their jobs. This has concrete implications. Such differences become more clearly evident on a task-based approach to the characterization of jobs. Similarly, the impacts on job substitution differ significantly from the perspective of jobs vs. tasks, with the identification of the substitutability of tasks not resulting necessarily in similar levels of job losses.

However, while Frey and [14, 13] followed a task-based approach, analysing task contents, they still considered the automation potential of occupations as a whole and not of the corresponding tasks directly. Different critiques emerged to Frey and Osborne's approach, reaching quite distinct conclusions regarding the potential impact on job losses in the economy. [3] used the same primary data characterizing the task content of different occupations in the US, but identified how automation was still dependent on human tasks, and reached a significantly different value of 9% of potential automation of jobs in the US economy.

Besides distinguishing in greater detail the concrete tasks that were automatable and their centrality to the corresponding occupations, thereby going beyond a simple threshold for automation, Arntz et al. took a more dynamic approach to the automation process. Technological innovation does not become implemented in a contextual void, as if it depended simply on the existence of the technology choice, almost in a deterministic model. On the contrary, technological innovation depends on a context of resources, financial, cultural or organizational, in addition to the concrete technological capabilities, that lead to its implementation. Such conditions represent differing assessment processes in different organisations. Criticising Frey and Osborne's approach, Artnz et al. note that the process of automation does not depend simply on technological feasibility but also "on the relative price of performing tasks by either humans or machines". Eurofound [11] followed a similar approach, further differentiating the task content, by work activity (e.g. physical, intellectual or more social activities) as well as regarding the tools and methods typically used, further thickening the characterisation of the milieu which will be subject to automation processes. Interestingly, they find that although there appears to be a small, but significant, decrease in routine jobs, there appears to be an increasing routinization of tasks, slightly higher, and also significant, and which reaches managers, professionals and clerical occupations more significantly, challenging received wisdom on the subject. They conclude that the polarization of the impacts of the automation process on skills is less clear from a task-based approach, pointing to new questions regarding the relationship between innovation, employment and skills.

What this analysis makes clear is that automation, through the increased use of robots or AI in the workplace, is not simply dependent on some isolated technological/economic potential of the available or envisioned technologies, but rather that the changing organisation of work is a process of coproduction of technology and social and economic orders. As [27] point out, new models of work organization framing human–robot interactions in industry are still in the making. In the processes through which robots and AI are developed and use, the reorganisation of jobs and work does not appear to amount to either the hyperbolic visions of the futurists, whereby technological change is arriving at a damning speed leaving none of us immune, nor to the catastrophic ones under which we will not simply experience change but should also be prepared to collect the fully fledged consequences, namely through significant job losses. In reviewing these different positions [35] points out, following John Urry, that "the social sciences must reclaim the terrain of future studies […] because future visions have enormously powerful consequences for society, carrying with them implicit ideas about public purposes and the common good." It is in these terrain that TA emerges as an opportunity to engage all actors, futurists, technologists, workers, entrepreneurs, regulators, citizens, to discuss our different visions of the future of work and to engage in constructing futures which align technological expectations with our understandings of what works in which contexts and what we consider undesirable outcomes.

3 Robotics, Automation and the Task of Technology Assessment

Robotics, automation and artificial intelligent systems [6] in general then progress not through a linear path towards self-awareness, general machine intelligence or "the singularity", but mostly through more domestic, service and industrial applications where the chance for a revolutionary breakthrough, affordability and market demand may justify their employment. Such is the case with carebots [1], neuroprosthetic limbs or autonomous vehicles, among others. By acknowledging its ambiguity, as in any other technology, one is led to believe that its good outcomes can be chosen and hazards avoided. While industrial robots do not elicit the same range of social and ethical concerns as their domestic counterparts, they have problems of their own. TA builds on these matters and attempts to structure the discussion around alternative technological innovations and designs.

The core notion of an ethics of technology and of the more thorough methodologies of TA is that man-made artifacts, either roads, buildings or gadgets such as smart watches or vacuum cleaners, are value-laden. This adds an insurmountable normative dimension to innovation, design, use and marketing of technologies. TA focuses not on the internal goods in the practices of makers, designers or engineers [26] and in the regulations to which such professions abide [39, 40], but on drawing attention to how technological artifacts and systems influence consumers in their actions and perceptions and alter how they relate with themselves and take up the world [5, 21]. In one word, TA acknowledges as philosophically and ethically relevant the hidden normativity of the contemporary technological way of being without taking it as necessarily helpful or harmful until further examination. It proceeds through several systematic methods of inquiry into the nature and consequences of technology for society 18 , encompassing many disciplines and methods related to risk assessment, decision theory, public policies, as well to social conflicts related to large scale projects like dams, nuclear power plants, highways or oil pipelines.

Since modern times, science has advanced based on the ontological assumption that facts and values are apart [23]. Theoretical endeavors are all about researching and establishing the lawfulness of natural phenomena considered as value free, while politics and ethics concern themselves with all that is persistent in the human condition. Discussion of basic values and rights in an open and favorable institutional background may provide collective and individual progress towards the good life, but such progress is always fragile as it depends on the long-term continuity of political regimes and must be enacted by each and every citizen. While scientific and technological advancements can be supportive, they should not be regarded as tantamount to actual moral and political progress. When they lose a connection to the ends of human flourishing, they can even spoil or forestall the process of internal human achievements.

Likewise, before paying more attention to the nature of technological phenomena, TA started five decades ago almost as one brand of economic analysis, balancing the impacts of technologies in terms of prediction and outcomes. Technology was mostly seen as neutral and its impacts were assessed in order to provide politicians with choices about what should be favored or avoided. This entailed a positivistic understanding of science, where society was taken as another, yet complex, mechanical system that one could tinker with and observe the desired or undesired effects through various and powerful quantification methods [38, 29]. TA would just report its objective predictions to the decision-makers.

In this regard, TA is the institutional and practical result of historical developments about the rising complexity of the interaction between science, technology and society. It is charged with the task of bridging both realms of facts and values, of searching for the adequate translations of theory into praxis. Instead of viewing the cumulative progress in scientific knowledge as having an intrinsically positive effect on the whole of society through the means of technology, criticism of the split between facts and values now sets itself the task of shaping technology according to values and the purported ends of social development. This task is of course heterogeneous, especially in pluralistic liberal-minded democracies. TA is thus found most likely in countries with a mature relation with scientific and technological development, usually within a post-industrial historical background where the road to growth is taken as more complex than just simply applying science and expecting good results. Providing opportunities for pausing, reflecting and assessing what kind of choice, if any, society has regarding technology is the key approach.

Overarching concerns about scientific and technical advancements started being raised with momentum due to some spectacular failures or disasters, usually connected to health and environmental harm. Accidents like Bhopal, Chernobyl or the Exxon Valdez oil spill broke up the fabric of an apparent seamless reality where technology always works according to what it was designed to [22]. TA does not concern itself with avoiding just such dramatic adverse effects, but with exposing the more invisible ones by which everyday reality is transformed, while dealing with recurrent optimistic or pessimistic metaphors that reify what for instance robots or automation will bring about.

TA as a reflective engagement with all the phases of technological development is also a departure from the earlier 19th and 20th metaphysical readings of technology as a force of either liberation or damnation. The empirical turn [33] and the influence of STS has given way to a more concrete analysis. Rather than seeking comprehensive structured readings about what technology is or does, TA provides greater attention to specific cases. While TA can be nevertheless operative, it should not be forgotten that, as a social engagement with technology, it should at least provide the spectra of the big philosophical questions and answers that are at stake. Those questions feed the narratives that frame public perceptions, worldviews and sense of self. Empirical analysis does not occur in a vacuum and it would be naive to consider them disconnected from any other system of beliefs.

TA has been evolving through welcoming some conceptual changes. Assessment has shifted from being settled on an outside perspective and tied to prediction to being more of an active inside engagement with developers, engineers and other social stakeholders [15]. On the one hand the increasing complexity of technological developments incurs in an artificial "techno-opacity" [32], while, on the other, moral agency has become largely emptied due to the ignorance that agents face when deciding about the outcomes of everyday banal actions [41], thus requiring renewed forms of engaging with technological innovation and moral responsibility.

The dilemma that modern societies face can then be shortly depicted as putting all the eggs in one basket. Research and innovation receive significant amounts of public funding in the hopes of discovering innovative ways of building a better future for all. Such efforts are nevertheless liable to backlashes, sometimes with significant unanticipated effects. An active and precautionary management of the risks and comeuppances that innovation may kindle is needed, which is simultaneously aware of its own epistemic limitations about establishing what the future will be.

TA now coexists with the notion that technology is socially constructed and although it never becomes a perfectly tamable and docile device, participatory processes can nudge it towards certain values by design in a more considerate way [30]. Establishing a dialog between the public and the stakeholders is the first step towards facing and identifying problems. In pluralist societies, where the meaning of flourishment is mostly left to individual choices and preferences, it is nevertheless hard to argue how society as a whole should roam forth. Only evident harms are set to be unanimously avoided, but these are often only identified following the resulting consequences. Setting ethical or acceptable limits in terms of risks is thus not a theoretical but an utterly practical matter of defining what is at stake while recognizing different values and readings of what ought to be done. Controversy surrounding different visions of the future is thus at the heart of the matter of assessing technological innovation.

4 Robotics and the Contextual Challenges of Technology Assessment

TA is thus not a way of solving once and for all the conundrums or quagmires of technological innovation, but a scientific and context-sensitive methodological exam of the complexities that overflow it. While it may help to problem-solving, it is mostly concerned with assisting the decision and development process, detailing the assumptions of the rationality involved in the technoscientific reasoning and pointing to epistemic opacities. It attempts to highlight procedures coupling forward-looking responsibility towards the unknown with a society that holds innovation as a key to its survival, keeping at bay the optimistic fallacy that only a high level of moral individual responsibility can chart the "best" course of technological progress [17]. The unexpected or unforeseeable effects are then of course related to social impacts.

In that sense, every impact seems to call back and reinforce the role of the subject: unemployment, accidents, poisoning and other environmental hazards are some of the most studied and measured due in part to their measurable side. "Soft" impacts like skills obsolescence, the sense of belonging or of being uprooted from one's world and the loss of meaning are far more difficult to account for and are usually deemed to belong to the private sphere and are considered not as serious or worthy of assessment as other "hard" impacts [38].

TA today has shifted towards a more comprehensive view, holding a guiding and advisory role to face the challenge of crossing the sea of the unknown technological effects. Participatory methods in general strengthen previous notions of deliberative democracy and emphasize how assessment of future consequences should not be just left to experts or politicians. Even if decisions later reveal themselves to be spurious and unwise, they have at least been made more legitimate and transparent [16]. Responsibility for decisions in this sense is at least shared and discussed and conflicts are allegedly represented, while civil society has a chance of asserting and renewing itself.

Constructive Technology Assessment (CTA), for instance, works as a real-time attempt to accompany all the various phases of technological development in order to bridge the asymmetry between our power to act and our power to predict the outcomes of technology. This promethean gap [2] was reinstated by Jonas [25] and according to Collingridge's dilemma it points at the difficulty to control technology once it is unleashed but ignoring how to adequately design it in the "laboratory". Theoretically, CTA aims at putting itself in the initial design phase in order to favor participation by various stakeholders in the co-shaping of a given technology. Rather than accepting the paradox that the more exact a prediction about the future is, the more likely it is that deviations occur [7], CTA opens up the process of construction of a technology, thus 'constructing predictions' along the way, about technology as well as about social order, whose fulfilment remains nevertheless an open question.

CTA is strongly dependent on an institutional and political context where stakeholders are able to cast doubts about a certain development or a successful market being on the brisk of being born. Questions about which technologies should be debated and analyzed are framed according to the stakeholders who raise them. Such a framework places epistemic challenges on what is likely to be asked and on criteria for understanding what is worth assessing beyond particular interests.

As seen above, the analysis of the impacts of automation processes on work and employment has been developed largely through an economic approach that underlines aggregate economic effects. Other approaches lead to distinct questions and issues. A sociological approach might highlight the extent to which changes in the work process in the shop floor due to the implementation of robots alienate workers or threaten their posts [9]. The nature of the problem is necessarily defined by perceptions and narratives of future gains and losses, interests, narratives and groups that push their framing of the issues and of how these are affected by technology. TA has the task of balancing such views and exposing their contributions as well as limitations, recognizing the conditions that frame the questions in the exercise, namely due to its own limited resources in workforce, time and funding. For example, monetizing the phenomena in terms of opportunities and threats, strengths and weakness at stake, as does the utilitarian ethics of cost–benefit analysis, leads to an easy and available numerical comparison that favors decision-making in terms of calculus, but has its own epistemic blind spot on the non- or less-quantifiable dimensions, such as moral values. This entails that TA is not taken from an outside or from a god's eye view, but rather is a methodological approach that expands the decision-making process and its reflective dimension. It is a step, albeit cautious, in the reinforcement of democratic procedures, attempting to see through the usual opacity of technologies to allow a more collective, shared, construction of technological futures.

Grunwald has recently argued that this expansion of TA should be taken as a broadening of responsibility towards the future with an emphasis on the NEST fields in early stages of development 19. By committing to the unpredictability of technological effects and how it usually thwarts our best expectations, the task dwells on the importance of interpreting and constructing meaning about the inchoative futures and the addressed ethical controversies [38]. It should nevertheless be noted that the institutional procedures under umbrella terms such as RRI (Responsible Research & Innovation) are proper to institutional contexts where funding and investment for breakthrough innovation and for a concurrent assessment is available and solidly grounded in a national scientific and technological system. Many national institutions bent on assessing more thoroughly certain products beyond the more obvious issues (safety, privacy, health and environmental) are very frequently powerless or out of pace to address all the development phases of technologies that are made outside their borders. Attention should be paid to how the dynamics between geopolitical centres and peripheries thus configure the institutional scope allowing TA to be applied before controversies are raised in the public sphere.

Correspondingly to the variety of contexts where technological innovation occurs, there are a variety of methods devised to acknowledge the most proper ways of constructing discussion and identifying conflicts [8]. The kind of stakeholders involved, their interests, power and backgrounds, along with the State as a kind of guardian of collective interests should help customize the methodological approach in order to deliver not a solution, but support to answers: knowledge, data and predictions about impacts and effects contribute to establish the different practical syllogisms that exhibit why and how such means should or should not be employed and according to how they promote what ends. Only then can communication establish the values in conflict according to a transparent structure of rationality. This entails and reinforces the normativity orientation of any TA as uncovering the layers of values involved in a conflict.

Such remarks illustrate how the contextual essence of TA prevents it to be considered as an algorithm-like process to be applied in every situation with only minor adjustments. Every assessment is an extension of practical reason, that is, a process by which moral agency is constituted in praxis, which cannot be reduced to a universal theory of decision awaiting to be applied and supplied through more and more theoretical knowledge [23]. The contextual details, the where, how, why, when and to whom of moral agency, not only matter but define the appropriateness between knowledge and ends.

TA is then about the expansion of democratic and participatory processes to the once more or less natural, automatic advancement of scientific and technological progress. Going outside the epistemic bubble of laboratories and industries can help scientists, engineers and developers to see that the meaning and the design of technology is not only a shapeable and open matter, but something that always constitutes the social lifeworld of meaning and should hence engage the public [20]. For industrial robots, for example, it means going beyond the question of how automation will impact jobs, employment and economic growth, in abstract, to ask also what does it mean for employees and the public in general to co-exist more and more with robots and with artifacts made by robots. This meaning is cross cultural and shaped not only by the real robots of factories and entertainment but also by an entire industry of science fiction literature and movies that help imagination run its course in the background of scenarios that will shape everyday future life.

Visions and images of a likely future condition how technologies are dealt with in the present. While the term fiction allots a figure of fantasy going wild and that "anything goes" about technology's effects, they actually contribute to structure the questions and debates involving the social consequences and impacts of technology. This kind of turn towards meaning is one of the shifts that TA went through and results from having to reconsider the connection between knowledge and praxis. The shortcomings of a predictable positivist reading of how society develops through technology do not lead to neglecting the connection between technology and a better society. It rather draws more attention to how the complexity of such relation must be actively and attentively built. It is through these meanings that decisions can be reached which are supported on a deliberative process and are at the same time not solid anticipations but visions of development.

5 Conclusion

In the face of considerable concerns regarding the impacts of increased use of robots and automation processes in industry as well as in the services, both in job substitution as well as in work organisation, TA can be an important resource to open up contemporary narratives about the future, explore ethical and social tensions and imagine collective futures which can engage actors in building joint visions centred on the public good. With the increased ubiquity of robotics and automation it is important to develop credible institutions for public engagement, which bring hand in hand both the images of its promises as well as the realities of its social situatedness. This is an important challenge at different levels. From the point of view of private organizations, workers unions and industrial relations, technology has largely been considered a given, which may be reacted upon, but not necessarily questioned or discussed in its form. As such, the analysis of the impacts of technology and employment are largely viewed from the economic perspective, reifying deterministic approaches. For TA practitioners, the assessment of the impacts of technologies on work has not been either a central object of analysis, being a particular subset of wider technological impacts on citizens. With many TA organisations linked to Parliaments, and hence focusing on citizen representation, the existence of concrete fora where industrial relations issues are discussed and negotiated may provide little incentive to such focus. In addition, TA organisations are often marred by limited funding and already existing institutional designs which limit the development of workplace oriented TA methods.

With this chapter we propose that, despite these constraints, the wider use of TA approaches in the context of institutions can contribute to better respond to both business needs for a favorable stakeholder environment that supports innovation and at the same time exploring their potential risks and impacts, and how these can be addressed, with the participation of the different actors. Public concerns over these matters will help build and shape new collective narratives that grant a technology their space of emergence and development and society the expectations to fulfil objectives for the common good.

References

1. AAVV (2018) Robotics in the care sector—challenges for society. TAB—office of technology assessment at the German Bundestag
2. Anders G ([1961] 1983) Commandments in the atomic age. In: Mitcham C, Mackey R (eds) Philosophy and technology. Readings in the philosophical problems of technology. The Free Press, Nova Iorque
3. Arntz M, Gregory T, Zierahn U (2016) The risk of automation for jobs in OECD Countries: a comparative analysis, OECD social, employment and migration working papers. OECD Publishing, Paris, p 189
4. Autor DH (2013) The 'task approach' to labor markets: an overview. J Labour Mark Res 46(3):185–199
5. Borgmann A (1984) Technology and the character of contemporary life. The University of Chicago Press, Chicago
6. Boucher P (2019) Why artificial intelligence matters. Briefing. European parliamentary research service
7. Collingridge D (1980) The social control of technology. New York: St. Martin's Press, Print
8. Decker M, Ladikas M (eds) (2004) Bridges between science, society and policy. Technology assessment—methods and impact. Springer
9. Decker M et al (2017) Service robotics and human labor: a first technology assessment of substitution and cooperation. Robot Auton Syst 87:348–354
10. Dreyfus HL (1997) What computers still can't do. A critique of artificial reason. The MIT Press, Cambridge, Massachusetts
11. Eurofound (2016) What do Europeans do at work? A task-based analysis: European jobs monitor 2016. Publications Office of the European Union, Luxembourg
12. Freeman C, Louçã F (2001) As time goes by: from the industrial revolutions to the information revolution. Oxford University Press, Oxford
13. Frey C, Osborne M (2017) The future of employment: How susceptible are jobs to computerisation? Technol Forecast Soc Chang 114:254–280
14. Frey C, Osborne M (2013) The future of employment: How susceptible are jobs to computerisation? Oxford martin programme on the future of work, Working Paper
15. Grunwald A (2000) Against over-estimating the role of ethics in technology development. Sci Eng Ethics 6(2):181–196

16. Grunwald A (2016) The hermeneutic side of responsible research and innovation, vol 5. ISTE and Wiley, London
17. Grunwald A (1999) Technology assessment or ethics of technology? Reflections on technology development between social sciences and philosophy. Eth Perspect 6
18. Grunwald A (2009) Technology assessment: concepts and methods. In: Anthonie W, Meijers M (eds) Handbook of the philosophy of science, pp 1103–1146
19. Grunwald A (2011) Responsible innovation: bringing together technology assessment, applied ethics, and STS research. Enterprise and work innovation studies, vol 7. IET, pp 9–31
20. Habermas J (1994) Técnica e ciência como ideologia. Edições 70: Lisboa
21. Ihde D (1990) Technology and the lifeworld: from garden to earth. Indiana University Press
22. Ihde D (2008) The designer fallacy and technological imagination. Philosophy and design. Springer, Dordrecht. DOI https://doi.org/10.1007/978-1-4020-6591-0_4
23. Jonas H (1983) The practical uses of theory. Soc Res 51:65–90
24. Jonas H (1984) The imperative of responsibility. In search of an ethics for the technological age. University of Chicago Press, Chicago
25. Kurzweil R (2014) The singularity is near. In: Sandler RL (eds) Ethics and emerging technologies. Palgrave Macmillan, London
26. MacIntyre A ([1981], 2007) After virtue: a study in moral theory. University of Notre Dame Press
27. Moniz AB, Krings, B-J (2016) Robots working with humans or humans working with robots? searching for social dimensions in new human-robot interaction in industry. Societies 6:23. https://doi.org/10.3390/soc6030023
28. Pianta M (2004) Innovation and employment. In Fagerberg J, Mowery D, Nelson R (eds) The oxford Handbook of Innovation, Oxford University Press, Oxford, pp 568–598
29. Porter TM (1996) Trust in numbers. The pursuit of objectivity in science and public life. Princeton: Princeton University Press
30. Rip A, Schot J (1997) The past and future of constructive TA. Technol Forecast Soc Chang 54:251–268
31. Rudolph A (2004) Technology assessment of autonomous intelligent bipedal and other legged robots. Final report, defense sciences office. Defense advanced research projects agency
32. Vallor S (2016) Technology and the virtues: a philosophical guide to a future worth wanting. Oxford University Press USA
33. Verbeek P-P (2011) Moralizing technology: understanding and designing the morality of things. University of Chicago Press, Chicago and London
34. Vivarelli M (2013) Technology, employment and skills: an interpretative framework. Eurasia Bus Rev 3:66–89
35. Wacjman J (2017) Automation: is it really different this time? Br J Sociol 68(1):119–127
36. Wallach W (2014) Ethics, law and governance in the development of robots. Ethics and emerging technologies. (Sandler, Ronald ed). Palgrave Macmillan, London, pp 363–379
37. van Est R, Gerritsen J (2017) Human rights in the robot age. Challenges arising from the use of robotics, artificial intelligence, and virtual and augmented reality. Rathenau Instituut
38. van Lente H, Swierstra, T, Joly P-B (2017) Responsible innovation as a critique of technology assessment. J Responsib Innov 4(2):254–261. https://doi.org/10.1080/23299460.2017.1326261.
39. van de Poel I (2006) Editorial: ethics and engineering design. Sci Technol Hum Value 31(3):223–236
40. van de Poel I, Goldberg DE (eds) (2010) Philosophy and engineering. An emerging agenda. Springer Science + Business Media B.V
41. van de Poel I, Fahlquist JN, Doorn N, Zwart S, Royakkers L (2011) The problem of many hands: climate change as an example. Sci Eng Eth 18(1):49–67

The manufacturer's authorised representative in the EU is Springer Nature Customer Service Centre GmbH, Europaplatz 3, 69115 Heidelberg, Germany. If you have any concerns regarding our products, please contact ProductSafety@springernature.com

Printed and bound by CPI Group (UK) Ltd, Croydon, CR0 4YY
30/09/2026
02225694-0009